W9-BFC-694

THE COSMOLOGICAL DISTANCE LADDER

THE COSMOLOGICAL DISTANCE LADDER

DISTANCE AND TIME IN THE UNIVERSE

MICHAEL ROWAN-ROBINSON

W.H. Freeman and Company
New York

Library of Congress Cataloging in Publication Data

Rowan-Robinson, Michael, 1942–
The cosmological distance ladder.

Includes bibliographies and index.
1. Cosmological distances. I. Title.
QB991.C66R68 1984 521 84-4088
ISBN 0-7167-1586-4

Printed in the United States of America

1 2 3 4 5 6 7 8 9 0 MP 3 2 1 0 8 9 8 7 6 5

CONTENTS

PREFACE

This book gives an introduction to modern astronomy from an unusual viewpoint, that of the measurement of distance, from the scale of the solar system up to the largest cosmological scales. Almost all types of astronomical object—stars, galaxies, clusters of galaxies—play a part in the cosmological distance ladder, and knowledge of distances is essential to all branches of astronomy. As we climb the rungs of the cosmological distance ladder, we follow the history of astronomy, from the geometrical arguments and speculations of the Greeks about the solar system to the first correct estimates of the relative distances of the planets from the sun by Copernicus; from the painstaking observations which led to the discoveries of the parallax and proper motion of the nearby stars to the opening up of Hubble's "realm of the nebulae". And this is also the history of mankind's expanding horizon.

In this century, there have been three great controversies about the cosmological distance scale. The first concerned the distances of the spiral nebulae, and culminated in the famous debate between Harlow Shapley and Heber Curtis in 1920; it was not conclusively settled till Edwin Hubble's work on cepheids in galaxies in the mid-20's. The second, on the nature of the redshifts of quasars, was at its peak in the early 70's, and it was at this time that my own interest in the cosmological distance ladder began. I noticed that all the arguments in favour of cosmological distances for quasars referred only to one class of quasar. It was therefore possible that, as with the nineteenth century controversy about the nebulae, there were two entirely distinct classes of object, one relatively local and one truly cosmological. I was disappointed to find that this reasonable argument made very little impression on the protagonists of the controversy, who seemed to know from some deep inner conviction that their hypothesis must prevail. As it happens, the supporters of the cosmological view have almost certainly been proved right by subsequent evidence.

The third controversy about the distance scale surfaced at a 1976 conference in Paris on the redshift and the expansion of the universe. There, Gerard de Vaucouleurs on the one hand, and Allan Sandage and Gustav Tammann on the other, arrived at estimates of the size of the universe, as measured by the Hubble constant, differing from each other by a factor of two. Moreover, when I asked the protagonists what was the range outside which they could not imagine the Hubble constant lying, these ranges did not even overlap. Given that they were studying more or less the same galaxies with rather similar methods, often using the same observational material, I found this incredible. It was to understand this disagreement that I set out to write this book, and I hope that the unprejudiced reader will find that I have at least partly succeeded.

After an introduction which gives the historical and philosophical background to the measurement of distances and times in the universe (Chapter I), the measurement of distances within our Galaxy is explored, and the structure and evolution of stars and galaxies are described (Chapter II). The main techniques of measuring distances to external galaxies are critically reviewed, both those primary methods which can be calibrated from theoretical arguments or from measurements in our Galaxy (Chapter III) and secondary methods which depend on already knowing the distances to some nearby galaxies (Chapter IV). An elementary introduction to the relativistic Big Bang (and other) cosmological models is given in Chapter V—the less mathematically minded reader might prefer to skip this chapter at a first reading. Finally, in Chapter VI, the value of the Hubble constant is reviewed and the origins of the current controversy over the distance scale are examined. The size and age of the universe are discussed and two questions are considered: Are there characteristic length-scales in the universe? Are we alive at a special time in the universe?

The book is aimed at undergraduate students in their final year of study. It could be the basis for a self-contained course for Physics or Astronomy students, or provide useful background reading for courses on stellar or galactic astronomy, or cosmology. The extensive tables and references should also make the book valuable for all those involved in astronomy, researchers, teachers and graduate students alike.

Throughout the text I have tried to present the results of others in an impartial way, and maintain a clear distinction between their quoted results and my own recalculations. In my discussion of the reliability of distance indicators, I tend to place more emphasis than an observational astronomer might on the theoretical justification for using the particular indicator.

My thanks are due to many astronomers who have answered my queries or given me copies of their work in advance of publication—Richard Fisher, David Hanes, Roberta Humphreys, Robert Kennicutt, Vera Rubin, Allan Sandage, Dave Schramm, Brent Tully and Don Vandenberg, in particular. I am especially grateful to Marc Aaronson, David Dewhurst, Paul Hodge, Barry Madore, G.C. McVittie, Gerry Neugebauer, Martin Rees, Joe Silk, Gustav Tammann and Sidney van den Bergh for reading parts or all of a draft of the book and providing many helpful comments and criticisms. I owe a special debt of gratitude to Gerard de Vaucouleurs, who read the whole manuscript closely and gave me hundreds of comments and corrections. These colleagues have generously given their time to try to keep someone who is not a noted expert on the subject on the straight and narrow. I have not been able to take all their advice: for one thing, one colleague's example of an appalling error is often another's example of an inalienable truth. Anyway, the errors and shortcomings that remain are entirely my own responsibility. I dedicate this book to all those who have toiled away in the building of the cosmological distance ladder.

Michael Rowan-Robinson

THE COSMOLOGICAL DISTANCE LADDER

CHAPTER ONE

EVOLUTION OF IDEAS ABOUT DISTANCE AND TIME

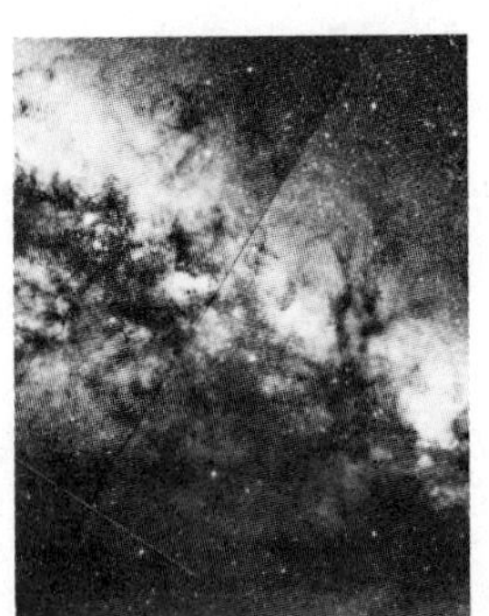

1.1 *Earliest ideas*

We shall probably never know whether the first men and women to walk on this planet a million or more years ago asked themselves questions about the sun, the moon, the stars. But from the time that the last ice age receded, 10,000 years ago, and human beings began to practice agriculture, we can be sure that they were interested in the seasons and the calendar. There is abundant evidence of astronomical purpose in the most ancient of the stone-age monuments of 3000 B.C. or so. We know that by 2000 B.C. the Babylonians were regularly observing Venus, and by the time of the height of the Babylonian civilization, around 600 B.C., we find sophisticated astronomical knowledge among both the Babylonians and the Chinese.

The intellectual horizon of the human race at any time has always been inextricably bound up with the perceived scale of the universe. Ancient myths and those of surviving primitive peoples in our own times suggest a conception of the heavens as not very far off at all. For example, in the most antique of the Chinese cosmologies, the Kai Thien or hemispherical dome theory, the heavens are seen as a rotating hemispherical cover, the earth as a bowl turned upside down, and the distance between the two as considerably less than the radius of the earth. Stories in which the sun or moon is carried in a chariot or boat suggest a distance scale measured on an even smaller, human scale. And there can be little doubt that a people's perceived scale of the universe must play a fundamental role in its culture and consciousness.

It is tempting to consider the growth of knowledge as a linear progress and to consider ourselves as therefore intellectually superior to the ancients. This fallacy is fueled by a picture of science, surprisingly widely held among scientists today, as data collection and analysis. The accumulation of data does indeed proceed monotonically, but science consists of the ideas that give meaning to these data, not the data themselves. When we look at the history of ideas about the scale of the universe, we find that the modern idea of an infinite universe through which countless stars and planets wander is not modern at all, but was held firmly by the Greek atomists and by the Chinese cosmologists of the Hsüan Yeh school. And even Aristotle, whom we have been taught to regard as the obstacle to scientific progress in the Middle Ages, summarized the views of his contemporaries in a way that could easily be adapted to modern cosmological predictions: "All thinkers agree that the world has a beginning, but some maintain that having begun, it is everlasting, others that it is perishable like any other formation of matter, and others again that it alternates, being at one time as it is now, and at another time changing and perishing, and that this process continues unremittingly."

1.2 Greek ideas about space and time

Western ideas of space and time have largely been molded by the thinking of the Greeks. Euclidean geometry is so much a part of our mental landscape that it is always hard at first to grasp the modern idea of curved space time. Euclid, who taught in Alexandria around 300 B.C., codified a geometry that had been built up piecemeal in scattered theorems by Greek mathematicians over the preceding three centuries, beginning with Thales and Pythagoras in the sixth century B.C. In an essay in *Concepts of Space and Time* [C1] Francis Cornford asks the question, "Did the Euclidean era, from which we are now emerging, stretch back, with no definable limits, through all recorded history into the darkness of the Stone Age?" His answer is that it did not and that there was a gradual and laborious transition from a pre-Euclidean common sense, in which the universe was seen as a finite sphere centered on the earth, to consistent Euclidean thinking with its concept of an infinite space. Pythagoras, for example, held that physical bodies are numbers, a number being defined as a plurality of units. Visible and tangible bodies are built up of these units, or monads, which are at the same time the units of arithmetic, the points of geometry, and the atoms of a body. To maintain the monads in a discrete plurality there have to be intervals of vacancy—a gap between terms in a numerical series, space between the boundaries of geometrical figures, and a void, or air, between the atoms of a body and between the surfaces of different bodies. But although the Pythagoreans asserted that this void extends beyond the heavens, they believed it did so only as a finite sphere. The Pythagorean philosophy was attacked by Parmenides, born at the end of the sixth century B.C., who denied the possibility of space without beings or of an internal void, and hence any plurality of separate things. Reality was the one, a unity without parts, and beyond a finite sphere containing all being, there must be an endless waste of nothing.

Plato and Aristotle still accepted the finite spherical world of pre-Euclidean common sense. Plato does seem to have recognized the implications of the emerging Euclidean geometry, for he located the finite spherical universe in an unlimited empty space. However, to Plato the space of geometry was not the same thing as physical space, being an object of thought rather than of the senses. This concept of an infinite void external to the finite world was vigorously pursued by the Stoic school. Aristotle, however, maintained that the cosmos was finite and boldly denied the infinite extension of space. But he did speak of the finite spherical world as "boundless" (thought of as a two-dimensional surface; a sphere does not have a boundary). Perhaps he would have felt at home with the modern picture of a finite universe of positive curvature.

The idea of an infinite physical frame first appears in Greek thought as the void of the atomists. Leucippus and Democritus, in the second half of the fifth century B.C., and Epicurus, a contemporary of Euclid, endowed the abstract space implied in Greek geometry with physical existence. Atoms must be illimitable in number and therefore demand an unlimited extent of space, through which innumerable worlds are scattered. The concept of a finite universe was refuted by Archytas of Tarentum, a friend of Plato's: "If I am at the extremity of the heaven of the stars, can I not stretch outward my hand or staff? It is absurd to suppose that I could not; and if I can, what is outside must be either body or space." The Roman atomist poet Lucretius used a similar image of a man at the boundary of the universe throwing a javelin. We now know this argument is fallacious, since there is no a priori reason why the universe should have a Euclidean geometry (though philosophers as eminent as Immanuel Kant and Bertrand Russell were to assert that it must). If the universe has a uniform positive curvature, then it can be finite, spherical, but unbounded (see Chap. 5).

It was the finite universe of Aristotle that prevailed in Western thought over the infinite universe of the atomists, and was to do so for almost two millennia until the time of Giordano Bruno. It is interesting that in Chinese cosmology there was a similar controversy between the Hun Thien or Celestial Sphere school and the Hsüan Yeh school, which taught the existence of an infinite, empty space. Again the former concept prevailed, becoming the official view by the second century A.D., though the latter idea was to persist for several centuries more (see Needham, *Science and Civilization in China* [N1]).

Turning now to Greek ideas about the nature of time, we find controversies which persist into the present century. The Pythagoreans identified time with the motion of the celestial sphere. Parmenides, on the other hand, denied the reality of succession or change: these are illusions generated by our mode of perception. The atomists attributed change to the motion of atoms and so, like the Pythagoreans, had a relational rather than an absolute theory of time. Time is a mere "appearance" (Democritus), an "accident of accidents" (Epicurus), it has "no being of itself" (Lucretius). For Aristotle, however, time had an absolute reality: time is everywhere alike simultaneously; time is present even if motion is absent, since it is a measure not only of motion but also of rest. But he did recognize that a prerequisite of time measurement is a periodic mechanism, the best being the revolution of the celestial sphere. The Stoics too were forced into a nonrelational theory of time, since they believed that the extramundane void coexists in time with the world. Time therefore flows in the void. (The Stoics had a cyclical cosmological theory, in fact: the universe is created from fire, and at the end of each cycle the universe is dissolved in the original fire. This coincides with the beginning of another cycle in which the events of the previous cycle are reconstituted in all their details and in the same order. The idea of cosmological cycles is a feature of several ancient philosophies. It was a strong element, for example, in the Meso-American cultures like those of the Maya and the Aztecs.)

The problem of whether time existed before the world was one that troubled medieval theologians, embarrassed by the concept of an infinite divine idleness. Saint Augustine resolved this brilliantly by asserting that the world was made not *in* time, but *with* time: "The world and time both had one beginning, and the one did not anticipate the other." I don't think there is any more convincing answer today to the perennial question "What was there before the big bang?"

1.3 The historical background to the cosmological distance scale

The Greeks were the first, as far as we know, to try to estimate astronomical distances. Aristarcos of Samos, who worked in Alexandria in the first half of the third century B.C., attempted the very difficult measurement of the angle between the sun and the moon at the point where the latter is exactly half full. From this he estimated that the distance of the sun is about 20 times that of the moon (the actual value is 390). By studying eclipses of the moon he correctly deduced that the moon's diameter is about one third of the earth's (near the true value of 0.27). He also recognized that the stars must be immeasurably distant. Another Alexandrine astronomer, Eratosthenes, estimated the circumference of the earth by measuring the angular distance of the sun from the zenith in Alexandria at midday on the summer solstice and combining this with the distance to Syene in upper Egypt (near modern Aswan) where the sun is directly overhead at this time. He found a value of 250,000 stadia which, if this refers to the famous stadium at Olympia, is 20% too great. Hipparcos, working at Rhodes in the middle of the second century B.C., found from the analysis of the geometry of eclipses of the moon that the distance of the moon was 59 times the radius of the earth, close to the correct value of 60.3. In the second century A.D. the last of the great Alexandrine astronomers, Ptolemy, confirmed this result by measuring the moon's parallax, the difference in the moon's apparent direction on the sky when observed from two different places on the earth at the same moment. Combining these estimates with the sun–moon distance ratio of Aristarcos, Ptolemy adopted as the sun's distance 1210 times the radius of the earth. This is a factor of about 20 below the correct value, because of the difficulty in deciding the exact moment when the moon is half full in Aristarcos's method. For the radius of the earth, Ptolemy adopted an estimate by Poseidonius of 180,000 stadia, about as much below the truth as Eratosthenes was above it.

Ptolemy had no way to measure the relative distances of the planets, and he simply assumed that the order of distance was one of increasing sidereal year (the time it takes a planet to complete its orbit through the fixed stars and return to the same place), namely, Mercury, Venus, Sun, Mars, Jupiter, Saturn. He then assigned relative distances on the basis of the nesting together of the crystalline spheres which were supposed to carry the planets around the sky. No significant advance was made in the knowledge of the distance scale until the time of Copernicus, though we do know that the Caliph Al Mumun, who built a magnificent observatory in Baghdad in 829 A.D., had Poseidonius's estimate of the size of the earth checked by his astronomers.

In the sixteenth century Nicolaus Copernicus made a very important advance in our knowledge of astronomical distances. His system gave for the first time nearly correct relative distances of the planets from the sun. Measured in terms of the distance of the earth from the sun, he found the distances from the sun of Mercury, Venus, Mars, Jupiter, and Saturn to be 0.36, 0.72, 1.5, 5, and 9, compared with the true values of 0.387, 0.723, 1.5, 5.2, and 9.5, respectively. He made a minor improvement in the estimate of the sun's distance to 1500 earth radii. From the lack of a measurable parallax for Mars, Johannes Kepler realized that this estimate should be increased by a factor of at least 3.

Copernicus is far better known, of course, for the indirect consequence of his

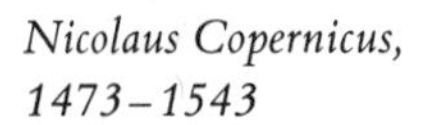

Nicolaus Copernicus, 1473–1543

system in shattering Aristotle's finite spherical world. Copernicus did not himself make this step, and his universe, with the sun in the center of a sphere of fixed stars, is decidedly Aristotelian in flavor. But because the stars now had to be very distant compared to the size of the earth's orbit, Copernicus's universe did represent a considerable enlargement of human horizons.

The natural and far more dramatic step of dethroning the sun, too, from the center of the universe was taken soon after by Giordano Bruno: "The stars are suns like our own and there are countless suns freely suspended in limitless space, all of them surrounded by planets like our own earth, peopled with living things. The sun is only one star among many, singled out because it is so close to us. The sun has no central position in the boundless infinite." With this return to the world view of the Greek atomists and the Chinese Hsüan Yeh school, the modern picture of the universe was ushered in. Not only was the universe infinite, but the distinction between the heaven and the sublunar world was eliminated forever.

The seventeenth century saw great advances in the accuracy with which astronomical distances were measured, mainly due to the work of the astronomers of the Paris Observatory. In 1671 Jean Picard measured the length of one degree of the meridian to an accuracy of a few meters. In 1671 – 1673 Jean Richer's expedition to Cayenne led to two important discoveries. First, gravity measurements established that the earth is not perfectly spherical. Second, Giovanni Domenico Cassini used Richer's observations of Mars to deduce that the distance of the sun was 140 million km, or 360 times the distance of the moon (only 7% below the true value). In 1675 Ole Römer, studying the time delay between eclipses of Jupiter's moons by Jupiter, made the first reliable measurement of the velocity of light, only about 25% below the correct value.

In the eighteenth century the English astronomer James Bradley embarked on a huge program of measurements of the positions of stars, from which he hoped to see the parallax, or angular displacement of nearby stars as they are viewed first from one side of the earth's orbit around the sun and then from the other. Instead he discovered aberration, the small change in apparent direction of a star due to the earth's velocity around the sun. The semimajor axis of the ellipse traced by stars on the sky, estimated by Bradley to be $20'' - 20''.5$ (true value $20''.47$), gave the ratio of the velocity of light to that of the earth and was consistent with Römer's result.

Edmond Halley had pointed out as early as 1679 the importance of observations of transits of Venus across the sun's disk for estimating the sun's distance accurately, and worldwide observations of the transits of 1761 and 1769 were organized. The results were of disappointingly low accuracy, and the final best estimate of the sun's distance was about 2.6% too high. It was not until the second half of the nineteenth century that measurements of the sun's distance close to the modern value of 150 million km (see Table 1.1) began to be made through observations of the parallax of Mars and asteroids.

Bradley's goal of measuring the parallax of the nearest stars, and hence their distances, was finally achieved by the German astronomer Friedrich Wilhelm Bessel in 1838 for the obscure star 61 Cygni (see Sec. 2.1). At last the distance scale of the stellar universe opened up by the astronomers and philosophers of the sixteenth century began to be measured.

1.4 The realm of the nebulae

By 1800 the possibility of a universe far grander in scale than that of the solar system and the stars had become a serious one. The suggestion that the nebulae might be "island universes" was made by Christopher Wren in the seventeenth century. To the peoples of the southern hemisphere the two fuzzy clouds of light in the night sky, first reported to Europe by Magellan's expedition, were familiar phenomena. In the northern hemisphere the first reliable record of a nebula was by the Arab astronomer Al-Sufi, who in his *Book of the Fixed Stars,* published in 964 A.D., called the Andromeda nebula a "little cloud." The invention of the telescope led to the discovery of many more nebulae, and in 1781 the French comet watcher Charles Messier completed a list of 103 nebulous objects which might be confused with comets.

The German philosopher Immanuel Kant elaborated Wren's island universe theory of nebulae, speculating that they were distant star systems like our own Milky Way. The latter he believed to be a disk-shaped distribution of stars. It was William Herschel who in the last decades of the eighteenth century inaugurated galactic and extragalactic astronomy as we know it today. He used counts of the

William Herschel, 1738–1822 (Courtesy of Yerkes Observatory.)

Henrietta Leavitt, 1868–1921 (From Popular Astronomy, *1920.)*

number of stars in different directions on the sky to try to estimate the size of the Milky Way and supported Kant's picture of a disk-shaped distribution of stars. Herschel's powerful telescopes resolved many nebulae into clusters of stars, and he argued that the unresolvable nebulae might be island universes similar to the Milky Way system. Herschel's faith in this idea was shaken by his observations in 1790 of the planetary nebula NGC 1574,* a star surrounded by a spherical halo of light. The spectroscopic studies of William Huggins in the 1860s showed that the halos of planetary nebulae, and a number of other unresolved nebulae like that in Orion, were due to hot gas and not to unresolved stars. This seemed fatal to the island universe theory, though the discovery by William Parsons (Lord Rosse) in the late

* NGC stands for *New General Catalogue of Nebulae and Clusters,* a very important compilation made by J.L.E. Dreyer in 1888, based on the earlier lists by William Herschel and his son, John Herschel.

1840s that some unresolved nebulae showed a remarkable spiral structure, reopened the possibility of a subclass of nebulae beyond the Milky Way. A difficulty was that these spiral nebulae appeared to be concentrated toward the poles of the Milky Way and therefore seemed to be part of the Milky Way system. At the turn of the twentieth century the island universe theory of the nebulae seemed to have been discredited.

The story of the reinstatement of the island universe theory over the succeeding 30 years is a fascinating and complex one (see Bibliography). The problem of the nature of the spiral nebulae was inextricably bound up with the determination of the scale of the Milky Way Galaxy. Important steps toward the latter were taken by Annie Jump Cannon and Antonia Maury, both working at the Harvard Observatory at the end of the nineteenth century. Harvard astronomers had embarked on a massive program of classification of stellar spectra, labeling them alphabetically according to the pattern of absorption or emission lines observed (see Chap. 2, Box 2.2). Annie Jump Cannon discovered the correct sequence of stellar spectral types O, B, A, F, G, K, M, now known to be one of decreasing surface temperature (see Sec. 2.1). Antonia Maury noticed that the lines of the spectra of stars of the same type varied in width from star to star and began to classify stars according to the widths of their spectral lines. In 1905 the Danish astronomer Ejnar Hertzsprung realized that the width of the spectral lines could be correlated with the luminosity of the stars, thereby discovering the distinction between giant and dwarf stars. Soon afterward Hertzsprung (1911) and the American Henry Norris Russell (1913) independently found that stars have a very characteristic distribution in a plot of stellar luminosity against color, or spectral type. This very important diagram is now called the Hertzsprung–Russell (HR) diagram.

The first and crucial step in establishing a distance scale for the spiral nebulae was taken by Henrietta Swan Leavitt, also working at the Harvard Observatory. In 1908 she noticed that in the Magellanic Clouds the variable stars (stars whose brightness changes, often with regular periodicity) showed a correlation between the brightness of the stars and the period of variation, and in 1912 she published the now famous period–luminosity relation. Hertzsprung recognized that these variable stars were in fact Cepheid variables, a class of variable star whose brightness varies extremely regularly on a time scale of 2–100 days. Cepheids, named after the prototype, δ Cephei, whose variations were first noticed in 1786 by John Goodricke, are also found relatively nearby in our own Galaxy. If the distances to the latter, closer stars could be reliably established, then Henrietta Leavitt's relation between period and the apparent brightness of the stars could be converted to one between period and luminosity (the amount of light emitted by the star). The intrinsic luminosity of the Magellanic Cloud Cepheids would then be known, and their distances could be found from their apparent brightnesses using the inverse-square law for radiation.

In 1901 the Dutch astronomer Jacobus Kapteyn had announced a model of the universe which was to be influential for more than 20 years. Using star counts Kapteyn showed that the density of stars falls off with their distance from the sun. He then estimated the scale of the stellar distribution from the parallax of nearby stars and from proper motions of stars, small changes in the relative motions of stars

FIGURE 1.1
Early evidence for a dust cloud in Taurus, photographed by Edward Barnard in 1907. The dark areas show where dust has extinguished the light from background stars. (From [B1].)

due to their motions through the Milky Way (see Sec. 2.3). Kapteyn's universe was a flattened system 30,000 light-years (lt-yr) in diameter and 10,000 lt-yr thick, with the sun near the center. Kapteyn considered the possibility that the apparent decline in the star density with distance from the sun was caused by extinction, the dimming of starlight due to particles of dust between the stars. But though Edward Barnard had shown conclusively the existence in the Milky Way of dark clouds of obscuring matter (Fig 1.1), there was at that time little evidence for a general distribution of dust, and Kapteyn concluded that the amount of extinction by dust must be too small to affect his model seriously. This view seemed to be confirmed by Harlow Shapley's observations in 1916–1917 of globular clusters, dense spherical agglomerations of 100,000 or more stars into a region no more than 30 lt-yr in diameter. Shapley's studies of globular clusters showed little evidence of extinction by intervening interstellar dust.

Globular clusters are found distributed all over the sky, but with an increased concentration toward the direction of the constellation of Sagittarius. In 1917 Shapley used them to derive a model for our Galaxy which conflicted dramatically with Kapteyn's. Shapley calibrated Leavitt's period–luminosity law using galactic Cepheids, and then established the distances of globular clusters via their Cepheid variables. Assuming that the globular clusters are part of our Galaxy, forming an almost spherical halo around it, Shapley concluded, correctly, that the center of the Galaxy must be in the direction of Sagittarius. He also estimated that the diameter of the Galaxy was 300,000 lt-yr, 10 times larger than Kapteyn's value.

In the same year Heber Curtis and George Ritchey found novae, a type of exploding star, in several spiral nebulae. Assuming that they were similar to the nova events seen in the Milky Way, Curtis deduced that the spiral nebulae lay far outside our Galaxy, supporting the island universe theory. He had to exclude specifically the novae S Andromedae (seen in 1885 in the Andromeda nebula) and Z Centauri (seen in 1895 in the nebula NGC 5253) as anomalous. These were later recognized as examples of the far more powerful supernova phenomenon. Similar conclusions were reached by Knut Lundmark in 1919, who used novae and the brightest nonvariable stars in the Andromeda nebula M31. In 1922 Ernst Öpik arrived at a distance of 1,300,000 lt-yr for M31 using an ingenious argument based on the assumption that it is similar to our own Galaxy and that the pattern of velocities seen in the nebula (through the Doppler shifting of lines in its spectrum) can be interpreted as a rotation of the whole nebula.

In 1920 a famous but inconclusive debate took place between Curtis and Shapley at a meeting of the American Association for the Advancement of Science on the subject of the scale of the universe. Curtis supported the Kapteyn model for our Galaxy and the island universe theory for spiral nebulae. Shapley confined himself mainly to the scale of our Galaxy which, as we have seen, he believed to be 10 times larger than the Kapteyn model. Against the idea that spiral nebulae were distant, he brought one apparently very convincing argument. In 1916 Adriaan van Maanen had claimed to have measured an apparent rotation of the spiral nebula M101 by comparison of photographic plates taken over a period of several years (Fig. 1.2). He had subsequently measured apparent rotations in several other spirals. At the distances proposed by Curtis, these angular rotations would have corresponded to rotation velocities close to the speed of light.

The question of the distance of the spiral nebulae was essentially settled in 1923 by Edwin Hubble, working at Mount Wilson Observatory. He found 12 Cepheids

Heber D. Curtis (Left), 1872–1942 (Courtesy of Lick Observatory.)
Harlow Shapley (Right), 1885–1972 (Courtesy of Yerkes Observatory.)

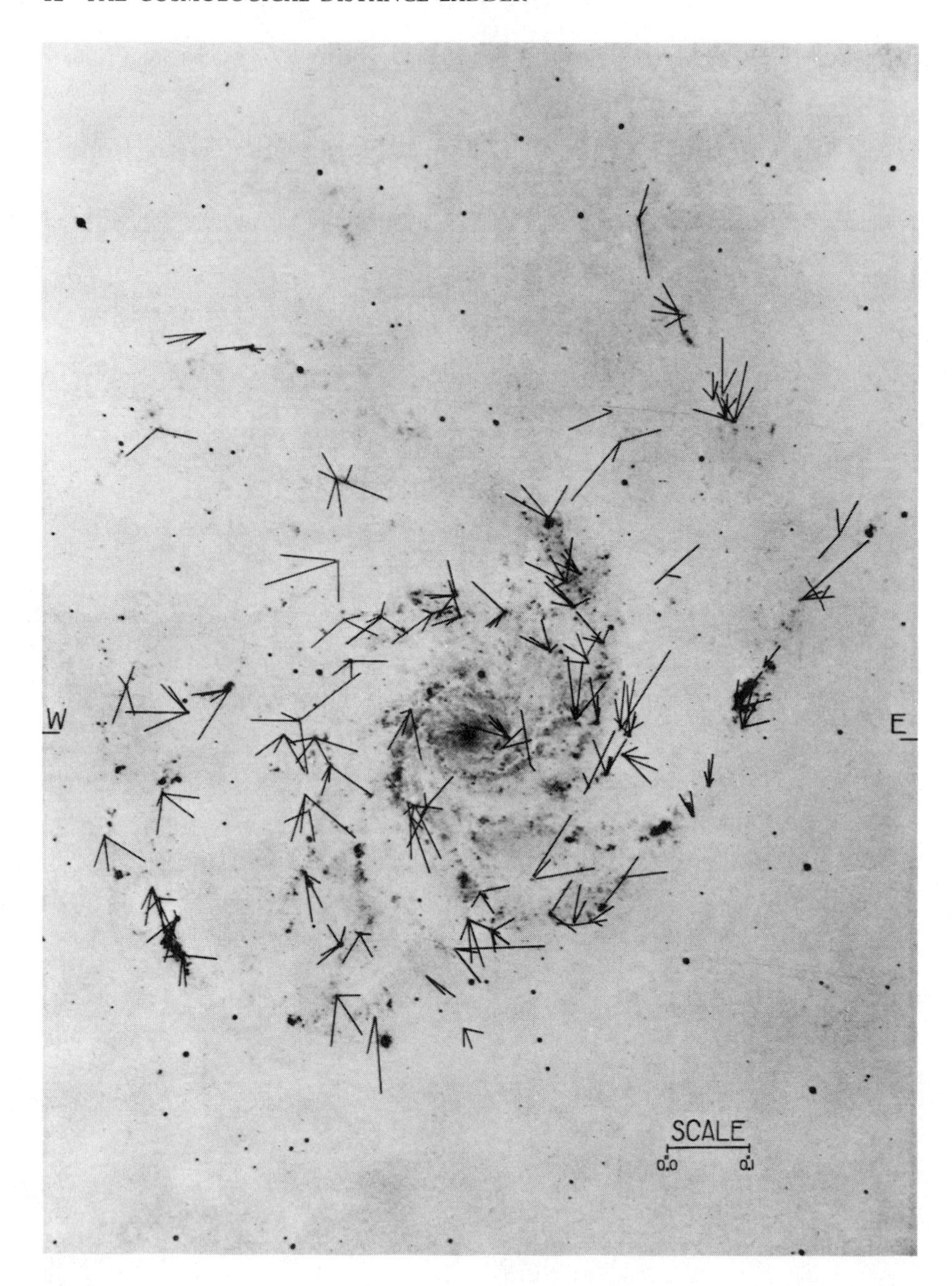

FIGURE 1.2

Adriaan van Maanen's evidence, later shown to be spurious, for rotation in M101. The lines show the direction of the motions he thought he had detected, the lengths of the lines being proportional to the magnitudes of motion (scale inset). (From [M1].)

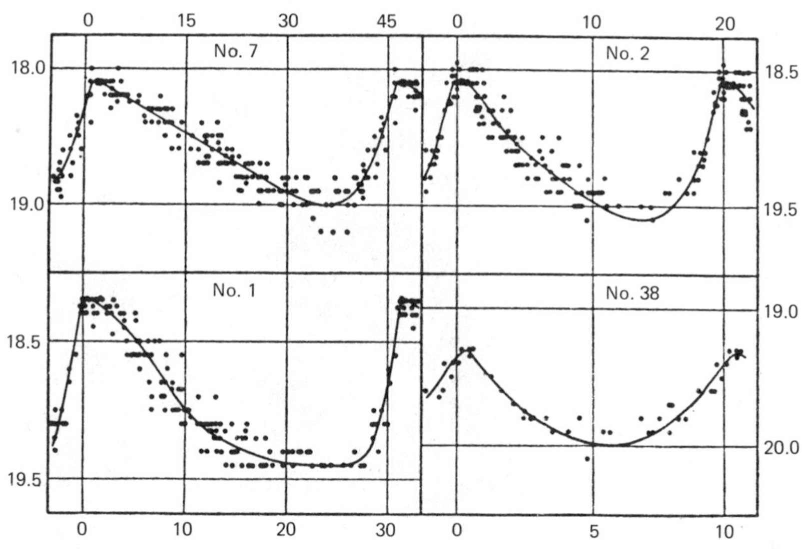

FIGURE 1.3

Hubble's observations of the light variations of four Cepheid variable stars in the Andromeda nebula M31. The vertical scale shows the brightness in magnitudes, and the horizontal scale gives the time in days. (From [H1].)

in M31 and 22 in the Triangulum nebula M33 (Fig. 1.3). Using Leavitt's period–luminosity relation as calibrated by Shapley, he deduced distances of 900,000 lt-yr for these two spirals, far outside our own Galaxy. In 1930 Robert Trumpler showed definitive evidence for extinction toward open star clusters by comparing the distances deduced from their apparent sizes with distances deduced from their brightnesses. The amount of extinction was just the right amount to reconcile the

FIGURE 1.4

Composite photograph of the Milky Way, from Sagittarius to Cassiopeia. The dark patches are due to extinction by large clouds of dust (and gas). (Courtesy of Mount Palomar Observatory.)

Edwin Hubble, 1889–1953 (Courtesy of Mount Wilson and Las Campanas observatories.)

Kapteyn and Shapley distance scales, the former being raised upward by a factor of 3 and the latter downward by a similar factor. Finally, in 1935 Hubble demonstrated that van Maanen's controversial apparent rotations did not exist and that they must have been some artifact of van Maanen's plate-measuring machine.

In the same years that the enormous scale of the universe was first being realized, the universe was also being shown to be expanding. In 1912 Vesto Slipher announced the results of his study at the Lowell Observatory of a sample of 12 spiral galaxies. Measuring their radial velocities from the Doppler shifts of their spectral lines, he had found that most showed large recession velocities. By 1925 he had measured the radial velocities of 40 galaxies, almost all of which were positive. In 1929 Hubble was able to compare these with his distance estimates for 18 galaxies, and he showed that there existed a clear correlation between the apparent velocity of recession, inferred from the red-shifting of the spectral lines, and the distance (Fig. 1.5). The possibility of an expanding universe had already been proposed by

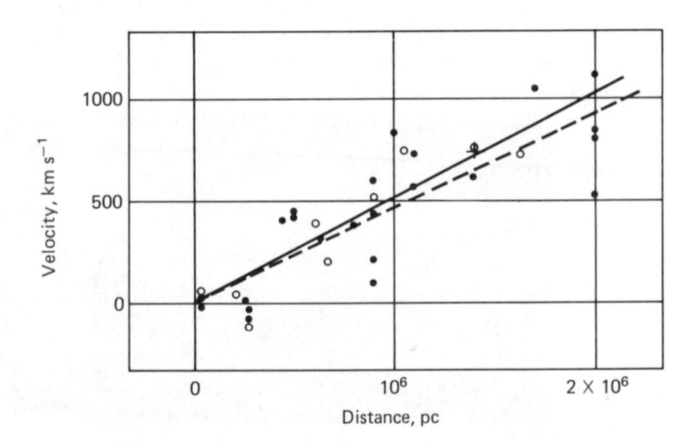

FIGURE 1.5

Hubble's velocity–distance law. The recession velocity of the galaxies is plotted versus their distance. (From [H1].)

Albert Einstein, 1879–1955. (Courtesy of Mount Wilson and Las Campanas observatories.)

theoretical cosmologists. When Albert Einstein published his general theory of relativity in 1916, one of the first problems he applied it to was the structure of the universe, for which he proposed a static model. Willem de Sitter responded in 1917 with an expanding universe model, also consistent with the general theory of relativity. The universe of de Sitter was devoid of matter, but more realistic expanding models were soon found by Einstein and de Sitter, by Aleksandr Friedmann in 1922 and 1924, and by Georges Lemaître in 1927.

An expanding universe must have been smaller in the past. Extrapolating this backward in time, we arrive at a time when the whole universe was concentrated together at infinite density, the big bang. Hubble's velocity–distance law therefore led him directly to an age for the universe, 1.8×10^9 years. This happened to be in good agreement with the age of the earth derived at that time from radioactive dating. However, the age of the earth was raised by the British geophysicist A. Holmes to 3.6×10^9 years, and subsequently to 4.3×10^9 years, embarrassingly long compared with Hubble's figure, particularly since the solar system is by no means one of the oldest objects in our Galaxy. In 1946 Lundmark pointed out that there was a discrepancy between the distance of the Andromeda nebula M31, deduced by the Cepheid method, and the larger distance estimates found by comparing novae or globular clusters in M31 with those in our Galaxy. It was not until 1952 that these discrepancies were resolved by Walter Baade, who found that there are two distinct populations of Cepheid variable stars in M31 and other spiral galaxies: those found in the disks of spiral galaxies are considerably more luminous than those found in globular clusters, in the halos of the galaxies. A further increase in the distance scale, and hence the age of the universe, took place in 1956 when Allan Sandage found that photographic images thought to have been the brightest stars in nearby galaxies were in fact groups of stars or clouds of ionized gas (called HII regions since their main constituent is ionized hydrogen). Sandage's 1956 value for the velocity–distance ratio, or Hubble constant, of 75 km s^{-1} Mpc^{-1} (megaparsec), represented an increase in the size of the universe of a factor of 6 from Hubble's first estimate (450 km s^{-1} Mpc^{-1}). The corresponding age of the universe was 1.3×10^{10} years.

Since 1956 most estimates of the Hubble constant have been in the range of 50–100 km s^{-1} Mpc^{-1} (Fig. 1.6), but there is fierce controversy about the actual value. The central figures of the controversy are Allan Sandage of the Mount Wilson Observatory, aided by Gustav Tammann of the University of Basel, on the one hand, advocating a value of 50 for the Hubble constant, and Gerard de Vaucouleurs of the University of Texas and his collaborators, advocating a value of 100. Sidney van den Berg of Dominion Observatory has also generally supported a value for the Hubble constant at the high end of the range. In the late 1970s and early 1980s many younger astronomers have been drawn into the controversy, refining traditional methods of measuring distance, and introducing new methods, for example, Marc Aaronson and Jeremy Mould of Steward Observatory, David Branch of the University of Oklahoma, David Hanes of the Anglo-Australian Observatory, Roberta Humphreys and Robert Kennicutt Jr. of the University of Minnesota, Barry Madore of the David Dunlap Observatory, and many others. The source of this controversy is one of the main themes of this book.

FIGURE 1.6

Measurements of the Hubble constant versus year of measurement, showing a change of a factor of 5–10 since Hubble's first estimate. Most measurements since 1960 lie in the range of 50–100 km s^{-1} Mpc^{-1}. (From [S1].)

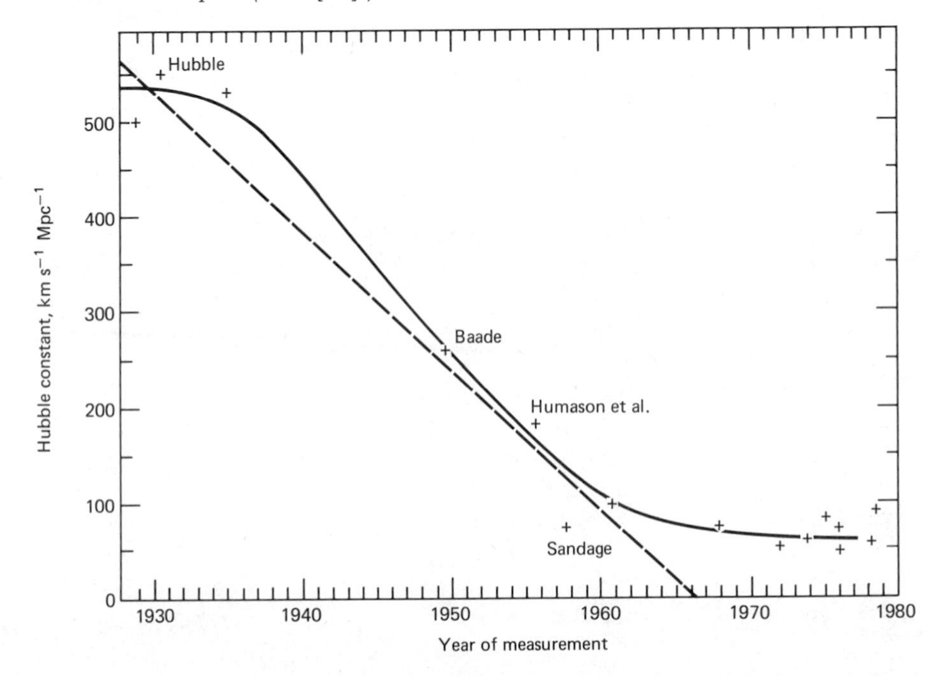

As the cosmological distance ladder extends outward, it also extends back in time. We have seen already that the age of the solar system has important implications for the cosmological distance scale since this age gives a lower limit to the age of the universe. Even more important are estimates of the age of our Galaxy and other galaxies since, while the solar system is not a particularly old object in the galaxy, galaxies are believed to be among the oldest structures in the universe. Later in the book (see Sec. 2.11 and 6.3) I shall be reviewing the evidence on these ages and their implications.

1.5 Fundamental concepts of space and time

As almost every philosopher throughout recorded history has discussed the nature of space and time, a brief account is bound to be superficial. In the bibliography of this chapter I give some suggestions for further reading.

The ideas of the Greeks, particularly those of Plato and Aristotle, exerted a powerful hold on European and Arabic thought until the fifteenth century and beyond. In the period from the Renaissance up to the time of Newton two important currents of thought can be detected. The first is that of the relativity of place, time, and motion. At the beginning of the fifteenth century Nicolas of Cusa asserted that the earth is not the center of the universe or of space and that "wherever in the heavens anyone may be placed, it would seem to him as if he were the center of the universe." Toward the end of that century Giordano Bruno argued that "since the horizon forms itself anew around every place occupied by the spectator as a central point, each determination of space must be relative, . . . since motion is only conceived in its relation to one fixed point, one and the same motion will present a different appearance according to whether I regard it from the earth or from the sun, and wherever I place myself in my thought, my own standpoint will always appear to be immovable." And from the relativity of motion followed, according to Bruno, the relativity of time. Since no absolutely regular motion can be discovered, we can find no absolute measure of time. Since motion appears different when regarded from different stars, there must be as many times in the universe as there are stars.

Later, René Descartes was to elaborate further the relational theory of space. Yet to make progress in understanding dynamics, he had to assert conservation of the quantity of motion, or momentum, mv (mass times velocity), and this assumes an absolute frame of reference.

The second philosophical trend after the Renaissance was toward the concepts of absolute space and absolute time, culminating in Sir Isaac Newton's famous definitions:

"Absolute space, in its own nature, without relation to anything external, remains always similar and immovable. Relative space is some movable dimension of the absolute space."

"Place is a part of space which a body takes up, and is according to the space, either absolute or relative."

". . . Absolute, true, and mathematical time, of itself, and from its own nature, flows equably without relation to anything external, and by another name is called duration: relative, apparent, and common time is some sensible and external (whether accurate or unequable) measure of duration by the means of motion, which is commonly used instead of true time; such as an hour, a month, a year."

These definitions gave a solid base to Newtonian mechanics, and even 200 years later James Clerk Maxwell, who believed that "all our knowledge of space and time is essentially relative," found it "more conducive to scientific progress to recognize, with Newton, the ideas of space and time as distinct, as least in thought, from that of the material system whose relations these ideas serve to coordinate."

Newton's concept of absolute motion was criticized by Bishop Berkeley, who argued that the concept was unnecessary for the formulation of the laws of motion and that it was impossible to tell whether a frame of reference is at rest or moving uniformly in a straight line. However, he was ignoring Newton's rotating-bucket experiment, in which a bucket of water suspended from a twisted string is released. After a while the surface of the water in the rotating bucket assumes a parabolic profile. Thus the laws of physics look different in a rotating frame of reference from those in a nonrotating frame. This illustrates that there are preferred frames of reference for the formulation of Newtonian mechanics, the inertial frames, which all move with respect to each other with a uniform velocity, without relative rotation.

In his correspondence with the English theologian Samuel Clarke, Gottfried Wilhelm Leibniz vigorously defended a purely relational theory of space and time, but was forced to concede the difference between relative and absolute motions because of the rotating-bucket experiment.

Toward the end of the nineteenth century Ernst Mach produced the intriguing suggestion that the inertia of a body is generated by the attraction of distant objects in the universe, and that any reference to motion in space must mean motion with respect to the entire universe (Mach's principle). This idea has been very influential in modern cosmological thinking, but it has been shown by Gödel that Einstein's general theory of relativity does not incorporate Mach's principle, in the sense that solutions can be found in which the whole universe has an arbitrary rotation velocity relative to the local inertial frame.

In 1905 Einstein published his special theory of relativity, which synthesized Newtonian mechanics and Maxwell's electromagnetic theory. Although the notion of absolute simultaneity was eliminated, the theory still required the existence of preferred frames, the inertial frames, and a Newtonian everflowing time in any particular inertial frame, manifested by an ideal clock assumed to be possessed by each inertial observer. The novelty was that the clocks of different observers appear to run at different rates, depending on their relative motions. Because the speed of light was postulated to be the same for each observer, spatial distances could be defined in terms of the light-travel time. The aspect of the theory emphasized by Hermann Minkowski in 1913, though, was the spatialization of time, with time becoming merely the fourth dimension in a four-dimensional space-time continuum. "Henceforth space by itself, and time by itself, are doomed to fade away into mere shadows, and only a kind of union of the two will preserve an independent reality." Our subjective perception of time becomes an illusion. As Hermann Weyl put it: "Subjectively there is an abyss between our modes of perception of space and time, but no trace of this difference remains in the objective universe which physics seeks to purge of immediate intuition. This universe is a four-dimensional continuum. There is neither space nor time but only consciousness, which, moving in an objective universe, records the section as it comes to it and leaves it behind as history, like a process which unfolds itself in space and opens out in time." And Gödel claimed that "from relativity . . . one obtains an unequivocal proof for the view of those philosophers who, like Parmenides, Kant, and the modern idealists, deny the objectivity of change as an illusion or appearance due to our special mode of perception."

At first sight this seems to be an extreme interpretation or at least an extreme emphasis, since it appears to lead directly to a deterministic and fatalistic view of the universe. "The world is, it does not become," in the words of Weyl. However, it is not true that relativity implies that the space and time dimensions are indistinguishable in status. Modern relativists emphasize, as Weyl did too in other writings, the $(3 + 1)$-dimensional nature of space-time, that is, that the four-dimensional space-time breaks up into three dimensions of space and one of time. As Einstein put it, "time and space are fused in one and the same continuum, but this continuum is not isotropic."

Nevertheless modern physics does not require the concept of a moving present, and several physicists have tried to give an explanation of the strong subjective sense of passage through time that we all have. A.S. Eddington introduced the idea of the arrow of time, defined by the direction of increasing entropy, or thermodynamic disorder, and suggested there might be a relation between entropy increase and the expansion of the universe. A thorough review of these questions is given by Paul Davies in *Physics of Time Asymmetry* [D1], in which he argues that it is preferable to speak of time asymmetry, thereby avoiding the subjective connotations of the arrow of time.

In his general theory of relativity, published in 1916, Einstein showed how the inertial frames of special relativity could be identified as the locally freely falling (and nonrotating) frames. He also showed how the laws of physics could be formulated with respect to an arbitrary, non-inertial frame of reference. The presence of matter results in a curvature of space-time, and an inertial (freely falling) observer moves on a geodesic of this curved space-time.

It would seem that in general relativity we reach the ultimate in a relational theory of space and time. Yet absolute space and time reemerge in modern relativistic cosmology. If the universe is homogeneous, that is, the same at every location in the universe, and isotropic, that is, looks the same in every direction, then a universal cosmic time exists for all observers at rest with respect to the matter in the universe. And, at every locality, a preferred frame exists, the one in which the expansion of the universe looks isotropic. This is even closer to absolute space than the inertial frame of Newtonian theory and special relativity, since the velocity of this cosmic frame is uniquely specified. Most current observations support the assumption that the universe is as homogeneous and isotropic on the large scale as we are capable of measuring. Our Galaxy's motion with respect to the local cosmic frame appears now to have been measured (see Chap. 5). How can this simple and absolute universe be reconciled with the relativistic philosophy? Several efforts have been made to explain how the universe might have come to have such a simple structure on the large scale. The best hopes may lie in processes in the very early universe when the basic forces of physics may have had a simpler and more unified structure than they do today.

1.6 *The basic forces of physics and their characteristic length scales*

The four forces of modern physics are gravitation, electromagnetism, and the strong and weak nuclear forces.

1 GRAVITATION. To a first approximation, gravitation obeys Newton's inverse-square law, and so there is no characteristic length scale associated with it. Except on the large scale, though, gravity is a very weak force. The mutual attraction generated by two 1-kg spheres 1 m apart is only about 10^{-11} times that due to the earth's gravity, so gravity is primarily an astronomical force. Einstein's general theory of relativity provides an important length scale associated with a spherically symmetric distribution of matter. This is the Schwarzschild radius, introduced by Karl Schwarzschild, one of the pioneers of relativity theory. The radius is defined as $r_s = 2GM/c^2 = 1.5(M/M_\odot)$km, where M is the mass of the object, $M_\odot$ is the mass of the sun, G is the gravitational constant, and c is the velocity of light. If the matter lies entirely within this radius, then $r = r_s$ defines an event horizon for external observers: no signals emitted inside this radius ever reach such an observer, and we have a black hole.

An important feature of Einstein's theory which has yet to be verified experimentally is that gravitational information is transmitted at the speed of light, and the acceleration of masses is therefore accompanied by the emission of gravitational radiation. However, indirect evidence for gravitational radiation may have been found in the binary pulsar, a double-star system in which both members have become neutron stars and are observable as pulsating radio sources or pulsars. A slowing down of the orbital period of the stars has been explained as loss of energy by gravitational radiation.

2 ELECTROMAGNETISM. Electromagnetism, the unification by Maxwell of electricity and magnetism, is also an inverse-square-law force. But it is a very much stronger force than gravity. The ratio of the electric force between a proton and an electron to the gravitational attraction between them is

$$\frac{e^2/r^2}{Gm_e m_p/r^2} = \frac{e^2}{Gm_e m_p} = 2.28 \times 10^{39}.$$

Electromagnetic forces therefore dominate over gravity on the atomic scale and are responsible for all chemical and biological phenomena. Electromagnetic forces also play a striking role in many astrophysical phenomena, for example, star spots and stellar flares. However, astronomical distributions of matter are, on average, electrically neutral so that over the large scale, positive and negative forces cancel out and are dominated by gravity. No satisfactory explanation of the charge neutrality of the universe has been found yet, and it seems we have to assume it as an initial condition.

3 STRONG NUCLEAR FORCE. The strong nuclear force is what binds neutrons and protons together in the nucleus of an atom. The range of this force is about 10^{-13} cm, the typical size of an atomic nucleus. The particles which are acted upon by the strong nuclear force, for example, protons, neutrons, mesons, and hyperons, are called hadrons. Particles which are not acted on by the strong nuclear force, like neutrinos, electrons, muons, tauons, and their antiparticles, are called leptons.

The current view of the nature of hadrons is that they are all made up of triplets of quarks, of which there are six different kinds or "flavors": up and down, strange and charmed, top and bottom. The proton is composed of two up quarks and one down quark, while the neutron is composed of one up and two down quarks. The attractive force, which holds the quarks together, is provided by particles known as gluons, and the detailed theory of this force is known as quantum chromodynamics. The name arises because another property of quarks has been given the name color (a quark can exist in one of three possible color states).

4 WEAK NUCLEAR FORCE. The weak nuclear force is responsible for the conversion of a proton to a neutron or vice versa, with the additional involvement of an electron or positron and a neutrino or antineutrino in the reaction. For example, a free neutron will decay in about 10 min into a proton, an electron, and an antineutrino. The range of the weak nuclear force is only about 10^{-15} cm.

In quantum-mechanical terms the four types of interaction (gravitational, electromagnetic, and strong and weak nuclear) can be thought of as arising from the exchange of a "virtual" elementary particle, one that exists for too short a time to be detectable in normal circumstances. Gravitation involves the exchange of gravitons, electromagnetism the exchange of photons, the weak nuclear force the exchange of intermediate vector bosons or weakons, and the strong nuclear interaction between neutrons and protons involves the exchange of mesons. The range of each force is then given by the Compton wavelength h/mc, where h is Planck's constant, m is the mass of the mediating particle, and c is the velocity of light. Because $m = 0$ for the graviton and the photon, gravity and electromagnetism have no length scale associated with them.

One of the goals of modern elementary particle physics is to unify these forces into a single theory. In 1971 Steven Weinberg, Abdus Salam, B.F.L. Ward, and G.'t Hooft successfully unified the weak and electromagnetic interactions.

J.C. Pati and Salam, Howard Georgi and Sheldon Glashow, and others have now proposed grand unified theories (GUTs), which seek to unify the strong and electroweak interactions. In these theories the leptons and quarks are grouped together into three generations:

1 Electron and its neutrino, up and down quark
2 Muon and its neutrino, strange and charmed quark
3 Tauon and its neutrino, top and bottom quark

Under conditions of very high energy [$> 10^{15}$ gigaelectron volts (GeV) per particle] the particles within each generation may be transformed into one another. The result is that baryons will no longer be conserved particles. For example, one of the quarks which make up a proton may be transformed into a positron, the other two making up a meson.

Such extreme conditions are found only in the very early phases of a big-bang universe, less than 10^{-35} s after the initial singularity. Now one of the puzzles about the universe today is that there is a clear excess of baryons (protons and neutrons) over antibaryons—galaxies are made of matter, not antimatter. This baryon excess can be quantified as the number of baryons in any volume of the universe minus the number of antibaryons (zero, in general, today) divided by the number of photons, since it turns out that this ratio is not altered as the universe expands: the observed number is about 10^{-9}. It is hoped that the baryon nonconserving processes occurring 10^{-35} s after the initial singularity can give an explanation of this value of the baryon excess. Incidentally these baryon nonconserving processes occur under normal conditions today, too, but extremely slowly. A proton is expected to decay into a Π° meson and a positron on a time scale of about 10^{31} years. Experiments are under way to measure this effect by studying very large masses of material. However, although examples of possible proton decay events have been published, the results are still inconclusive.

We can make an estimate of the epoch at which, extrapolating the expansion of the universe backward toward ever higher densities, general relativity must break down because of quantum effects. Quantum theory predicts that associated with a particle of mass m there is an uncertainty in its position of amount h/mc, the Compton wavelength. Now in general relativity the event horizon for a mass m has a radius of order Gm/c^2 (neglecting the factor of 2 which appears in the definition of the Schwarzschild radius). Equating these two gives a mass $m_p = (hc/G)^{1/2} = 5.46 \times 10^{-5}$ grams (g), known as the Planck mass, and the corresponding Planck length $\lambda_p = h/m_p c = (Gh/c^3)^{1/2} = 4.05 \times 10^{-33}$ cm. We can associate with this length a time, the time for light to travel this distance,

$$t_p = \frac{\lambda_p}{c} = \left(\frac{Gh}{c^5}\right)^{1/2} = 1.35 \times 10^{-43} \text{ s}$$

the Planck time. General relativistic big-bang cosmological models cannot be extrapolated to times earlier than this, and a quantum theory of gravity is required.

1.7 International Standards of Length and Time

How are distances and times defined and measured today? During the past century, standards of length and time have been regulated by the Comité Général des Poids et Mésures (CGPM), which meets every few years. In 1884 Greenwich mean time (GMT), based on the length of the mean solar day, was adopted as the international basis for timekeeping. Toward the end of the nineteenth century S.C. Chandler and Friedrich Küstner discovered that the earth wobbles, roughly periodically, relative to its axis of rotation, resulting in variations of about 1 millisecond (ms) in the length of the day at Greenwich. This phenomenon, with a period of 427 days, is superposed on the much larger wobble in the earth's axis found in 1748 by James Bradley and known as nutation. The latter has a period of 19 years and is caused by the action of the moon's gravity on the earth's equatorial bulge. In 1926 H. Spencer Jones showed, through observations of the Sun, Mercury, Venus, and Mars, that unexplained irregularities in the motion of the moon, first found by Simon Newcomb in 1878, were due to small fluctuations in the rate of rotation of the earth on a time-scale of decades. In 1935 additional small seasonal fluctuations in the rate of rotation due to worldwide meteorological effects were found. Finally, there appears to be a long-term, secular slowing-down of the earth, amounting to about 1 s per 50,000 years, believed to be due to tidal friction between the moon and earth.

During the 1930s quartz clocks, in which a crystal is set vibrating by an electric circuit that provides an oscillating voltage, began to replace pendulum clocks for accurate work. In 1948 ephemeris time, based in theory on the earth's annual orbit about the sun, but in practice on observations of the moon against the background of stars, was introduced. After the 1955 general assembly of the International Astronomical Union, corrections to GMT, by then known as universal time (UT), began to be published. UTO is the observed universal time (that is, mean solar time); UT1, needed by navigators, surveyors, and astronomers, is UTO corrected for the effects of the irregular rotation and secular slowing down of the earth; and UT2 is UT1 corrected for seasonal fluctuations.

However, events were beginning to overtake the astronomers. In 1947 Lyons and colleagues at the U.S. National Bureau of Standards succeeded in using the frequency of an absorption line in ammonia gas to stabilize a quartz crystal oscillator, and later produced a microwave frequency standard based on the transition between two energy levels in an atom of cesium-133, the cesium atomic clock. Improvements in cesium frequency standards soon followed, and in 1957 the frequency of this standard line was determined jointly by the U.S. Naval Observatory and the Royal Greenwich Observatory as 9,192,631,770 hertz (Hz) (1 Hz = 1 cycle per second). In 1967 the CGPM finally abandoned ephemeris time and defined the second as "the duration of 9,192,631,770 periods of rotation corresponding to the transition between two hyperfine levels of the ground state of the cesium-133 atom." International atomic time is now based on atomic clocks at several locations, and international time signals conform to a system called universal time (coordinated) (UTC). This corresponds exactly in rate with international atomic time, but is adjusted by inserting or deleting seconds—positive or negative leap seconds—to ensure that it agrees approximately with UT1. Leap seconds are applied when necessary as the last seconds of a month, preferably at the end of December or June. As an example of the accuracy of modern atomic clocks, the U.S. National Bureau of Standards' primary cesium frequency standard, known as NBS-6 and placed in service in 1975, has an accuracy of one part in 10^{13} and a long-term stability of one part in 10^{14}.

Meanwhile, in 1960 the CGPM had replaced the definition of the meter, which had been in force since 1889, based on an international prototype made of platinum-iridium, by the following definition: "The meter is the length equal to

165,073.73 wavelengths in vacuum of the radiation corresponding to the transition between the levels $2p_{10}$ and $5d_5$ of the krypton-86 atom."

However, the implementation of this standard was limited to an accuracy of a few parts in 10^9, and most laboratories preferred to use the wavelength of a stabilized optical laser as their basic unit of length. For example, the reproducibility between laboratories of the 633-nm (0.633-μm) helium-neon laser stabilized with intercavity iodine is better than four parts in 10^{11}. In 1983 a team at the U.S. National Bureau of Standards succeeded in establishing a link from the microwave region of the spectrum (where the cesium time standard is located) to the optical region via a chain of intermediate frequencies, thereby measuring the frequency of an optical laser to one part in 10^{10}. They were able to combine this accurate frequency measurement with wavelength measurements by teams in the United States, Great Britain, and Canada, using the relationship

$$\text{frequency} \times \text{wavelength} = \text{speed of light}$$

to arrive at a value for the speed of light in vacuo of

$$c = 299{,}792{,}458.6 \pm 0.3 \text{ m s}^{-1}.$$

In October, 1983, the CGPM therefore decided to change the definition of the meter. From now on the velocity of light will be defined to be 299,792,458 m s^{-1}, and the meter will be defined in terms of the distance traveled by light in 1 s.

The great accuracy of laser-wavelength measurements is not immediately transferable to normal length measurements. Thermal expansion of engineering materials is a practical limit for the measurement of large objects, and problems in defining the end points for lengths of 1 m or less remain a limitation. Terrestrial geodetic measurements are nearly all limited in accuracy by the refractive index of air.

1.8 Distances within the solar system

We saw that Copernicus, using pretelescopic observations of very limited accuracy, was able to estimate the relative distances of the planets then known to an accuracy of about 5%. By the end of the nineteenth century the shapes and relative sizes of the planetary orbits had been measured to very high accuracy. The absolute scale of the solar system, characterized, for example, by the mean radius of the earth's orbit or astronomical unit (A.U.), was however known only to an accuracy of 0.1%. This had been improved by a factor of 10 by the 1940s.

An enormous improvement in the accuracy with which this absolute scale can be measured, by a factor of 10,000, took place when Soviet, U.S., and British astronomers succeeded in reflecting radar signals off Venus in 1961, to be followed soon after by radar detections of Mercury (1962) and Mars (1965). Table 1.1 summarizes the mean distances of the planets from the sun, though it should be emphasized that several parameters are needed to specify each planetary orbit precisely.

TABLE 1.1
Distances of the planets from the sun.

Planet	*Semimajor axis of orbit** (A.U.)
Mercury	0.387099
Venus	0.723332
Earth	1.000000
Mars	1.523691
Jupiter	5.202803
Saturn	9.53884
Uranus	19.1819
Neptune	30.0578
Pluto	39.44

* 1 A.U. = 149,600,000 km (International Astronomical Union, 1963)
=149,597,892.9 ± 5.0 km (Goldstone, Jet Propulsian Lab.: Muhleman, D.O., *Month. Not. R. Astron. Soc.*, 1969, **144,** pp. 151–157)
=149,597,892.3 ± 1.5 km (Lincoln Lab., U.S. Naval Observatory: Ash, M.E., Shapiro, I.I., and Smith, W.B., *Astron. J.*, 1961, **72,** pp. 338–350)

SOURCE: From [A1]. See also [C2] for a discussion of the orbital elements of the planets.

Solar-system studies contribute to our understanding of the very largest scales in the universe in several ways. For example, to understand the motions of the planets, account has to be taken not only of the Newtonian gravitational effect of each planet on the others, but also of the effects of the general theory of relativity. In fact an unexplained discrepancy in the rate of precession of Mercury's orbit (the gradual rotation of the major axis of the elliptical orbit) led to many searches during the nineteenth century for a new planet, Vulcan, nearer to the sun than Mercury. The discrepancy turned out to be a consequence of general relativity. Also in general relativity it is predicted that light is bent by concentrations of matter, and this phenomenon has been found during eclipses of the sun, though the accuracy of the measurements is not very high.

The advent of spacecraft led to a new precision test of general relativity. By monitoring the time delay between transmission and reception of a radar signal reflected from the spacecraft as it passed behind the sun, the predictions of general relativity have been shown to be confirmed to high accuracy. Most rival theories of gravity have been eliminated unless they happen to give identical solar-system predictions to general relativity, in which case general relativity triumphs for the moment on the grounds of simplicity. The cosmological models we shall be using in Chap. 5 to describe the universe are all based on general relativity.

There are several other ways that the solar system plays a significant role in building up a picture of the very largest scales in the universe. First, we can use the extent of the earth's orbit to measure the distances to the nearest stars by the surveyor's technique of triangulation, the method of trigonometric parallaxes (see Sec. 2.1). Second, the abundances of the elements in the solar system have a universal significance. Carbon, nitrogen, oxygen, and heavier elements (referred to collectively as heavy elements by astronomers) were made in stellar interiors and tell us about the history of chemical evolution in our Galaxy up to the time that the solar system formed. Most of the light elements, lithium, beryllium, and boron, were made by cosmic rays colliding with the nuclei of heavier elements, a process known as spallation. And most dramatically of all, the majority of the helium, deuterium (the isotope of hydrogen, with a neutron in the nucleus in addition to the usual proton), and lithium-7 in the solar system are now believed to have been made during the early "fireball" phase of a big-bang universe (see Chap. 5).

Finally, we can even estimate the age of the universe from measurements in the solar system alone. The oldest solar-system material, notably meterorites and parts of the lunar surface, can be dated as having solidified 4.55×10^9 years ago. Long-lived radioactivities, for example, certain isotopes of thorium, uranium, and osmium, can be used to give the age of the Galaxy (see Sec. 2.11). Since the time it took our Galaxy to form is believed to have been brief compared with its present age, the age of the universe is likely to be not much greater than that of our Galaxy.

BIBLIOGRAPHY

[A1] Allen, C.W., *Astrophysical Quantities,* 3rd ed., Athlone Press, 1973.

[B1] Barnard, E.E., "On a Nebulous Groundwork in the Constellation of Taurus," *Astrophys. J.,* 1907, **25,** pp. 218–225.

[B2] Berendzen, R., Hart, R., and Seeley, D., *Man Discovers the Galaxies,* Science History Publ., 1976.

[B3] Berry, A., *A Short History of Astronomy,* 1898; reprinted by Dover, 1961.

[C1] Capek, M., Ed., *Concepts of Space and Time,* Reidel, 1976.

[C2] Clemence, G.M., and Brouwer, D., The accuracy of the coordinates of the five outer planets and the invariable plane, *Astron. J.,* 1955, **60,** p. 118–126.

[D1] Davies, P.C.W., *Physics of Time Asymmetry,* Surrey University Press, 1974.

[F1] Fernie, J.D., "The Historical Quest for the Nature of the Spiral Nebulae," *Publ. Astron. Soc. Pacific,* 1970, **82,** pp. 1189–1230.

[H1] Hoskin, M.A., *Stellar Astronomy: Historical Studies,* Science History Publ., 1982.

[H2] Hubble, E.P., *The Realm of the Nebulae,* Dover, 1936.

[M1] Maanen, A., van, "Preliminary Evidence of Internal Motion in the Spiral Nebula Messier 101," *Astrophys. J.,* 1916, **44,** pp. 210–228.

[M2] Munitz, M.K., Ed., *Theories of the Universe,* Free Press Paperback, 1957.

[N1] Needham, J., *Science and Civilization in China,* vol. 3, Cambridge University Press, 1959.

[P1] Pannekoek, A., *History of Astronomy,* Allen and Unwin, 1961.

[S1] Schramm, D.N., "Nuclear Constraints on the Age of the Universe," in *Highlights of Astronomy,* vol. 6, West, R.M., Ed., Reidel, 1983, pp. 241–253.

[S2] Shu, F.H., *The Physical Universe,* University Science Books, 1982.

[S3] Smith, H., "The Steady March of Atomic Time," *New Scientist,* 1982, **93,** p. 382.

[S4] Smith, R.W., *The Expanding Universe, Astronomy's Great Debate,* Cambridge University Press, 1982.

[S5] Struve, O., and Zebergs, V., *Astronomy of the 20th Century,* Macmillan, 1962.

[V1] de Vaucouleurs, G., *The Discovery of the Universe,* Faber, 1957.

[W1] Waterfield, R.L., *A Hundred Years of Astronomy,* Duckworth, 1938.

[W2] Wilkie, T., "Time to Remeasure the Metre," *New Scientist,* 1983, **100,** pp. 258–263.

CHAPTER TWO

DISTANCES WITHIN THE GALAXY

Understanding distances in our Galaxy is a crucial first step before we can attempt to measure distances to other galaxies. Most of the confusion about the distances of the extragalactic nebulae in the early part of this century arose from mistaken estimates of distances within the Milky Way. For example, neglect of extinction by interstellar dust in our Galaxy was one of the key factors delaying recognition of the huge scale of the "realm of the nebulae." In this chapter we study how the distances to stars and other types of object in the Galaxy are measured and build up a picture of the structure and evolution of stars and galaxies.

It is sometimes hard to realize how very far away even the nearest star is. From the radius of Pluto's orbit to the distance of the nearest star, α Centauri, is a factor of more than 6000 up the cosmological distance ladder. The distances of several thousand stars have been measured by the method of trigonometric parallax, that is, the change in direction of the star as seen from the earth when the latter is on opposite sides of the sun (Sec. 2.1). The proper motions of stars across the sky have allowed us to find the distances of many more thousands of stars (Sec. 2.3) and of the nearest star clusters (Sec. 2.4). The majority of nearby stars are in binary or multiple systems (Sec. 2.2), and many other stars in the Galaxy are to be found in clusters of different types (Sec. 2.6). Sec. 2.7 includes an outline of the evolution of stars of different masses, with particular reference to the Hertzsprung–Russell diagram. Spectroscopic and photometric studies of stars of different types and at different evolutionary stages extend the distance scale for stars to the edge of the Galaxy (Sec. 2.5).

Spread between the stars are clouds of interstellar dust and gas (Sec. 2.8). Kinematical distances to these gas clouds are described in Sec. 2.9. The gas and the associated young stellar objects trace out the spiral arms of our Galaxy. The classification of galaxies as spirals, ellipticals, or irregulars is introduced in Sec. 2.10. Finally, in Sec. 2.11 the age of our Galaxy is discussed, which gives a lower limit to the age of the universe.

During the next few years we have the exciting prospect of space projects like the Hipparcos satellite and the Hubble space telescope, which will enormously enhance our knowledge of the distance scale both within our Galaxy and on the larger, cosmological scale.

2.1 *Trigonometric parallax*

The most fundamental of our distance measurement techniques outside the solar system is the method of trigonometric parallax. As the earth goes around the sun, the apparent directions, or positions, of the stars on the sky change with the season in a way that depends on the star's distance and on its direction on the sky.* This is the phenomenon of parallax. If the direction of the star lies in the plane of the earth's orbit (that is, the ecliptic), the star will appear to move backward and forward on a line in this plane by an amount of $\pm\pi$, where $\pi = r_E/d$ rad. Here d is the distance of the star and r_E the radius of the earth's orbit around the sun, both measured in, say, meters. If the star lies perpendicular to the plane of the earth's orbit, at the ecliptic pole, the star will appear to move in a circle of angular radius $\pi = r_E/d$. If the star lies at ecliptic latitude b^E, then it will appear to move in an ellipse of semimajor axis $\pi = r_E/d$ and of semiminor axis $\pi \sin b^E$ (Fig. 2.1). The quantity π is called the trigonometric parallax of the star, and is usually measured in seconds of arc(″); so $\pi = 2.06 \times 10^5 r_E/d''$.

* The positions of the stars have to be corrected automatically for the effects of aberration (see Sec. 1.3, p. 6), the precession of the equinoxes, and nutation (see Sec. 1.7, p. 23), which affect all stars, regardless of their distance.

FIGURE 2.1

Geometry of trigonometric parallax. (a) As the star is observed from different positions E, E′, E″, E‴ *on the earth's orbit around the sun* S, *the apparent direction of the star changes, and the star traces out an ellipse on the sky (c). (b) View from the direction* EE″. r_E—*radius of earth's orbit;* d—*distance of star;* π—*trigonometric parallax;* b^E—*ecliptic latitude.*

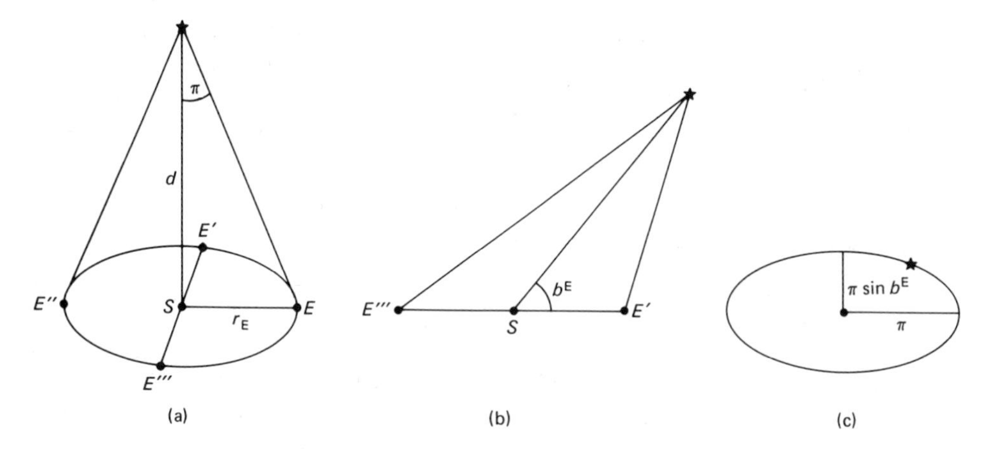

The parallactic motion of stars was predicted by Copernicus's theory. From the absence of motions obvious to the naked eye he concluded that the stars must be very distant. The greatest of the pretelescopic observers in Europe, Tycho Brahe, looked very hard for this effect, but it was too small for him to see with the naked eye, even with the aid of his beautifully constructed instruments. As he seriously underestimated the distances of the stars, since due to a physiological effect they look slightly fuzzy to the naked eye, he concluded that Copernicus was wrong about the earth's motion, though he accepted that the other planets went around the sun.

It was not until 1838–1939 that the first accurate measurements of the parallax of stars were made. First, in 1838, Friedrich Wilhelm Bessel at the Königsberg Observatory in Prussia found $\pi = 0''.31 \pm 0''.02$ for the star 61 Cygni, close to the modern value of $0''.292 \pm 0''.004$. Then, in 1939, Thomas Henderson found $1''.16 \pm 0''.11$ for α Centauri (modern value $0''.760$). Friedrich Struve had found $0''.125$ for the star Vega (α Lyrae) in 1937, close to the modern value of $0''.123$, but he did not stick to this value, and in 1939 he published the result $\pi = 0''.2619 \pm 0''.254$, not accurate enough to constitute a significant measurement.

Work on measuring parallaxes proceeded slowly. In 1878 17 stellar parallaxes were known, and by 1908 about 100. α Centauri has the largest parallax known, and is the nearest star system to the sun. It is a triple system, in fact, and the nearest member to the sun at present is the dwarf star Proxima Centauri. Only 39 star systems are known with $\pi \geq 0''.2$ (see App. A.1). The *Yale Parallax Catalog,* published in 1952 [J1], gives parallaxes for 5822 stars, and about a further 1000 stars have been measured since then. However, for the majority of these stars the parallaxes are too small and not known accurately enough to give reliable distances. There are 1180 stars in the Yale catalog which have parallaxes at least 5 times the uncertainty quoted in the catalog, and this corresponds roughly to those stars with $\pi > 0''.04$ [H4]. Parallaxes smaller than this do not yet give reliable distances to an individual star. However, if we average the results for many stars, we can find a reliable *mean* distance for them. This is useful for stars in a cluster, for example, or for stars of one particular type. More recently, in 1978, Walter Gleise has estimated that there are in all over 2000 stars with $\pi > 0''.044$, of which 1500 have been measured [G4].

Astronomers like to measure the distance to stars in terms of the parallax they would give. 1 *parsec* is defined as the distance at which the star would give a parallax of 1 arc sec. (The term is a contraction of *par*allax of 1 *sec*ond of arc.) The distance d in parsecs is therefore $d = 1/\pi('')$. In more physical units,

$$\begin{aligned} 1 \text{ pc} &= 206{,}265 \text{ A.U.} \\ &= 3.08568 \times 10^{16} \text{ m} \\ &= 3.2615 \text{ lt-yr.} \end{aligned}$$

Of course the parsec, the meter, and the light-year are all highly earth-centered measures of distance, but the most satisfactory is the light-year, which is based on light-travel time. It would have been better for astronomy if the kilo-light-year, mega-light-year, and giga-light-year had come into use in preference to the kiloparsec (10^3 pc), megaparsec (10^6 pc) and gigaparsec (10^9 pc). But an advantage of the parsec was that it gave distances which did not depend on the knowledge of the astronomical unit, a quantity determined with high precision only during the past 20 years.

Once we have measured the parallax of a star and thus know its distance, we can convert the observed brightness of the star to the luminosity, or total energy output, using the inverse-square law for radiation,

$$L = 4\pi d^2 S$$

where L is the luminosity in watts, d is the distance in meters, and S is the radiant flux, or energy received per second per unit area, in watts per square meter (W m^{-2}). In practice we can never measure the total radiant flux at all wavelengths with a single detector. Instead we study the radiation within quite a narrow band of wavelengths, or frequencies. Ground-based astronomers are able to study the radiation from astronomical sources in the radio, near infrared, visible, and near ultraviolet bands. To study sources in the far infrared, far ultraviolet, or X-ray bands, airborne or satellite-borne telescopes are needed. The range of wavelengths is often deliberately narrowed down, using a filter, to gain extra spectral resolution.

The distribution of radiation from a source with wavelength or frequency, called the spectrum of the source, can tell us a great deal about the nature of the source and the physical processes going on in it. From the spectrum of a star, for example, we can measure the temperature of its surface layers. The pattern of absorption and emission lines in the spectrum depends strongly on the temperature of the star (see p. 36) and also, to a lesser extent, on its surface composition. Let $S_\nu \, d\nu$ be the flux of radiation received in the frequency range $(\nu, \nu + d\nu)$ and let $L_\nu \, d\nu$ be the energy emitted by the star per second in the same frequency range. S_ν and L_ν are called the monochromatic flux, or flux density, and the monochromatic luminosity, respectively. Then it follows from the inverse-square law that

$$L_\nu = 4\pi d^2 S_\nu$$

where $d\nu$ has been dropped from both sides of the equation.

The unit of brightness used by optical and infrared astronomers is the *magnitude* m, defined by

$$m = A - 2.5 \log S \tag{2.2}$$

where A is a constant that depends on the wavelength band being used. (It simply gives the zero point for the magnitude scale in the band.) This choice of units seems peculiar until we understand its historical origin. In his star catalog, Hipparcos subdivided the stars visible to the naked eye according to their brightness, first magnitude for the brightest to sixth magnitude for the faintest. When quantitative measurement techniques were developed in the nineteenth century, it was found that each magnitude corresponds to a decrease of about 0.4 in the logarithm of the flux, so the magnitude scale was defined to agree approximately with the traditional estimates. One advantage of the magnitude is that it can be used as a relative unit by assigning the magnitude of a standard bright star (usually Vega).

For observations using a standard filter, the V (visual, or yellow-green) filter, say,

$$m_V = A_V - 2.5 \log \left\{ \int \phi_\nu S_\nu \, d\nu \right\} \tag{2.3}$$

where ϕ_ν is the fraction of the incident light transmitted by the atmosphere, the telescope, and the filter to the detector.

Radio astronomers use a more physically based unit for S_ν, the *Jansky*, named after the pioneer radio astronomer Karl Jansky. This is defined as 10^{-26} W m^{-2} Hz^{-1}.

The unit of luminosity used by optical and infrared astronomers seems even more peculiar, the *absolute magnitude M*, which is the magnitude the star would have at a distance of 10 pc. (It would have been more convenient to have used a distance of 1 pc.) Thus

$$\begin{aligned} M &= m - 5 \log \left(\frac{d}{10 \text{ pc}} \right) \\ &= m - 5 \log \left(\frac{0''.1}{\pi} \right). \end{aligned} \tag{2.4}$$

The above equation shows that the difference between the apparent magnitude of a star and its absolute magnitude, $m - M$, is a measure of its distance. A logarithmic measure of distance is extremely convenient in astronomy, where such enormous ranges of distance are found. The quantity $m - M$ is called the distance modulus and is written as

$$\mu_0 = m - M = 5 \log \left(\frac{d}{10 \text{ pc}} \right).$$

I shall give distances in terms of parsecs and distance moduli in this book.

The *bolometric flux* (and the corresponding *bolometric magnitude* m_{bol}) is defined to be the total radiant flux received from a source integrated over all frequencies. For sources with a standard spectrum, for example, normal stars of a particular type, knowledge of this spectrum allows us to convert the magnitude in a single frequency band to the bolometric magnitude. The quantity $m_{\text{bol}} - m$, the difference between the bolometric and apparent magnitudes, is called the *bolometric correction*, appropriate for a particular type of source and for a particular frequency band (the visual band if not specified).

Having found the intrinsic luminosities of the nearest stars, it is natural to see whether a star's luminosity is correlated with any other property. The simplest property to observe is the color of the star. Astronomers define color in terms of the ratio of the fluxes at two different wavelengths, say, S_B/S_V, where S_B and S_V are the fluxes at the blue and visual (yellow-green) wavelengths. If we take the logarithms of this ratio, we can see from Eq. (2.2) that the color is given by the difference of the magnitudes at the two wavelengths. The most widely used color index in optical astronomy is the B-V index, the difference in the magnitudes in the blue and visual standard photometric bands (Box 2.1).

Gleise has made a plot of the absolute visual magnitude M_V against the B-V color for nearly 500 stars within 22 pc (Fig. 2.2). These stars are all those with accurate trigonometric parallaxes greater than 0″.044, for which the B and V magnitudes are known. We see that there is a strong concentration of stars in a band running from the upper left to the lower right (the *main sequence*), a second parallel band to the lower left (the *white dwarf* or *degenerate* branch), and a few scattered stars at the top center of the diagram (the *giant* branch). Now the color of the star gives an indication of its surface temperature: the bluer the star, the hotter the temperature. The correlation would be exact if stars had perfect blackbody spectra, since the flux

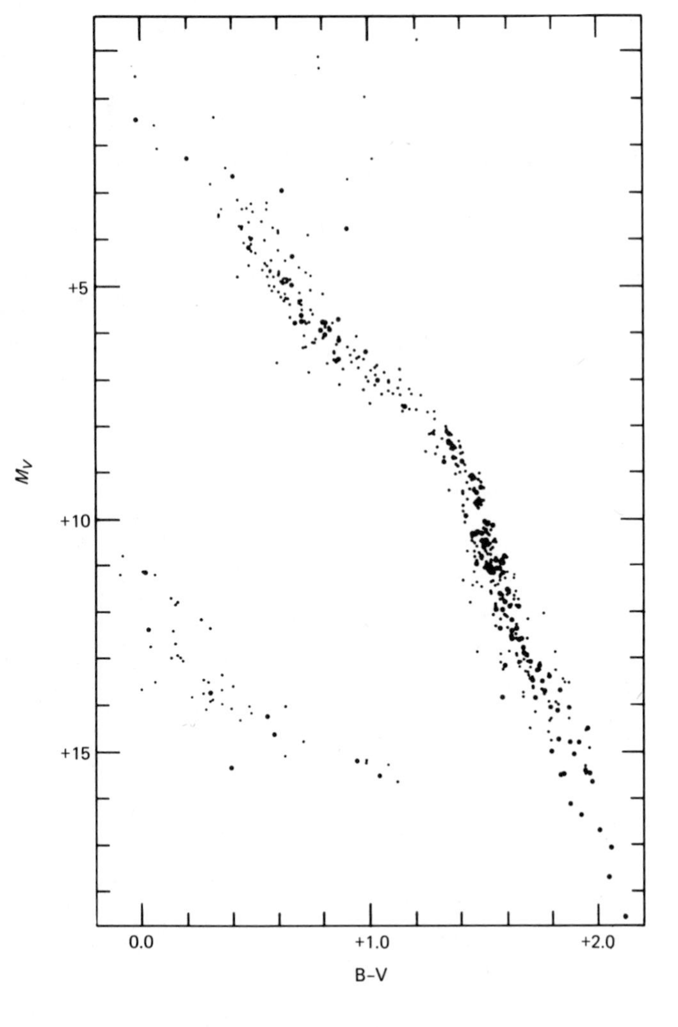

FIGURE 2.2

HR diagram. Absolute visual magnitude M_V *versus B-V color for stars with trigonometric parallax* $\leq 0''.044$*. Note the main sequence running from upper left to lower right, the white dwarf locus to the lower left, and a number of stars on the giant branch in the upper center. Smaller dots denote less accurate data. (From [G4].)*

BOX 2.1 MAGNITUDES AND COLORS OF STARS

Until the 1950s most astronomical photometry was done photographically with two systems: the blue-violet-sensitive international photographic system m_{pg} and the photovisual system m_{pv} whose wavelength sensitivity simulates that of the eye. Unfortunately the photographic plate has a nonlinear response to different levels of light intensity, and each plate has to be calibrated empirically by comparison with standard stars. Furthermore the dynamic range of a plate is only about a factor of 20 in intensity.

Astronomical photometry was revolutionized with the advent of photoelectric photometers. These instruments are strictly linear, have enormous dynamic range, and are capable of a factor of 10 better precision than photographic plates, yielding magnitudes accurate to ±0.01 mag and magnitude differences often accurate to ±0.002 mag. The standard photometric system today is the ultraviolet-blue-visual (UBV) system of H.L. Johnson and W.W. Morgan, and the characteristics of this and other widely used systems are given below.

Filter characteristics of astronomical photometry systems.

System	*Filter*	*Wavelength of peak transmission* λ_0 (Å)	*Full-width to half intensity* $\Delta\lambda_{1/2}$ (Å)
UBV (Johnson–Morgan)	U	3650	700
	B	4400	1000
	V	5500	900
Six colors (Stebbins–Whitford–Kron)	U	3550	500
	V	4200	800
	B	4900	800
	G	5700	800
	R	7200	1800
	I	10,300	1800
Infrared (Johnson)	R	7000	2200
	I	8800	2400
	J	12,000	3800
	K	22,000	4800
	L	34,000	7000
	M	50,000	12,000
	N	104,000	57,000
uvbyβ (Strömgren–Crawford)	u	3500	340
	v	4100	200
	b	4700	160
	y	5500	240
	β	4860	30, 150

SOURCE: From [M1].

ratios for any two wavelengths would then be the corresponding ratios for the Planck blackbody spectrum at the appropriate temperature. The deviations from blackbody spectra are not enormous, though, except for very hot and very cool stars, and we can say that stars with B-V = 0, 1, and 2 have surface temperatures of about 10,000, 4000, and 2000 degrees Kelvin (K), respectively. We can immediately deduce two important facts about the stars of Fig. 2.2:

1 The stars on the white dwarf branch, being 10^4 times less luminous than the stars of the same color (and, hence, temperature) on the main sequence, are about 100 times smaller.

2 The stars above the main sequence (giants) must be bigger than the stars with the same color on the main sequence.

We shall see in Sec. 2.7 that stars on the main sequence are supplying their energy by converting hydrogen to helium, that the giants are stars in which the hydrogen has begun to be exhausted, and that the white dwarfs are dying stars with no nuclear energy sources left.

Before accurate photometric methods were developed, astronomers had found another way to classify stars and estimate their surface temperatures. When the spectrum of a star is examined with high spectral resolution, many spectral lines are found in either absorption or emission superposed on the star's continuous radiation, or *continuum.* Since the excitation of the atoms responsible for the spectral lines depends strongly on temperature, it is natural that the spectra of hot and cool stars should look very different. The different spectral types used by astronomers are described in Box 2.2. The strange order of the letter codes (remembered by Henry Norris Russell's mnemonic: Oh Be A Fine Girl/Guy, Kiss Me Right Now, Sweetheart) arose because the astronomers of the Harvard Observatory who invented the classification in the 1890s did not then know what the temperatures of the stars were and had arranged the types alphabetically according to the complexity of the spectra.

Fig. 2.3 shows a plot of the absolute visual magnitude M_V against the spectral type for stars with trigonometric parallaxes greater than $0''.044$ ($d < 22$ pc). Again we see the main sequence and a few giants. The surface temperatures are about 10,000 K for type A0, 6000 K for type G0, 3500 K for type M0, and 2500 K for type M6. The significance of Figs. 2.2 and 2.3, which are known as the Hertzsprung–Russell (HR) diagram, is more fully discussed in Sec. 2.7.

What of the future prospects for the method of trigonometric parallax? Comparison of the trigonometric parallaxes in the 1952 Yale catalog for the same stars measured by different observatories (Allegheny, McCormick, Yale, Cape, Greenwich, Mount Wilson, Yerkes, and Sproul) showed that there were small systematic differences from one observatory to another, on the order of $\pm 0''.005$ (see the review by A. Heck [H4]). A big program of parallax measurements has been undertaken during the 1970s at the U.S. Naval Observatory using a new telescope built especially for the purpose, and these parallaxes are of higher accuracy than the older measurements. Accurate programs are also being pursued at several other observatories (Lick, Yerkes, and Greenwich). The planned Hipparcos satellite (an acronym for *h*igh-*p*recision *par*allax *co*llecting *s*atellite), due to be launched in the

BOX 2.2 SPECTRAL TYPES OF STARS

The MK system of spectral types was developed by W.W. Morgan and P.C. Keenan from the earlier Harvard system. The principal characteristics of the spectral classes are as follows:

Spectral class	*Spectral features*
O	**He II lines visible; lines from highly ionized species, for example, C III, N III, O III, Si IV; H lines relatively weak; strong ultraviolet continuum**
B	**He I lines strong, attain maximum at B2; He II lines absent; H lines stronger; lower ions, for example, C II, O II, Si III**
A	**H lines attain maximum strength at A0 and decrease toward later types; Mg II, Si II strong; Ca II weak and increasing in strength**
F	**H weaker, Ca II stronger; lines of neutral atoms and first ions of metals appear prominently**
G	**Solar-type spectra; Ca II lines extremely strong; neutral metals prominent, ions weaker; G band (CH) strong; H lines weakening**
K	**Neutral metallic lines dominate; H quite weak; molecular bands (CH, CN) developing; continuum weak in blue**
M	**Strong molecular bands, particularly TiO; some neutral lines, for example, Ca I, quite strong; red continua**
C(R,N)	**Carbon stars; strong bands of carbon compounds C_2, CN, CO; TiO absent; temperatures in range of classes K and M**
S	**Heavy-element stars; bands of ZrO, YO, LaO; neutral atoms strong as in classes K and M; overlaps these classes in temperature range**

late 1980s, is designed to determine accurate parallaxes as small as 0″.01 and will represent an enormous advance in our measurement of stellar distances. It will measure the trigonometric parallaxes, proper motions (see Sec. 2.3), and positions of about 100,000 stars, and some of the results expected from the mission are:

1 At least five or six "open" star clusters will now fall within the range of the trigonometric parallax method (see Sec. 2.4).

2 HR diagram positions of many O and B stars and red giants, not accurately known at present.

3 Accurate information on the age and chemical composition of clusters and groups of stars from their positions in the HR diagram (see Sec. 2.7).

4 Detection and measurement of a few thousand presently unknown binary star systems (see Sec. 2.2).

The classes are further subdivided into decimal subclasses, running from 0 at the hot end to 9 at the cool end; for example, B0, B1, B2, . . . , B9. In addition, stars are assigned to luminosity classes, mainly on the basis of the widths of the spectral lines, as follows:

Designations	*MK luminosity class*
Ia-0	**Most extreme supergiants**
Ia	**Luminous supergiants**
Iab	**Moderate supergiants**
Ib	**Less luminous supergiants**
II	**Bright giants**
III	**Normal giants**
IV	**Subgiants**
V	**Dwarfs**
VI	**Subdwarfs**

Extreme metal-poor stars lying below the main sequence in the HR diagram are called subdwarfs, denoted by the prefix sd. White dwarfs are denoted by the prefix w (for example, wA, wG) or by the symbol D followed by subdivisions A, B, or C.

The suffices p, e, f, m denote stars which are peculiar, stars with unexpected emission lines, O stars with He II and N III emission, and A stars with metallic lines, respectively. The symbols WC and WN (or sometimes WR) denote Wolf–Rayet stars, which are O, B, or A stars that show exceptionally strong, broad emission lines.

SOURCE: From [M1].

Hipparcos can be expected to make a major impact on our knowledge of the first rungs of the cosmological distance ladder. It is hoped that even more accurate parallax measurements, perhaps with errors as small as 0″.0005, can be made with the Hubble space telescope, also due to be launched in the 1980s. The number of stars that can be measured will be limited by the many other demands on the space telescope, but the higher accuracy and fainter magnitudes that can be reached make this an important complement to Hipparcos. Table 2.1 summarizes some present and future parallax programs.

Radio astronomers have already achieved positional accuracies comparable with the best ground-based optical results, and have the potential to achieve even higher accuracy (see the review by C.C. Counselman [C6]). It is hoped to link radio and optical methods by using Hipparcos to measure the few known stars which are detected as radio emitters and to measure stars near the radio reference sources on the sky.

TABLE 2.1

Summary of past, present, and future parallax programs.

Status	*Name*	*Limiting apparent magnitude*	*Number of stars*	*Measurement period*	*Precision*	*Mean distance reached with* $\sigma/\pi_0 < 0.15$	*Remarks*
Published	L.F. Jenkins (1952, 1963)	Most ≤ 13	5822 + 600		On the average 0″.009	15 pc	760 stars with $\sigma/\pi_0 < 0.15$
In progress	USNO Flagstaff	75% ≥ 13		300 stars in 10 years	0″.004 for 40 plates	40 pc	5 stars fainter than $m_V = 16$; UBV photometry also measured
	Herstmonceux	$16 \leq m \leq 18$	20 projected	2 stars in 3 years			
Projects	Palomar	20	A large number planned	4 stars in 6 months	0″.023 for 20 plates	6 pc	Precision twice as good for small areas
	U.K. Schmidt	19	A large number planned	1 plate of 30,000 stars in 3 months with GALAXY measuring machine	0″.01	15 pc	Planned measurement of all stars in some selected areas with large *b*
	Hipparcos	13	100,000	2 years	0″.002	80 pc	Planned for 1986
	Space telescope	20			0″.0005	300 pc	Only relative parallaxes, projected for 1985

SOURCE: Compiled from [H4].

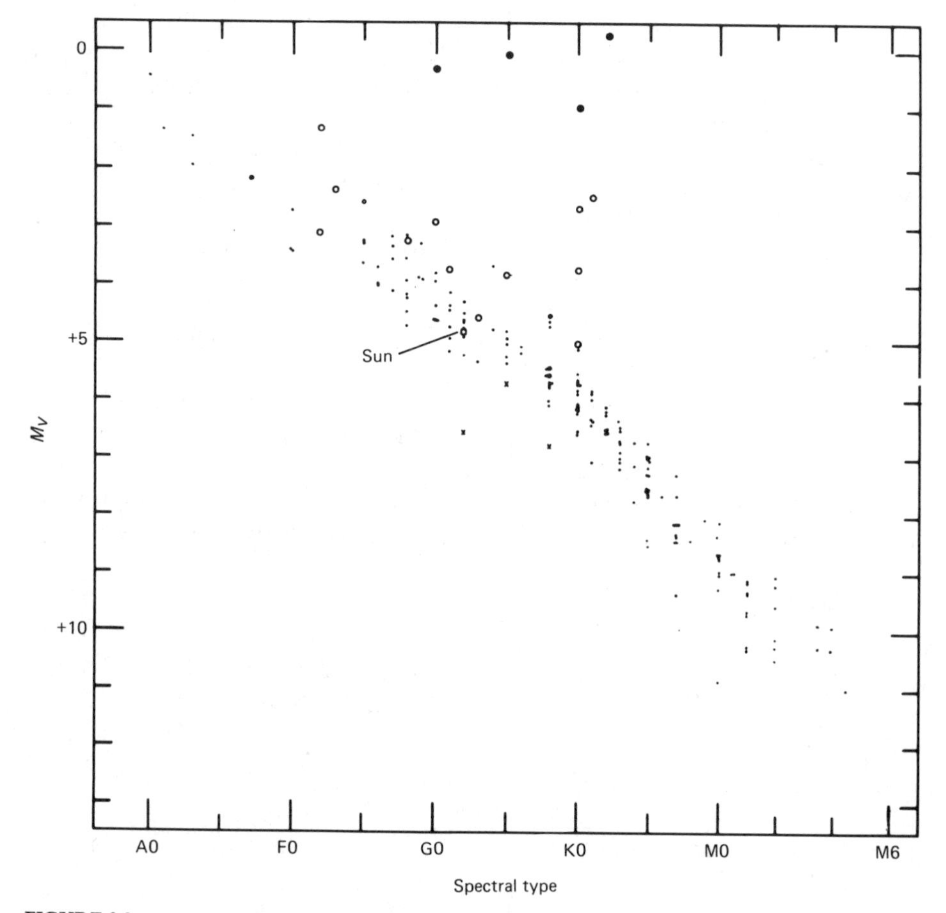

FIGURE 2.3

HR diagram. Absolute visual magnitude M_V *versus spectral type for stars with trigonometric parallax* $\leq 0''.044$. *The white dwarfs have been excluded from this figure. Different symbols denote different luminosity classes: filled circles—III, open circles—IV, dots—V, crosses—VI. (From [G4].)*

2.2 *Binary and multiple stars*

One of the most striking facts about the stars within 5 pc of the sun (App. A.1) is the large proportion of double or triple systems. Half of the 39 systems are known multiple systems, and other stars may have companions too small to be observed. Of the 60 known stars within 5 pc, at least 60% are in binary or multiple systems. We have already seen that the nearest star system to the sun, α Centauri, is a triple system.

There are three ways in which double or multiple systems are discovered. The first is by simply looking through a telescope and seeing that what to the naked eye looked like one star is in fact two or more. These are *visual binaries.* The first double star to be discovered was Mizar, or ζ Ursae Majoris, the central star in the handle of the Plow (or Big Dipper), studied through the telescope by Gianbattista Riccioli in 1650. In 1782 William Herschel published the first catalog of 269 double stars found in this way, in which the separations between the stars were smaller than 2 minutes of arc. Of course some of these were simply chance line-of-sight coincidences. With stars separated by a few seconds of arc or less, however, there cannot be much doubt that a real physical association is involved. This can be demonstrated by painstaking observations over many years, during which time the stars can actually be seen to orbit around each other.

One of the first double stars to be discovered, though its significance was not then understood, was Algol, the demon star, noticed by Geminiano Montanari in 1669. Regularly every 68 hours the star dims to one third of its normal brightness for a few hours and then brightens up again. This is an example of an *eclipsing binary,* the second way that double stars are found. Here the dimming occurs because the smaller and brighter of the two stars is eclipsed by its larger but fainter companion. To see eclipses, our line of sight must be close to the plane of the stars' orbit (Fig. 2.4).

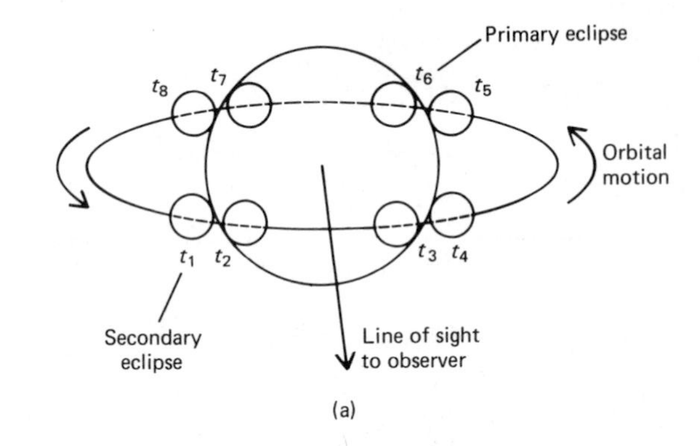

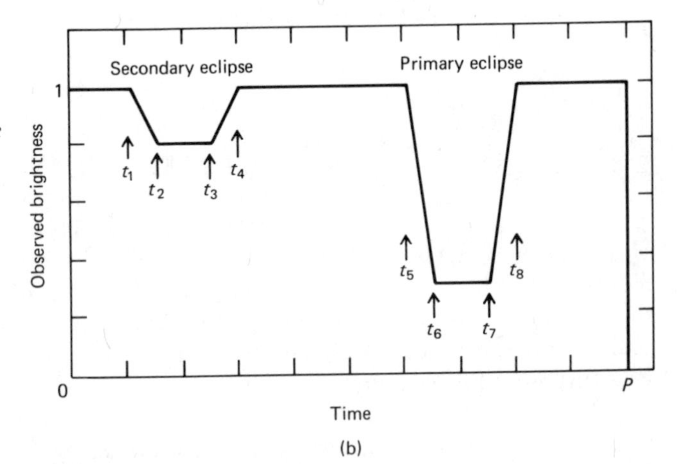

FIGURE 2.4

Eclipsing binary system. If we view a binary system from a direction close to the plane of its orbit, eclipses will occur as the two stars pass in front of each other. (a) Geometry of the system and definition of times t_1–t_8. *(b) Schematic light curve for the system, that is, brightness of the system (in units of the total brightness of both stars) as a function of time over one orbital period* P. *In the case illustrated the smaller star has a greater surface brightness and contributes most of the light. (From [M1].)*

The third method involves studying the spectrum of a binary unresolved by the telescope. Sometimes two distinct sets of spectral lines can be seen, moving backward and forward in wavelength due to the Doppler shift, as the orbital motion of the stars moves them first toward us and then away from us. This phenomenon was first noticed by Edward Pickering in 1889 in the spectrum of the binary star Mizar, already mentioned above. Such stars are called *spectroscopic binaries.* Even if only one set of lines is seen in the spectrum, their oscillatory motion back and forth in wavelength shows that the star is orbiting around an unseen companion. App. A.2 lists the properties of a few of the best studied visual, spectroscopic, and eclipsing binaries.

Binary systems provide the only good estimates of stellar masses. To see how the distances of binary systems are measured and the masses of the stars are inferred, let us now consider the dynamics of a double-star system. In general each star moves in an elliptical orbit around the common center of mass of the system. In fact, our solar system can be thought of, dynamically, as a binary system, because the mass of Jupiter is so much greater than that of the other planets. The center of mass of the solar system lies not at the center of the sun, but near the surface of the sun close to the line joining the center of the sun to Jupiter. The sun and Jupiter each orbit around this point once every 11.86 years. As with the planets of the solar system, most stars in binary systems move in approximately circular orbits. (This becomes a poorer approximation for binaries with periods longer than 100 days.) For illustrative purposes we shall assume the orbits are exactly circular, which simplifies the analysis greatly.

How much we can learn about a binary system depends on whether it is a visual, eclipsing, or spectroscopic binary (it can be all three). Suppose A and B are two stars of mass M_1 and M_2, orbiting about their center of mass C, where $AC = r_1$, and $AC = r_2$. Then

$$\frac{r_1}{r_2} = \frac{M_1}{M_2}. \tag{2.5}$$

If the inclination angle i [Fig. 2.5(a)] is zero, so that we are looking from a direction normal to the plane of the orbit, the motion of the stars on the sky will look circular [Fig. 2.5(b)]. Then if d is the distance to the stars from the earth, the angular radii of the orbits θ_1 and θ_2 are given by

$$\theta_1 = \frac{r_1}{d}, \qquad \theta_2 = \frac{r_2}{d}. \tag{2.6}$$

If the inclination angle i is not zero, then the orbits on the sky will look elliptical [Fig. 2.5(c)] due to the tilting of the plane of the orbit away from an orientation normal to our line of sight. However, the angular radii of the major axes of the orbits will still be θ_1 and θ_2, as above. Thus for a visual binary we can determine the ratio

$$\frac{M_1}{M_2} = \frac{\theta_1}{\theta_2}. \tag{2.7}$$

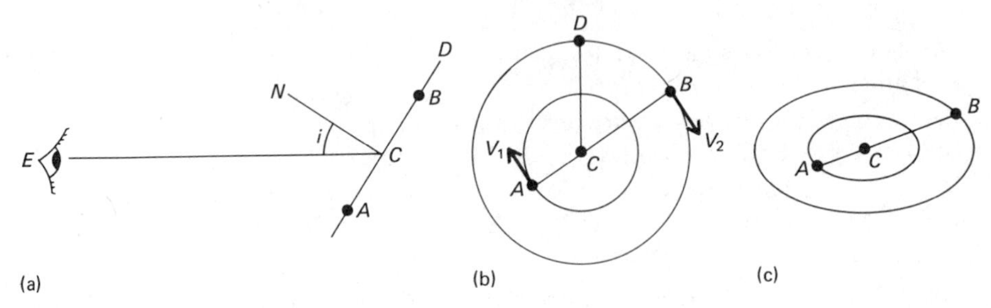

FIGURE 2.5

Geometry of a noneclipsing binary system. (a) As seen from a point in the plane of the stars' orbit. (b) As seen from a point N *on the line normal to the plane of the stars' orbit, through their center of mass* C. *(c) As seen from the earth* E. A, B*—the two stars;* v_1, v_2*—their velocities;* i*—inclination angle of orbit.*

But since we can also measure the period of the orbit P, we can use Kepler's law to obtain, in this case,

$$P^2 = \frac{4\pi^2(r_1 + r_2)^3}{G(M_1 + M_2)}. \tag{2.8}$$

From Eqs. (2.6)–(2.8) we can now determine the individual masses M_1 and M_2 for the two stars, provided we know the distance d.

For a spectroscopic binary we can determine, from the maximum red shift and blue shift of the spectral lines in each star, the quantities $V_1 \sin i$ and $V_2 \sin i$. The velocities are related to the orbital radii r_1 and r_2 and the period P by

$$V_1 = \frac{2\pi r_1}{P}, \qquad V_2 = \frac{2\pi r_2}{P}. \tag{2.9}$$

Eq. (2.8) can then be written

$$(M_1 + M_2) \sin^3 i = \frac{P(V_1 \sin i + V_2 \sin i)^3}{2\pi G} \tag{2.10}$$

where the quantities on the right-hand side can be determined entirely from spectroscopic observations. If the orbit is elliptical, with eccentricity e, an additional factor $(1 - e^2)^{3/2}$ appears on the right-hand side. Since $M_1/M_2 = r_1/r_2 = V_1/V_2$, we can determine $M_1 \sin^3 i$ and $M_2 \sin^3 i$, but since i is in general unknown, we obtain only lower limits on M_1 and M_2. However, if the binary is also an eclipsing system, then we know that i is close to 90°, and we can obtain good estimates of M_1 and M_2. From the duration and depth of the eclipses we can also deduce the radii of the stars relative to the orbital radii and get an accurate estimate of i.

Altogether 40,000 visual binaries are known, but only 300 of these have well-determined orbits. About 600 spectroscopic binaries and 100 eclipsing binaries with well-determined orbits are known. In his 1980 review on stellar masses, D.M. Popper gives accurate masses for about 140 stars, most of them within 300 pc of the sun [P2]. If we know the distances and masses for a number of stars, we can calculate the number of stars of different masses per cubic parsec in the neighborhood of the sun. The mass function $\eta(M)$ is defined by letting $\eta(M)\,dM$ be the number of stars per cubic parsec in the mass range M to $M + dM$. The function $\eta(M)$ is found to increase steeply as we go from higher masses to lower masses, and from the steepness of this distribution we can immediately deduce that low-mass stars ($M < M_\odot$, the mass of the sun) are not only the most numerous, but they also contribute most of the average local mass density due to stars.

For nearby binary stars for which we can deduce the distance, and thus their masses, we can make a plot of luminosity against mass. This is shown in Fig. 2.6 for stars on the main sequence. For these stars there is quite a tight correlation between mass and luminosity of the form

$$L \propto M^4, \qquad \text{for } 0.4 \lesssim \frac{M}{M_\odot} \lesssim 10. \tag{2.11}$$

The largest accurately determined stellar mass is 26.9 $M_\odot$ for V382 Cyg, where $M_\odot$ is the mass of the sun, but by extrapolating Fig. 2.6 to more distant, luminous stars it can be inferred that stars of up to 60 – 100 $M_\odot$ exist. It has even been suggested that the giant 30 Doradus HII region in the Large Magellanic Cloud may be illuminated not by a cluster of stars, but by a single star of 1000 $M_\odot$ [C3]. The lowest accurately measured stellar mass is 0.11 $M_\odot$ for UV Cet. From theoretical calculations the lowest mass star that can fuel itself by fusing hydrogen to helium is 0.07 $M_\odot$. Objects in the range of 0.01 – 0.07 $M_\odot$, though they do not become true hydrogen-burning stars, do have a brief episode in which they burn up the deuterium made in the big bang (see Sec. 5.1). Objects of mass less than 0.01 $M_\odot$ never become stars in any sense, though they will radiate the energy stored in their radioactive elements and also the heat generated as a result of their gravitational contraction. Jupiter, at 0.001 $M_\odot$, can be thought of as a failed star.

In recent years considerable effort has gone into trying to find whether nearby stars have low-mass companions in the range of 0.001 – 0.01 $M_\odot$. If a star could be shown to have a companion at the lower end of this mass range, then we could say that other stars beside the sun have planetary systems. The technique is to look for the tiny wobbles in the position of the star as it orbits around the center of mass of itself and its companion. Several cases of companions of 5 – 10 Jupiter masses are known, but companions on the scale of the solar-system planets have not yet been conclusively demonstrated (see the review by Peter van de Kamp [K1]).

For visual binaries for which we also have spectroscopic determinations of the orbital velocities, we can deduce both the masses of the star and their distance. Alternatively if we use the mass – luminosity relation to estimate the masses of the stars in a binary system, then the distance of the stars can be estimated. Distances determined in this way are called *dynamic parallaxes.* A comparison of dynamic (π_d)

FIGURE 2.6

Mass–luminosity relation for stars with well-determined masses. $L_\odot$, $M_\odot$*—luminosity and mass of sun. The straight line corresponds to* $L \propto M^4$. *(Data from [P2].)*

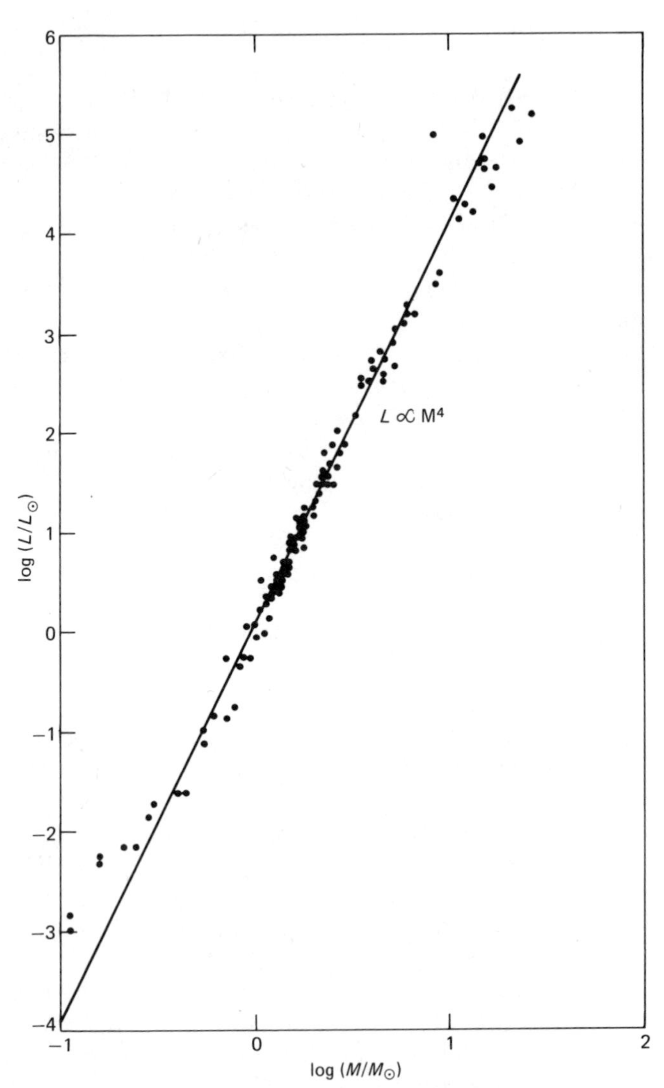

and trigonometric (π_{tr}) parallaxes shows that, over the range for which trigonometric parallaxes are reliable ($\pi_{tr} > 0''.05$), there is good agreement between the two [H4]. Dynamic parallaxes can give reliable distances up to about 200 pc.

2.3 The motion of nearby stars

Looking up at the sky from a mountain top on a clear night it is easy to visualize the fact that the fainter and more distant stars are distributed in a thin disk, delineated by the Milky Way. The basic motion of the stars in this disk is in a circle about the center of the Galaxy, rather like the planets of the solar system. However, each star experiences the gravitational pull of every other star, so that there is an additional random motion on the order of 10 km s^{-1} superposed on the circular motion. The net result for an observer on earth is that the stars near the sun do not stay in a fixed direction, or position, on the sky, but appear to move across the sky relative to more distant stars. This relative motion is called the *proper motion* of a star. Such motions were first noticed by Edmond Halley in 1718, when he compared the positions of the stars Sirius, Procyon, and Arcturus with the positions measured by the Greeks. Sirius even seemed to have changed its position since the time of Tycho Brahe. Modern measurements show that these three stars move relative to more distant stars by 1″.324, 1″.247, and 2″.285 per year, respectively. Only one other of the 100 brightest stars has a proper motion μ greater than 1″ per year, namely, the nearest star, α Centauri, with $\mu = 3''.678$ per year, though several other fainter stars have higher values than this.

If we know the distance of a star as well as its proper motion, then we can calculate its velocity tangential to the line of sight,

$$V_t = 4.74\,\mu d \tag{2.12}$$

where μ is in seconds of arc per year, V_t is in kilometers per second, and d is in parsecs. The factor 4.74 appears because of the units used. From spectroscopic observations we can deduce the radial component of the star's velocity, V_r by measuring the Doppler shift of the spectral lines. We then know the speed and direction of the star's motion relative to the sun.

The motions of the nearby stars relative to the sun arise from two causes. First, the stars move around the Galaxy as just described. Second, the sun itself has some velocity with respect to the local average motion, and this velocity of the sun is reflected in an opposite motion in every other star. By taking a large well-defined sample of nearby stars and measuring their average velocities with respect to the sun, the random motions of other stars relative to the sun average out, and we are left with the velocity of the local average motion.The frame of reference found in this way is called the *Local Standard of Rest* (LSR). Let us use the cylindrical polar coordinate system (ϖ, θ, z) in which ϖ is the radial distance in the Galactic plane from the center of the Galaxy, θ is the longitude of the star measured in a frame centered on the Galaxy with the galactic plane as equator, and z is the star's height above the galactic plane. Then at the sun this defines the three directions shown in Fig. 2.7. Let (Π, Θ, Z) be the components of a star's velocity in this frame, and let

FIGURE 2.7

Geometry of stellar motion. Definition of the velocity components of a star (Π, Θ, Z): Π—directed along radius vector from galactic center; Z—perpendicular to galactic plane; Θ—in galactic plane perpendicular to Π and in direction of galactic rotation.

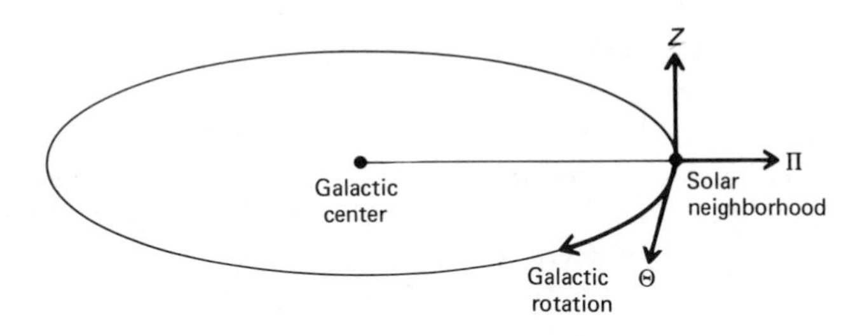

the velocity of the local standard of rest LSR be $(0, \Theta_0, 0)$. Then the star's *peculiar* velocity (that is, its velocity relative to the LSR) has components $u = \Pi$, $v = \Theta - \Theta_0$, $w = Z$, and we expect that the average peculiar velocity for many stars gives $\langle u, v, w \rangle = (0, 0, 0)$. Let the peculiar velocity of the sun with respect to the LSR be (u_0, v_0, w_0). The peculiar velocity of a star relative to the sun is then (u', v', w'), where

$$(u', v', w') = (u - u_0, v - v_0, w - w_0).$$

The mean of this for nearby stars is [D1]

$$\boldsymbol{v}_{\mathbf{LSR}} = \langle u', v', w' \rangle = (10.4, -14.8, -7.3) \text{ km s}^{-1}. \tag{2.13}$$

Hence the velocity of the sun relative to the LSR is

$$u_0 = -10.4 \text{ km s}^{-1}, \qquad v_0 = 14.8 \text{ km s}^{-1}, \qquad w_0 = 7.3 \text{ km s}^{-1}. \tag{2.14}$$

The direction of the sun's motion with respect to the LSR is called the *solar apex*.

We shall be going further into the subject of the large-scale motion of stars in the Galaxy and into the structure of the Galaxy in Sec. 2.9. What we want to pursue now is the possibility of using the proper motions of stars to determine their distances. We saw in Eq. (2.12) that the proper motion expected for a given velocity tangential to the line of sight varies inversely with distance, $\mu \propto V_t/d$. If all stars were at rest with respect to the LSR except the sun, then we would be able to determine their distances precisely by measuring their proper motion μ and radial velocity V_r, since we can work out what radial and tangential velocities a star in a particular direction ought to have if $\boldsymbol{v} = \boldsymbol{v}_{\mathbf{LSR}}$. But in fact stars move randomly with respect to the LSR, so this method can only be used statistically to measure the average distance of groups of stars. This is the method of *statistical* or *secular parallaxes,* which can be used reliably up to distances of 500 pc if care is taken that the sample selected is not biased in some way.

Proper motions have other applications too. Most searches for parallax are concentrated on stars with high proper motion. On the average the parallax π, in seconds of arc, tends to be about one-tenth of the proper motion μ, in seconds of arc per year, for stars selected for their high proper motion. By using surveys of the proper motion of hundreds of thousands of stars over large areas of the sky, W.J. Luyten [L3] has measured the average luminosities of different types of stars by simply assuming that their distance is inversely proportional to their proper motion, $d \propto 1/\mu$. (This neglects the variation of the tangential velocity V_t from one star to another.)

Finally it should be emphasized that not all stars near the sun partake of the basic circular motion around the galactic disk which defines the LSR. A certain fraction of stars have much higher than average velocities relative to the sun, up to hundreds of kilometers per second. These are called the *high-velocity* stars and are believed to come from the *halo* of our Galaxy, an extended, roughly spherical region surrounding the Galactic disk.

2.4 Distances to nearby clusters of stars

Even a modest telescope reveals that the Pleiades cluster contains many more stars than the six or seven visible to the naked eye. And toward the end of the eighteenth century William Herschel demonstrated the existence of many such clusters of stars. To establish the distance of a cluster is immensely valuable for it gives us immediately the absolute magnitude of every star in the cluster. It is also natural to suppose that the stars of a cluster were all born at the same time, so that a cluster gives us a snapshot of the evolution of stars of different masses.

One of the nearest clusters, and one of the key rungs on the cosmological distance ladder, is the Hyades cluster (Fig. 2.8). This is an example of an *open* or *galactic* cluster, a loose aggregate of a few hundred stars found in the disk of the Galaxy. The distance of the Hyades can be measured by a variety of methods, and it then serves as the benchmark for finding the distance to hundreds of other open clusters.

FIGURE 2.8

Composite photograph of central portion of Hyades cluster. The bright star on the left is Aldebaran, not in fact a cluster member. (Courtesy of Peter Malin and the Mount Palomar Observatory.)

Although the Hyades cluster, at 40–50 pc away, is rather beyond the range where the parallax to an individual star is reliable, a reasonable estimate of the distance to the cluster can be found by taking the average for many cluster stars. Dynamical parallaxes can also be found for a number of stars in the cluster which are binaries. Distances obtained in this way are summarized in Table 2.2

TABLE 2.2

Hyades distance modulus.

Method	*Authors*	$\mu_0 = m - M$
1 Proper motions, convergent point	R.B. Hanson (1975)	3.42 ± 0.20
	T.E. Corbin et al. (1975)	3.19 ± 0.04
	R.B. Hanson (1977)	3.32 ± 0.06
	M.S. Carpenter (1977)	3.27 ± 0.1
	S.C. Morris (1979)	3.66 ± 0.2
2 Trigonometric parallaxes	L.F. Jenkins (1963)	3.26 ± 0.14
	O.J. Eggen (1967)	3.23 ± 0.19
	A.R. Upgren (1974)	3.28 ± 0.18
	A.R. Klemola et al. (1975)	3.23 ± 0.15
	B.G.F. Scales (1979)	3.10 ± 0.31
3 Dynamic parallaxes	G. Wallerstein and P.W. Hodge (1967)	3.25 ± 0.12
4 K-line absolute magnitudes	H.L. Helfer (1969)	3.25 ± 0.20
	T.E. Lutz and D.H. Kelker (1975)	3.80
5 Photometric parallaxes		
UBV	E.K.L. Upton (1971)	3.19 ± 0.06
R-I	O.J. Eggen (1969)	3.10 ± 0.06
	A.R. Upgren (1974)	3.22
Stebbins–Whitford system	R.L. Sears and A.E. Whitford (1969)	3.16 ± 0.07
Geneva system	M. Golay (1972)	3.28 ± 0.20
BVr	E.J. Mannery and G. Wallerstein (1971)	3.25 ± 0.10
ubvy (white dwarfs)	J.A. Graham (1972)	3.25 ± 0.08
6 Stellar evolution models	I. Iben (1967)	3.40
	I. Iben and R. S. Tuggles (1972)	3.30
	D. Koester and V. Weidemann (1973)	3.24
Summary	**R.B. Hanson (1980)[H1]**	$\mathbf{3.30 \pm 0.04}$

SOURCE: Compiled from [H1], [H4].

The proper motions of the stars in the cluster can be used in an ingenious technique known as the *convergent-point* method. Suppose the cluster as a whole has a mean velocity relative to the sun which is large compared with the random motions of the stars in the cluster. Then exactly analagously to the effect of perspective, whereby parallel lines appear to meet on the horizon, the proper motions of the stars will all be directed toward a point on the sky, the convergent point, which is the direction parallel to that of the cluster's velocity. Lewis Boss applied this technique for the first time in 1908 to the Hyades, and Fig. 2.9 shows a more recent application of the method by H.G. van Bueren. If λ is the angle between the direction of the cluster and that of the convergent point (Fig. 2.10), then

$$\tan\lambda = \frac{V_t}{V_r} = \frac{\mu d}{V_r}$$

or, in the more usual units,

$$d = V_r \frac{\tan\lambda}{4.74\mu} \quad \text{[pc]} \tag{2.15}$$

where V_r is in kilometers per second and μ is in seconds of arc per year.

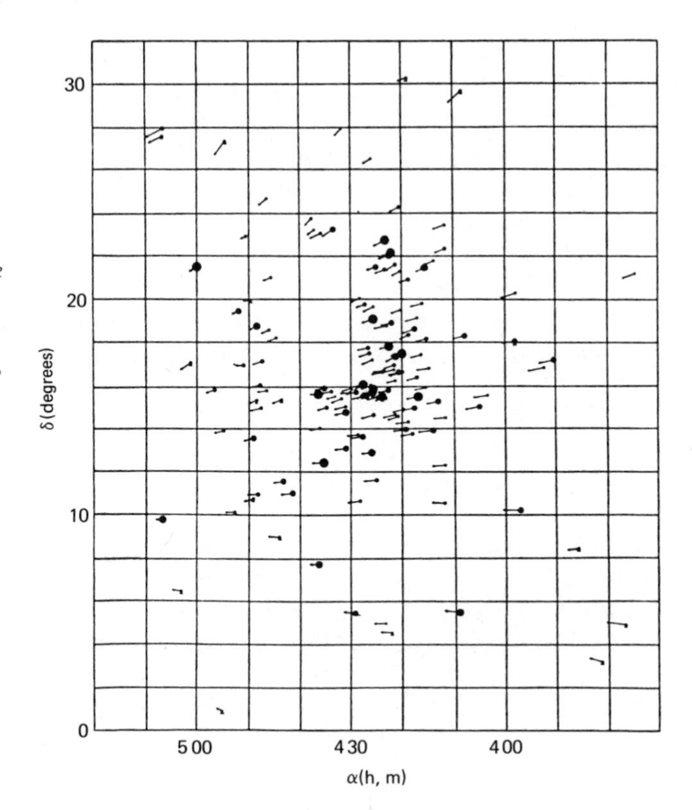

FIGURE 2.9

Determination of Hyades convergent point by van Bueren in 1952. δ—declination; α—right ascension. The size of the dots is a measure of the brightness of the stars. The lines denote the direction and magnitude of the proper motions of the stars, which converge to a point off to the left of the figure. (From [H4].)

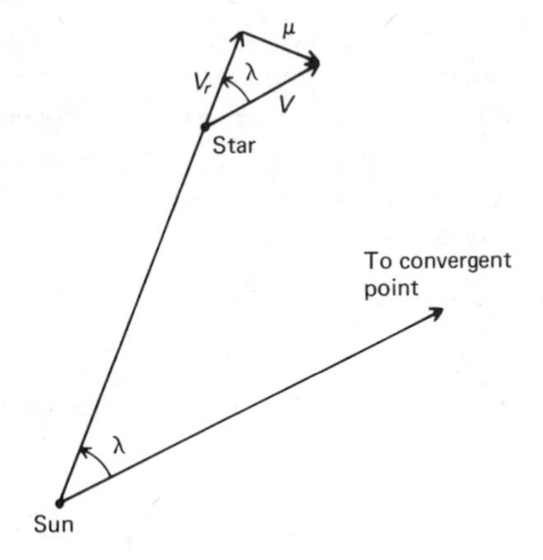

FIGURE 2.10

Definition of angle λ in convergent-point method. μ—(tangential) proper motion of star; V_r*—its radial velocity;* V*—its total velocity. The direction of the convergent point is parallel to* V.

This can be determined for each individual star for which a radial velocity and proper motion are known and then averaged to find the distance of the cluster (Table 2.2).

Unfortunately in practice it turns out to be hard to determine the convergent point with much precision. An alternative way of using the proper motions, known as the *moving-cluster* method, is based on the rate of variation with time $\dot{\theta}$ of the angular diameter θ of the cluster, due to its radial motion relative to the sun. Then it follows that

$$\dot{\theta}d = \theta V_{\mathrm{r}}. \tag{2.16}$$

From the observed proper motions we can deduce

$$\frac{\dot{\theta}}{\theta} = \frac{d\mu_\alpha}{d\alpha} = \frac{d\mu_\delta}{d\delta} \tag{2.17}$$

where μ_α and μ_δ are the components of proper motion in the directions of right ascension α and declination δ. (These are the coordinates of longitude and latitude projected onto the sky.) Then

$$d = V_{\mathrm{r}} \frac{\theta}{\dot{\theta}}.$$

Some of the photometric methods of the next section can also be used to measure the Hyades distance.

In a recent review of all the different methods available, Robert B. Hanson [H1] obtained as the most probable Hyades distance modulus

$$\mu_0 = 3.30 \pm 0.04 \text{ magnitudes (mag)} \qquad (d = 45.7 \pm 0.8 \text{ pc}). \tag{2.18}$$

In light of the measurements summarized in Table 2.2, we can say that the distance modulus of the Hyades almost certainly lies in the range of 3.2–3.4 mag, and the value of 3.3 will be adopted in the remainder of this book. It is worth noting, though, that between 1939 and 1980 the distance of the Hyades cluster has been revised from 35 to 46 pc, an increase of 30%. Such a large change in so basic a rung on the distance ladder gives some idea of how difficult this field is and how uncertain distances on the largest scale must be.

To obtain distances to other open clusters like the Hyades, we can use the method of *main-sequence fitting*. Since we now know the distance of the Hyades, we can plot its HR diagram, the absolute magnitude M_V versus spectral type or color. For a more distant cluster, for example, the Pleiades, we can make a plot of the apparent magnitude V versus spectral type or color. The similarity of the two clusters means that these two distributions should look very similar, apart from the labeling on the vertical (magnitude) scale. In particular the concentration of stars to the main sequence should define two lines of the same shape. If we now superpose the two distributions so that the horizontal (spectral type or color) coordinates are aligned, but displace them vertically until the main sequence for the Pleiades fits onto that for the Hyades, we can then read off $\mu_V = V - m_V$, the apparent visual distance modulus for the Pleiades. We continue successively with the clusters NGC 2362, α Persei, III Cephei, and NGC 6611 to obtain the composite main sequence shown in Fig. 2.11. If the stars in the cluster are dimmed by extinction due to dust, say, by A_V magnitudes at visual wavelengths (see Sec. 2.8), then μ_V has to be corrected by this amount to yield the true distance modulus,

$$\mu_0 = \mu_V - A_V. \tag{2.19}$$

Another important effect, which has to be taken into account in comparing the

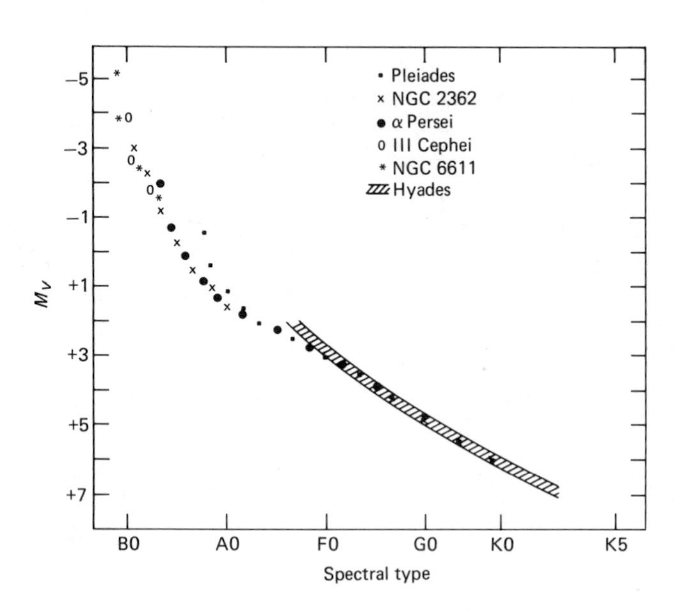

FIGURE 2.11

Zero-age main sequence. A plot of absolute visual magnitude M_V *against spectral type, as defined by the unevolved, lower parts of the main sequences of the Hyades and five other open clusters. The other clusters have been fitted to the Hyades by the method of main-sequence fitting. (From [H4].)*

spectra of stars in distant clusters with those of the Hyades, is *line blanketing,* the absorbing effect on the continuum of a star due to the presence of myriads of spectral lines. These absorption lines are caused by the great variety of elements present in the surface layers of a star, and the strength of the line blanketing depends on both the surface temperature of the star and its composition. By comparing stars of the same spectral type or color, we can ensure that the surface temperature is approximately the same in both cases. However, the abundance of the heavy elements* (carbon, nitrogen, oxygen, and elements heavier than these), which are synthesized in stellar interiors, depends strongly on the age of the star. The stars in a cluster with lower heavy-element abundance than the Hyades will suffer less line blanketing than those in the Hyades, and the colors of the stars will be shifted. Because line blanketing increases toward the ultraviolet end of the spectrum, the effect is that stars of lower heavy-element abundance appear relatively brighter in the ultraviolet than similar stars in the Hyades. The heavy-element abundance of a star can therefore be measured through the ultraviolet excess in the spectrum of the star compared to a similar star in the Hyades (Fig. 2.12). The Hyades have in fact an unusually high heavy-element abundance, about 1.6 times more than the solar abundance (see Table 2.9). Most nearby young clusters have a heavy-element abundance similar to that of the sun. In 1977 Sidney van den Bergh [B5] suggested that neglect of this effect has led to a considerable error in the distance estimates for galactic Cepheid variable stars, in that the distance moduli to the open clusters containing Cepheids have been overestimated by about 0.15 mag. I have adopted this correction in this book.

* Astronomers also use the term "metals" to refer to all elements except hydrogen and helium. Thus the metals consist of the heavy elements plus the light elements lithium, beryllium, and boron.

FIGURE 2.12

Correlation of ultraviolet excess due to line blanketing δ(U-B), measured with respect to the Hyades sequence, with heavy-element abundance [Fe/H] relative to the sun. δ(U-B) is the difference between the U-B color for a star and the U-B color for a star of the same spectral type in the Hyades cluster. $[Fe/H] = \log\{[N(Fe)/N(H)]_{star}/[N(Fe)/N(H)]_{sun}\}$, *where N(Fe) and N(H) are the abundances of iron and hydrogen, respectively. As the metal abundance decreases, the effect of line blanketing is reduced and the star has a more ultraviolet color. (From [M1].)*

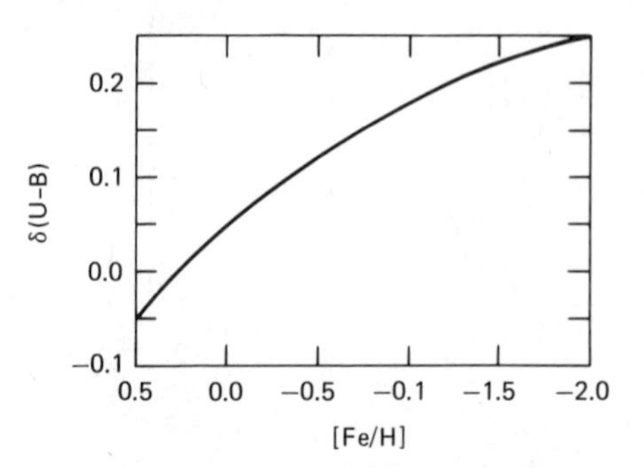

Distances to open clusters within 0.75 kpc are given in App. A.3a. An important result is the distance of the double cluster *h* and χ Persei, which contains many very luminous stars (supergiants) and several Cepheids. At this distance (230 pc) the cumulative uncertainty in using this method is estimated to be 0.23 mag in μ_0, or 12% in *d* (see, for example, the reviews by A. Blaauw [B7] and A. Heck [H4]). Recently R.F. Fenkart and B. Binggeli [F1] have given the distances of 190 open clusters by this method, with distances of up to 7 kpc.

Care has to be taken to use only stars which actually are on the main sequence. In the older and more evolved open clusters, stars that would have been on the upper parts of the main sequence have already begun to evolve away toward the giant branch (see Sec. 2.7). The method of main-sequence fitting can even be used, as we shall see in Sec. 2.6, to estimate the distance of the much older globular clusters, though with less reliability.

2.5 Distances to stars by spectroscopic and photometric methods

With the method of statistical parallaxes we reach out to stars at distances of 500 pc or so. This is still only a short step compared with the size of our Galaxy, which has an overall radius of 30 kpc or more. For the young massive stars, which are found in open clusters, we can reach to distances of 7 kpc or more with the method of main-sequence fitting, as we have seen. But to push further for other types of stars we have to rely on our ability to classify and recognize particular kinds of stars and to calibrate their absolute magnitudes.

We saw in Sec. 2.1 that stars may have the same spectral type but very different luminosities, according to whether they are giants or main-sequence stars. The spectra of the stars are not identical, though, and from small differences (mainly in the widths of the spectral lines) a *luminosity classification* has been introduced. The most luminous supergiants are classified as class I stars (subdivided into Ia, Iab, Ib) or class II; giants are class III; stars intermediate between giants and the main sequence (subgiants) are class IV; the main sequence is class V; and stars below the main sequence but not as far below as the white dwarfs (subdwarfs) are class VI (Box 2.2). Pioneering work on this method was carried out by W.S. Adams and co-workers at Mount Wilson Observatory during the early part of the century. Fig. 2.13 shows where the mean loci for some of these luminosity classes are found in the HR diagram. The zones of applicability of different ground-based distance methods are illustrated in Fig. 2.14a. Trigonometric parallaxes can be used to calibrate the main sequence from F5V to M5V and for a few G and K giants. Statistical parallaxes extend the calibration to B8 stars on the main sequence, to M giants and to M and K supergiants of class II. For 0 and B stars and for all other classes of supergiant we have to use the method of main-sequence fitting for clusters. App. A.4 summarizes the mean absolute visual magnitude M_V as a function of spectral type and luminosity class determined by different methods. Fig. 2.14b shows the impact the

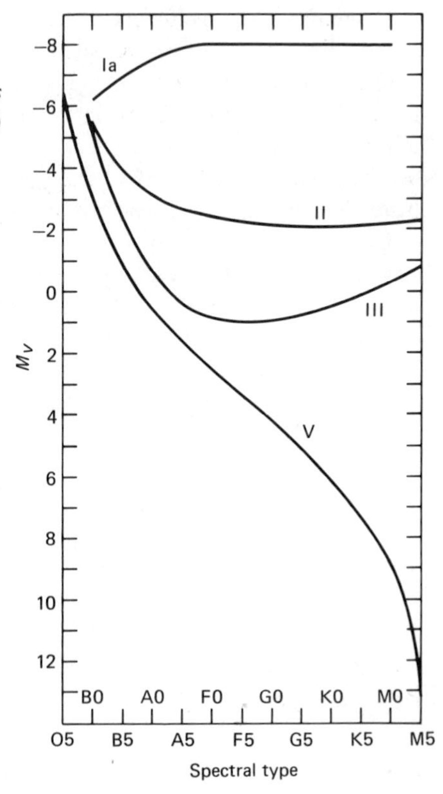

FIGURE 2.13

Absolute visual magnitude as a function of spectral type for selected luminosity classes. Individual curves are labeled with the luminosity class. (From [M1].)

Hipparcos satellite (p. 35) will have on the distance scale for stars.

Naturally it is desirable to look for other indicators of stellar luminosity. The widths of certain spectral lines turn out to be especially good luminosity indicators. O.C. Wilson and M.K.V. Bappu [W1] showed in 1957 that the width of a particular emission line, the K line from the H and K doublet of ionized calcium at a wavelength of 3933.7 Å, is well correlated with the stellar luminosity of stars of spectral type G, K, and M over a wide range in luminosity (Fig. 2.15). This spectral line arises in a zone known as the *chromosphere,* which in the sun lies just outside the visible disk (the latter defines the *photosphere).* On the sun the K-line emission is associated with a fine motley of bright points located within the granular cell structure seen on the surface due to the convection taking place just below the surface.

For hotter stars of type O, B, and A the intensities of the Balmer lines of hydrogen H_α, H_β, and H_γ are correlated with luminosity. App. A.3b gives distances to a number of clusters determined by David Crampton using the H_γ intensity and by D.L. Crawford and co-workers using the H_β intensity.

A very important method of estimating the distances of pulsating stars like RR Lyrae and Cepheid variable stars was introduced by Walter Baade in 1926 and modified by A.J. Wesselink in 1947, and is known as the *Baade–Wesselink* method. The general idea of the method is as follows. If we assume that the star has a blackbody spectrum, then a measurement of the color of the star gives its surface temperature. The surface brightness of the star, in the visual band, say, is then given

FIGURE 2.14

HR diagram. Zone of applicability of different methods of measuring distances to stars. (a) With ground-based techniques. (b) Planned performance of Hipparcos satellite due to be launched in the late 1980s.

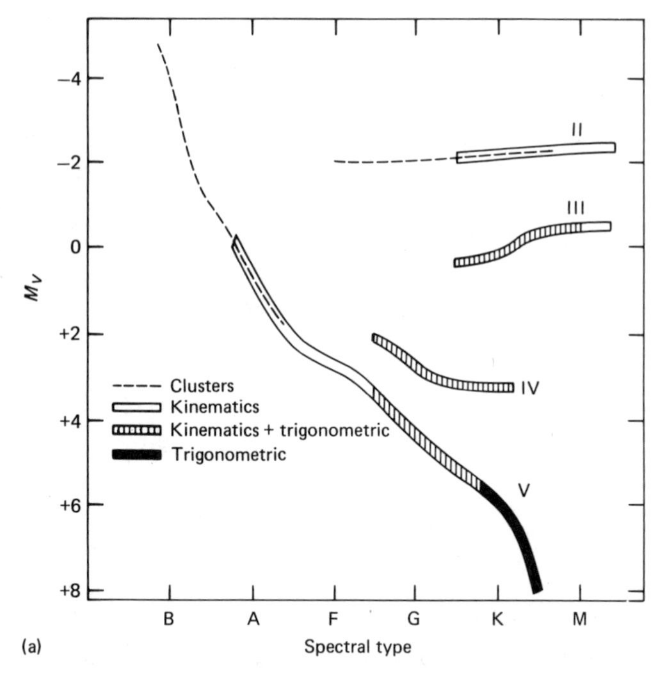

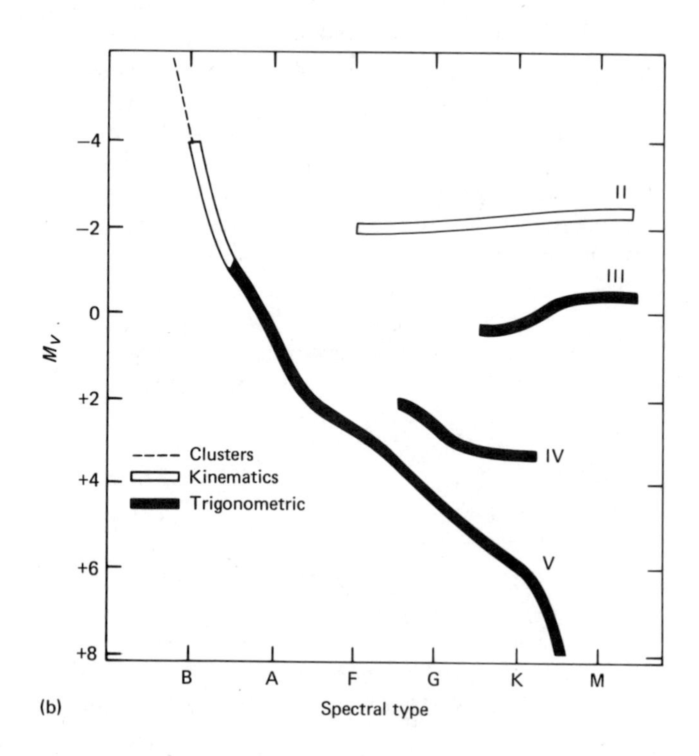

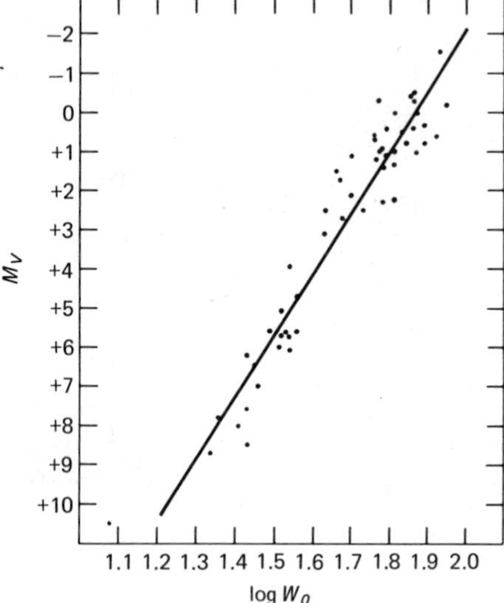

FIGURE 2.15

Basis of the Wilson–Bappu K-line method. Absolute visual magnitude M_V against logarithm of the width W_0 of the K line of calcium. (From [W1].)

by Planck's radiation law, which gives the radiant intensity of a blackbody at a particular temperature as a function of wavelength. Measurements of the brightness in the same band then give the angular radius of the star as a function of time. Integrating the observed radial velocity of the star with respect to time over the variability cycle of the star yields the displacement of the star's radius about its mean position. Combining these two sets of measurements, assuming that they are in phase with each other, allows both the mean radius of the star and its distance to be determined.

Wesselink's modification is to measure the brightness of the star at two phases at which the color of the star is the same. This eliminates the need to know how the surface brightness depends on color. L.A. Balona has shown [B2], that the effect of observational errors leads to a biased estimate of the star's radius. He gives a version of the method based on the principle of maximum likelihood (see App. A.17) and applies it to the determination of the radius of Cepheids.

T.G. Barnes and co-workers [B3] have applied the Baade–Wesselink method to Cepheids by finding an empirical correlation between visual surface brightness and the visual-to-red (V – R) color index for stars which have had their angular radii measured by interferometric methods.

2.6 Star clusters and associations in the Galaxy

We have already encountered the clusters of young stars found in the disk of our Galaxy, like the Hyades, which are known as open or galactic clusters. These contain, typically, a few hundred stars within a diameter of

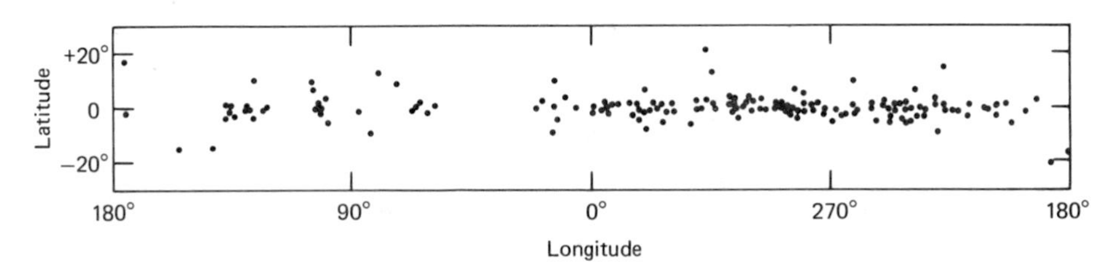

FIGURE 2.16

*Distribution on the sky, in galactic coordinates (*l, b*), of open clusters with distance* d ≥ *400 pc. Note the strong concentration toward the galactic plane. (Data from [F1], [M2].)*

10 pc or so, and some 1039 such clusters are listed in the catalog of G. Alter et al [A2]. Open clusters are, apart from a few of the nearest ones, confined to the plane of the Milky Way (Fig. 2.16).

OB associations are less concentrated but well-defined groupings of luminous young main-sequence 0 and B stars. They contain from a few tens to several hundred stars within a diameter of 10 – 100 pc. Like the open clusters, they are confined to the plane of the Milky Way, and about 70 are known, mostly within 4 kpc of the sun, because interstellar dust makes more distant associations hard to identify. When the proper motions and radial velocities of the stars in an OB association are examined, it is found that the whole association is expanding outward and will disperse in a time of a few million years.

The third type of star cluster in the Galaxy, the *globular cluster,* is very different from the other two types. First, globular clusters have a far more concentrated and regular, spherical appearance (Fig. 2.17). They contain about 100,000 stars concentrated into a region no more than 10 pc or so in diameter. Second, their light is dominated by red giants and solar-type stars. The blue, luminous O and B stars, which dominate the light of open clusters, are completely absent. Third, they are found distributed all over the sky at high galactic latitudes, though as we shall see below, a higher density of clusters is seen near the Milky Way, in particular toward the galactic center. Altogether about 125 globular clusters are known in our Galaxy [A2].

The HR diagrams of globular clusters are very different from those we have seen for open clusters (such as Fig. 2.11). Fig. 2.18 shows the HR diagram for the globular cluster M92, the appearance of which is illustrated in Fig. 2.17. Only the lower part of the main sequence is present, the red giant branch is very prominent, and a new concentration of stars appears, the *horizontal branch.* These turn out to be stars which have completely exhausted the hydrogen in their central regions and are now fusing helium into heavier elements (see Sec. 2.7). Finally, stars in globular clusters have much lower heavy-element abundances than younger stars like the sun.

How can distances to these three types of cluster be found? We have already seen in Sec. 2.4 how trigonometric parallaxes and proper motions allow us to measure the distance of the Hyades open cluster and that the method of main-sequence fitting can reach out to open clusters at distances of 7 kpc or so. For more distant clusters we must rely on spectroscopic and photometric parallaxes to individual stars in the cluster.

The smaller number of stars in an OB association means that the main sequence is not very well defined, and therefore the method of main-sequence fitting cannot be used with great certainty. Roberta Humphreys has estimated the distances to 53 OB associations using spectroscopic parallaxes (see App. A.5). Allowance has to be made for interstellar extinction once distances of 100 pc or more are involved. This will be discussed in Sec. 2.8.

FIGURE 2.17
Globular cluster M92. (Courtesy Lowell Observatory.)

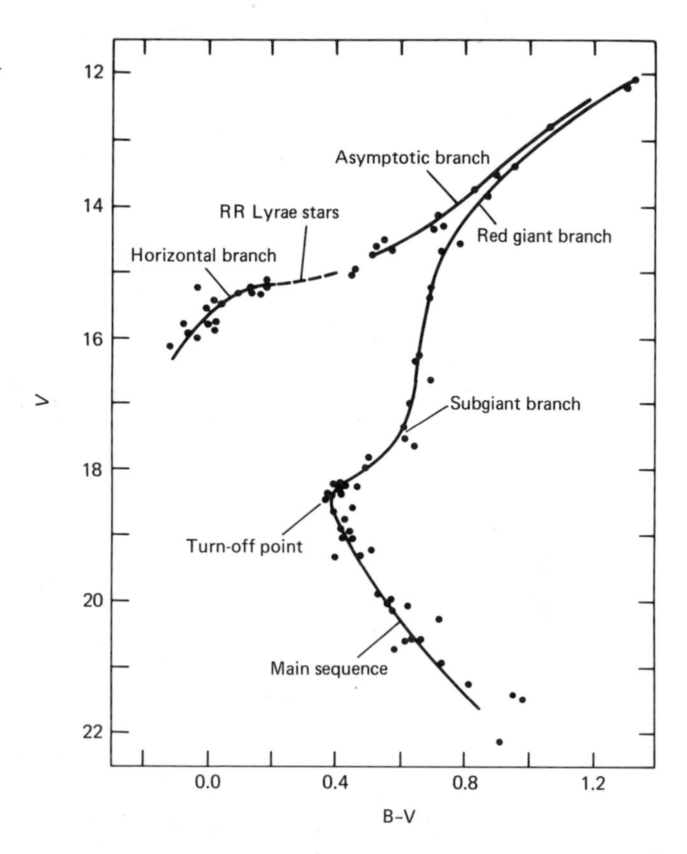

FIGURE 2.18
Plot of visual magnitude V *against B-V color for stars in the metal-poor globular cluster M92. Shown are the main sequence, the turnoff point, the subgiant, red giant, asymptotic, and horizontal branches, and the zone of RR Lyrae variable stars (broken line). (From [I2].)*

For some of the nearer globular clusters, the method of main-sequence fitting can give an estimate of the distance. However, the main sequence must be compared not with the normal "solar" heavy-element-abundance main sequence of open clusters, but with the main sequence determined for subdwarfs in the solar neighborhood with low heavy-element abundance (see, for example, Sandage [S1]). As we shall see in Sec. 2.7, this is because the position of the main sequence in the HR diagram depends quite sensitively on chemical composition. The distances of these subdwarfs are estimated by trigonometric parallaxes, and the method is limited in accuracy by the small number of subdwarfs for which the trigonometric parallax is known. The most accurate globular cluster distances are obtained by making a detailed comparison of the observed HR diagram for the cluster with theoretical calculations of the locus where stars with the same age should lie (isochrones, see Sec. 2.7). The results of a recent calculation of this kind by Bruce Carney are given in App. A.6a.

For more distant globular clusters, in which only the more luminous stars can be observed, it is possible to estimate distances by assuming that the horizontal branch occurs at the same absolute magnitude in all clusters, though this does in fact depend on chemical composition. The accuracy of these estimates is probably about 5 – 10%. Distances to a number of globular clusters determined in this way are given in App. A.6b.

A particular type of horizontal branch star that can easily be recognized in globular clusters is the *RR Lyrae* variable star, named after a prominent variable in the constellation of the Lyre, a star that varies its light output regularly on a time scale of 0.4 – 1 day. Table 2.3 lists some recent determinations of the mean absolute magnitude of RR Lyrae stars, determined by the main-sequence fitting of globular clusters, by the method of statistical parallax, and by the Baade – Wesselink method. Distances to globular clusters can be estimated by assuming that all RR Lyrae stars have the same absolute magnitude, though there is evidence from both theory and observation that the metal-rich RR Lyrae stars (for example, those in the globular cluster 47 Tucanae) are fainter than the metal-poor ones.

Another useful indicator of the distance of globular clusters is the angular size θ

TABLE 2.3

Mean absolute magnitude of RR Lyrae variables.

Location of stars	*Method*	*Authors*	M_V(RR)
M3	Main-sequence fitting	A.R. Sandage (1970)	0.60
M13	Main-sequence fitting	A.R. Sandage (1970)	0.05
M15	Main-sequence fitting	A.R. Sandage (1970)	0.98
M92	Main-sequence fitting	A.R. Sandage (1970)	0.91
NGC 6838	Main-sequence fitting	F.D.A. Hartwick and D.A. VandenBergh (1973)	0.35
47 Tuc	Main-sequence fitting	F.D.A. Hartwick and J.E. Hesser (1974)	0.90
5 clusters	Main-sequence fitting	A. Sandage (1982)	0.87 ± 0.16
Galactic field	Statistical parallax	G. van Herk (1965)	0.7 ± 0.2
Galactic field	Statistical parallax	R. Wooley and A. Savage (1971)	0.6 ± 0.2
Galactic field	Statistical parallax	A. Heck (1973)	0.5 ± 0.2
Galactic field	Statistical parallax	M.K. Hemenway (1975)	0.5 ± 0.4
Galactic field	Statistical parallax	A. Heck and J.M. Lakaye (1978)	0.63 ± 0.30
Galactic field	Statistical parallax	S.V.M. Clube and J.A. Dawe (1980)	1.0 ± 0.25
Metal-weak stars	Baade – Wesselink	L.H. McDonald (1977)	0.55 ± 0.2
Metal-strong stars	Baade – Wesselink	L.H. McDonald (1977)	0.85 ± 0.2
Metal-strong stars	Baade – Wesselink	D.H. McNamara and K.A. Feltz (1977)	0.9 ± 0.2
Metal-weak stars	Baade – Wesselink	R. Wooley and E. Davies (1977)	0.46 ± 0.23
Metal-weak stars	Baade – Wesselink	G. Wallerstein and G.W. Brugel (1979)	0.6 ± 0.2
Metal-weak stars	Baade – Wesselink	A. Manduca *et al* (1981)	0.6 ± 0.2
Metal-weak stars	Baade – Wesselink	M.J. Siegel (1982)	0.42 ± 0.07
adopted mean for metal-weak RR Lyrae stars			**0.6**

SOURCE: Compiled from [G5], [M1], [C4].

of the cluster, which is related to the linear diameter D by $\theta = D/d$, where d is the distance of the cluster. Here we have to assume that all globular clusters have the same linear diameter. However, it is hard to define the edge of a cluster, and there is considerable spread in the value of the linear diameter, so the method is not very accurate. Distances to additional globular clusters using RR Lyrae stars or cluster diameters are included in App. A.6b.

Having measured the distances to the globular clusters, we can study their distribution in the Galaxy. The space distribution is illustrated in Fig. 2.19. The contrast with the distribution of open clusters (see Fig. 2.16) and associations is striking. The globular clusters are found to occupy a spherical region centered on a point about 8.5 kpc in the direction of Sagittarius. This is now identified by several different means as the center of our Galaxy. Table 2.4 summarizes recent estimates of the distance to the center of the Galaxy from the sun. The official value adopted by the International Astronomical Union in 1964 was $R = 10$ kpc, which is probably too large by about 10 – 15%, though there is not yet a consensus on what the true value should be.

While studying the nearby Andromeda galaxy M31 at Mount Wilson Observatory during the exceptionally good conditions of a Los Angeles blackout in 1944, Walter Baade was able to resolve many individual stars in this galaxy and determine their color and luminosity. He noticed a striking difference between the stars in the disk of M31 and those in the central region of this galaxy (the 'nucleus'). The stars in the disk resembled the stars of an open cluster, with the brightest stars being blue O and B stars, while the stars in the central region of the Andromeda galaxy (and also the stars in M31's dwarf elliptical companions NGC 205 and 221) resembled the

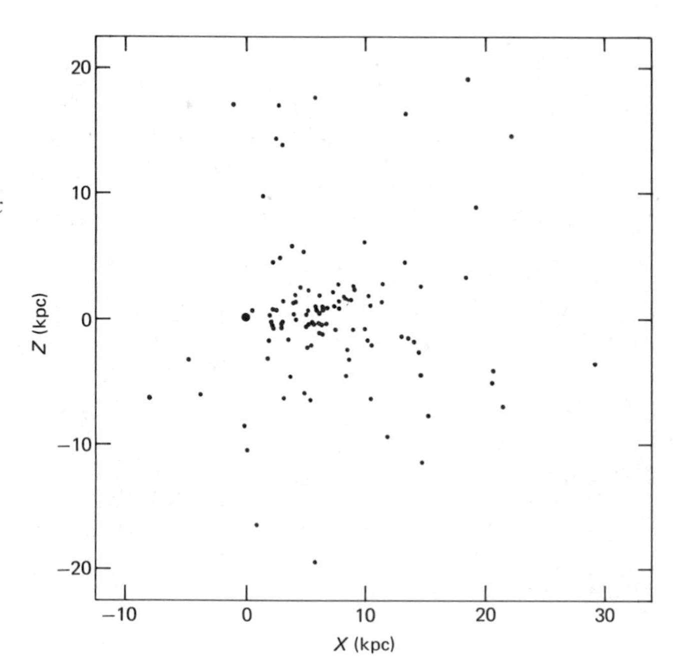

FIGURE 2.19
Space distribution of globular clusters, projected on the plane through the sun and the galactic center, perpendicular to the galactic plane. Z—distance from galactic plane; X—distance in plane. The sun is the circled dot, and the galactic center lies at Z = 0, X ≃ 8 – 10. The apparent deficiency of globular clusters close to the plane (Z ≤ 1) at X > 10 is due to the effect of extinction by dust. (From [H2].)

TABLE 2.4

Distance to galactic center.

Method	Authors	R_0
OB stars	D. Crampton et al. (1976)	8.4 ± 1
Globular clusters	W.E. Harris (1976)	8.5 ± 1.6
RR Lyrae stars	J.H. Oort and L. Plaut (1975)	8.7 ± 0.6
Galactic disk	A. Toomre (1972) and G. Rybicki et al. (1974)	9.0 ± 1
Average		**8.7 ± 1**

SOURCE: Compiled from [G5].

stars of a globular cluster (Fig. 2.20). He called these two types of stars Population I and II, respectively, and found that the stars in the disk of our Galaxy are also Population I, while those in the nucleus and halo of our Galaxy are Population II. Dust, gas, young stars, and regions of star formation are all associated with Population I, while Population II is composed of old stars, so the distinction is primarily one of the mean age of the populations. This age difference accounts for another important difference between the two populations, namely, that the average abundance of heavy elements is much lower in Population II than in population I. As

FIGURE 2.20

Stellar Populations I and II. (Left) Andromeda nebula, photographed in blue light, shows giant and supergiant stars of Population I in spiral arms. The hazy patch at the upper left is composed of unresolved Population II stars. (Right) NGC205, companion of Andromeda nebula, photographed in yellow light, shows stars of Population II. The brightest stars are red and 100 times fainter than the blue stars of Population I. The very bright, uniformly distributed stars in both pictures are foreground stars belonging to our own Milky Way system. (Courtesy of Mount Palomar Observatory.)

time goes on, the gas from which new stars form becomes steadily enriched with the heavy elements synthesized in the cores of older stars and ejected by them at their death.

It is realized today that the situation is more complex. First, it is now recognized that the boundary between the two populations is not a sharp one and that several intermediate populations can be defined. Table 2.5 summarizes the modern picture of stellar populations in the Galaxy. A discussion of these differences in terms of the evolution of the Galaxy will be given in Sec. 2.11. Second, stars in the nuclei of galaxies have higher heavy-element abundances than the stars in the halo of our Galaxy, suggesting that there have been several generations of stars in the nuclei.

TABLE 2.5

Stellar Populations.

	Population I				*Population II*
		Disk population			
Property	*Spiral-arm population*	*Young*	*Intermediate*	*Oldest*	*Halo population*
Representative objects	Interstellar gas and dust; Open clusters and stellar associations; OB stars; Supergiants; Classical Cepheids; T Tauri stars; *Some* stars of type A and later	A stars; F stars; A-K giants; Me dwarfs; *Some* G, K, and M dwarfs and white dwarfs	Sun; Most G dwarfs; *Some* K and M dwarfs and white dwarfs; *Some* subgiants and red giants; Planetary nebulae	*Some* K and M dwarfs and white dwarfs; *Some* subgiants and red giants; Moderately metal-poor (weak-lined) stars; Long-period variables; RR Lyrae stars with $\Delta S < 5$, $P < 0.5$ day	Subdwarfs; Globular clusters; RR Lyrae stars with $P \geq 0.5$ days
Age, 10^9 years	$\lesssim 0.1$	~ 1	~ 5	$\lesssim 10(?)$	9–20
$\langle \lvert z \rvert \rangle$, pc	120	200	400	700	3000
$Z/Z_\odot$	1–2	1–2	0.5–1	0.2 or 0.3–0.5	0.005–0.5
HR diagram	h and χ Persei	Hyades	M67	NGC 188	M92

SOURCE: Compiled from [M1], [B6].

2.7 *Stellar evolution and the Hertzsprung–Russell diagram*

We now try to understand the different types of stars we have encountered — main sequence, giants, horizontal branch, white dwarfs, and so on — in terms of the evolution of stars of different masses. We consider the structure of a main-sequence star, its evolution to the giant branch, the later stages of evolution and stellar remnants, and how stars form and reach the main sequence. Many different types of stars at different stages of their evolution, for example, Cepheids, RR Lyrae variables, novae, supernovae, red giants, and supergiants, all play a part in establishing the cosmological distance ladder.

MAIN SEQUENCE

When a star forms it can be supposed that its composition is roughly homogeneous. For a star forming out of the interstellar gas today, this composition would be something like 73% hydrogen, 25% helium, and 2% heavier elements by mass. About 90% of the stars near the sun lie on the main sequence (see Fig. 2.2), and it can be presumed that stars spend the major part of their lives in this region of the HR diagram. Provided the star is not rotating strongly, it can be assumed to be spherically symmetrical and supported against gravity by a steady increase in pressure as we go from the surface to the center. The increase in pressure toward the center of the star implies, in a normal gas, an increase in temperature. There is therefore a flow of energy from the central regions of the star to the outside. This flow of energy has to be supplied by thermonuclear reactions in the central regions of the star, where the temperature exceeds one million degrees Kelvin.

The three main modes of energy transport through matter are radiation, convection, and conduction. Conduction turns out to be unimportant in stellar interiors. Radiation would be capable of carrying all the energy from the center of the star to the surface, were it not for the fact that the material of the star continually absorbs the radiation and reemits it. The capacity of the material to restrict the flow of radiation is described by the *opacity* of the material, which measures the rate of absorption of radiation per gram of material. The rate of energy transport by radiation is therefore limited to a maximum value at any given location in the star. If the star needs to transport energy at a greater rate than this in order to maintain its pressure support against gravity, then the gas in the star will become convectively unstable, forced into a turmoil of rising and falling currents. This turns out to be a very difficult process to model.

The main complexity of stellar models arises from the detailed dependence on density, temperature, and composition of the opacity of the stellar material and the rate of generation of energy by thermonuclear reactions, and from the difficulty in finding a good mathematical description of convection. An added complication is that some stars rotate quite rapidly, and this can affect their internal structure and

TABLE 2.6

Convective cores and envelopes of star of different mass.

	$M/M_\odot$						
	0.5	1.0	1.5	3.0	5.0	9.0	15.0
M_{cc}/M	0.01	0.00	0.06	0.17	0.21	0.26	0.38
M_{ce}/M	0.42	0.01	0.00	0.00	0.00	0.00	0.00

M_{cc}—mass in form of convective core; M_{ce}—mass in form of convective envelope.

SOURCE: Compiled from [T1].

their shape quite dramatically. Higher mass stars ($M > 1.5M_\odot$) are found to be convectively unstable in their central regions, while stars of mass less than or equal to that of the sun have convective zones just below their surface (Table 2.6). It is because of these outer convective zones that the difficulties in constructing a good physical and mathematical model lead to considerable uncertainty in our knowledge of stellar structure.

The bulk of the energy available to a star from thermonuclear reactions (more than 70%) derives from the fusion of hydrogen into helium. This can take place via a number of different reaction chains, as shown in Box 2.3. The two main types of reaction chain are the proton–proton (PP) chain, in which protons are fused directly into deuterium (heavy hydrogen) and then to helium, and the carbon–nitrogen (CN) cycle, in which carbon and nitrogen act as catalysts. The rate of energy generation by the PP chain is approximately proportional to the density of the gas, to the square of the hydrogen abundance, and to a steeply rising function of temperature (Fig. 2.21). For the CN cycle the rate of energy generation is proportional to the gas density, to the hydrogen abundance, to the carbon (and nitrogen) abundance, and to an even more steeply rising function of the temperature. Thus the PP chain dominates the energy production when the central temperature of the star is $< 2 \times 10^7$ K, and the CN cycle dominates when the central temperature is higher than this.

The opacity of the stellar material is caused by a great variety of atomic and molecular processes. In the surface layers of cool stars ($T < 3000$ K) molecular band absorption plays the major role. At very high temperatures ($T > 10^7$ K) scattering by free electrons is the main source of opacity. At intermediate temperatures the opacity is dominated by bound–free and free–free atomic transitions, with bound–bound transitions playing an important role in the surface layers of hotter stars in producing the spectral lines observed in a star's spectrum (where bound and free refer to whether or not the electron taking part in the transition is bound to an atom).

A difficulty with comparing theoretical and observed main sequences is that the vertical (magnitude) axis has to be converted to the total or bolometric luminosity,

BOX 2.3 HYDROGEN BURNING REACTIONS IN STARS

1 PP I CHAIN

$p(p, e^+ + \nu)d$ (1)

$d(p, \gamma)^3\mathrm{He}$ (2)

$^3\mathrm{He}(^3\mathrm{He}, p + p)^4\mathrm{He}.$ (3)

2 PP II CHAIN.
This starts with reactions (1) and (2). Then

$^3\mathrm{He}(^4\mathrm{He}, \gamma)^7\mathrm{Be}$ (3′)

$^7\mathrm{Be}(e^-, \nu)^7\mathrm{Li}$ (4′)

$^7\mathrm{Li}(p, \alpha)^4\mathrm{He}.$ (5′)

3 PP III CHAIN.
This starts with reactions (1), (2), and (3′). Then

$^7\mathrm{Be}(p, \gamma)^8\mathrm{B}$ (4″)

$^8\mathrm{B}(, e^+ + \nu)^8\mathrm{Be}$ (5″)

$^8\mathrm{Be} \rightarrow 2\ ^4\mathrm{He}.$ (6″)

4 CN CYCLE

$^{12}\mathrm{C}(p, \gamma)^{13}\mathrm{N}$ (1)

$^{13}\mathrm{N}(, e^+ + \nu)^{13}\mathrm{C}$ (2)

$^{13}\mathrm{C}(p, \gamma)^{14}\mathrm{N}$ (3)

$^{14}\mathrm{N}(p, \gamma)^{15}\mathrm{O}$ (4)

$^{15}\mathrm{O}(, e^+ + \nu)^{15}\mathrm{N}$ (5)

$^{15}\mathrm{N}(p, ^4\mathrm{He})^{12}\mathrm{C}.$ (6)

SOURCE: From [T2].

and the horizontal axis has to be converted from the relatively easily observable spectral type, or color, to the effective temperature T_e defined by

$$L = 4\pi r_s^2 \sigma T_e^4 \tag{2.20}$$

where L and r_s are the luminosity and the radius of the star and σ is the Stefan–Boltzmann constant (which relates the total energy per unit area emitted by a blackbody to the fourth power of the temperature—Stefan's law). The radius of a star r_s is not a sharply defined quantity. For example, although the sun appears to have a well-defined visible disk, the density does not suddenly drop to zero at this surface. The effective temperature is defined to be the temperature on the surface at which the *optical depth* to the outside world is unity. This agrees quite well with the surface defined by the visible disk of the sun. The optical depth τ of a slab of material which absorbs or scatters radiation is defined in terms of the dimming effect on the radiation, $\tau = \ln(I_0/I)$, where I_0 is the intensity of the radiation entering the slab and I is that emerging, so $I = I_0 e^{-\tau}$.

The angular diameters of a number of stars have been measured by interferometric methods (in which the light from the star is collected by two separate mirrors and made to undergo interference), by lunar occultation studies (in which the signal from the star is carefully observed as the moon passes in front of it), and from analysis of the light curves of eclipsing binaries (see Sec. 2.2). If the distance of the star is known, then T_e can be obtained from Eq. (2.20).

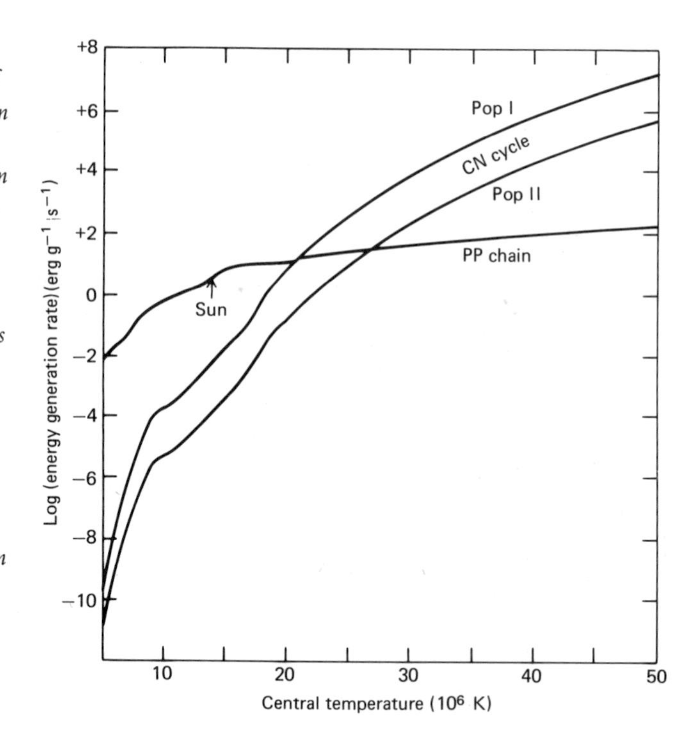

FIGURE 2.21
Temperature dependence of the rate of energy generation for the main hydrogen-burning nuclear reactions in stars. PP chain—proton–proton chain, which dominates for stars with central temperatures less than 20 million K, such as the sun. Two curves are shown for the CN cycle, since this depends on the abundance of carbon and nitrogen: Pop I—young stars of Population I; Pop II—old stars of Population II. (Adapted from [F3].)

If stars had blackbody spectra, then T_e could be deduced from their colors. But this is a poor approximation, especially for cool stars and for very hot stars. App. A.7 summarizes the colors, spectral types, luminosities, masses, effective temperatures, and radii, which have been determined for main-sequence stars.

Fig. 2.22(a) shows theoretical main sequences calculated for several different helium abundances. We see that as the helium abundance by mass Y increases, the position of a star in the HR diagram shifts upward and to the left. Fig. 2.22(b) shows the observed HR diagram for nearby well-studied stars compared with theoretical curves appropriate to Populations I and II. The overall agreement of theory and observation is quite good. Fig. 2.23 shows a theoretical mass–luminosity relationship based on calculations by Icko Iben for helium abundance $Y = 0.27$and heavy-element abundance by mass $Z = 0.02$, compared with some stars of well-determined mass and luminosity, including the sun. We see that we can be reasonably confident that we have qualitatively understood main-sequence stars, though we must bear in mind the considerable uncertainties in the models (especially the effects of convection and rotation) and in the observations. A nagging doubt that remains is why the observed flux of neutrinos from the sun is well below that predicted by theory. Neutrinos are a by-product of some of the nuclear reactions in the sun's interior (see Box 2.3). They escape freely from the sun and have been detected in a subtle experiment designed by Ray Davis and located at the bottom of a deep gold mine (see the review by John Bahcall [B1]).

How long do stars spend on the main sequence? If all the hydrogen in the star

FIGURE 2.22

(a) Theoretical absolute bolometric magnitude M_{bol} versus effective surface temperature T_{eff} for unevolved stars, showing the effect of varying the helium abundance. The heavy line is for Y = 0.28, Z = 0.02, and the thin lines show the effect of varying the helium abundance to Y = 0.18 and 0.38. The filled circles are loci corresponding to a fixed stellar mass.
(b) Absolute bolometric magnitude M_{bol} of well-studied stars as a function of effective surface temperature T_{eff}.
×—Population I stars;
●—old disk stars;
□—halo stars (Population II). The curves are theoretical, the heavy curve (Y = 0.28, Z = 0.02) fitting Population I and old disk stars, and the broken curve (Y = 0.30, Z = 0.0004) fitting the more extreme Population II subdwarfs. (Adapted from [G6].)

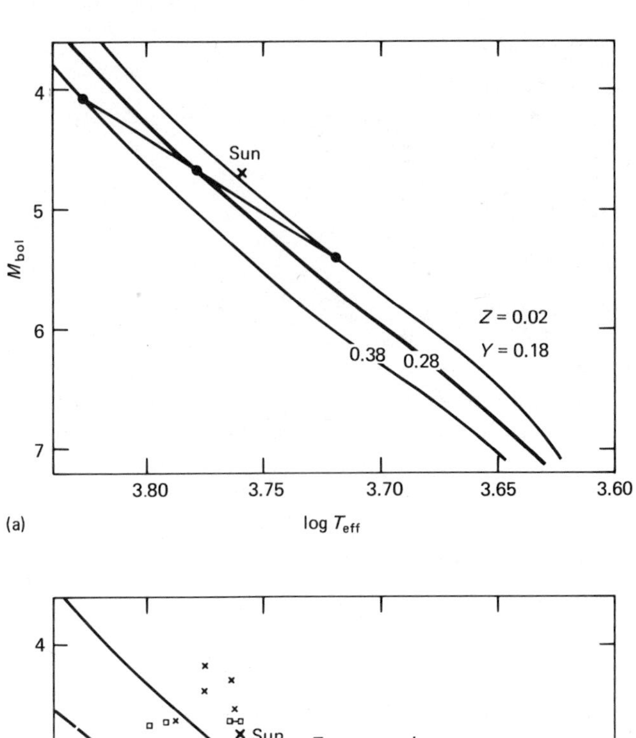

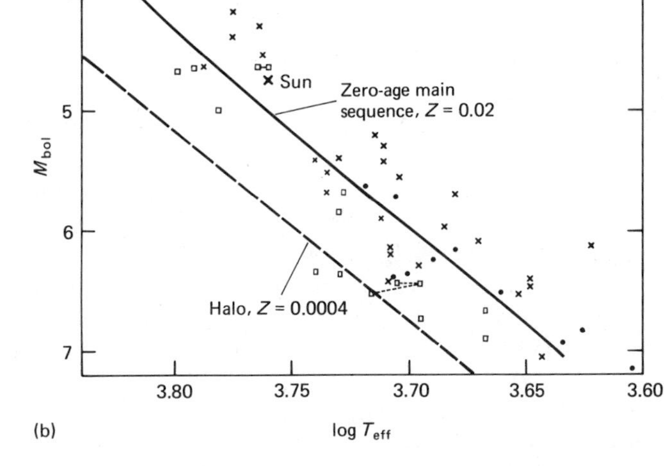

were available to be converted to helium, then since the conversion of hydrogen to helium releases 0.7% of the original rest mass in the form of energy, the time that the star could be fueled for would be

$$\tau_{ms} = \frac{0.007 M c^2}{L} \quad [\mathrm{s}] \tag{2.21}$$

where M is in grams, c is in centimeters per second, and L is in ergs per second. Since the luminosity L increases with mass M as a high power of M (between 3 and 4, see Figs. 2.6 and 2.23), the main-sequence lifetime must decrease sharply as mass

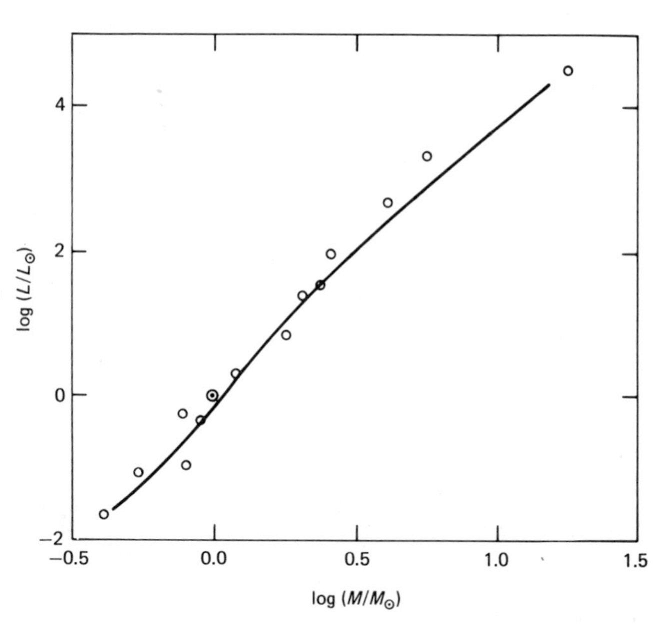

FIGURE 2.23

Mass–luminosity curve for main-sequence stars deduced from sequences of theoretical models for stars of different masses compared with some well-studied nearby stars, including the sun (circled dot). $L_\odot$, $M_\odot$*—luminosity and mass of sun. See Fig. 2.6 for data on a larger number of stars. (From [T1].)*

increases. Table 2.7 gives the results of detailed calculations of main-sequence lifetimes. These are smaller than would be given by Eq. (2.21) because not all hydrogen in the star is converted to helium before the main-sequence phase ends. There is also a small energy wastage in the form of neutrinos, which escape freely from the stars.

POST-MAIN-SEQUENCE EVOLUTION OF LOW-MASS STARS ($0.5-4M_\odot$)

There is no convection taking place at the center of a low-mass star, and hence no mixing of the stellar material. Thus when hydrogen is exhausted in the center of such a star, hydrogen burning ceases there and a roughly constant-temperature core of helium forms in which no nuclear reactions occur. Hydrogen burning continues in a shell surrounding this core, which gradually moves outward as hydrogen is exhausted (Fig. 2.24). Meanwhile the helium core contracts and heats up and the outer layers of the star readjust their structure too. The star slowly leaves the main sequence.

The increasing central density of the star means that the material there soon becomes *degenerate.* The atoms are so closely packed together that there is no space for the electrons to travel freely, and an effect of quantum theory, the Pauli exclusion principle (roughly, no two electrons can occupy the same place at the same time), becomes important. The main consequence is that, in these circumstances, the pressure of the gas depends only on the density of the gas and not on its temperature. Fig. 2.25 shows the results of detailed calculations by Iben of the actual post-main-sequence evolution of stars in the range of $0.25-15M_\odot$. Table 2.8 gives

TABLE 2.7

Main-sequence lifetimes for stars of different mass.

$M/M_\odot$	1.0	1.25	1.5	2.25	3.0	5.0	9.0	15.0
Lifetime, years	8.2×10^9	3.0×10^9	1.7×10^9	5×10^8	2.3×10^8	6.8×10^7	2.2×10^7	1.0×10^7

SOURCE: Compiled from [T1].

TABLE 2.8

Stellar evolution times (years).

	Mass (in units of $M_\odot$)							
Evolution-track interval (numbered points in Fig. 2.25)	*1.0*	*1.25*	*1.5*	*2.25*	*3*	*5*	*9*	*15*
1–2	7×10^9	2.8×10^9	1.5×10^9	4.8×10^8	2.2×10^8	6.5×10^7	2.1×10^7	1.0×10^7
2–3	2×10^9	1.8×10^8	8.1×10^7	1.6×10^7	1.0×10^7	2.2×10^6	6.1×10^5	2.3×10^5
3–4	1.2×10^9	1.0×10^9	3.5×10^8	3.7×10^7	1.0×10^7	1.4×10^6	9.1×10^4	
4–5	1.6×10^8	1.5×10^8	1.0×10^8	1.3×10^7	4.5×10^6	7.5×10^5	1.5×10^5	7.5×10^4
5–6	$\geq 10^9$	$\geq 4 \times 10^8$	$\geq 2 \times 10^8$	3.8×10^7	4.2×10^6	4.9×10^5	6.6×10^4	
6–7	–	–	–	–	2.5×10^7	6.1×10^6	4.9×10^5	7.2×10^5
7–8	–	–	–	–		1.0×10^6	9.5×10^4	6.2×10^5
					4.1×10^7			
8–9	–	–	–	–		9.0×10^6	3.3×10^6	1.9×10^5
9–10	–	–	–	–	6.0×10^6	9.3×10^5	1.6×10^5	3.5×10^4

SOURCE: From [I1]. Reproduced, with permission, from *Annual Review of Astronomy and Astrophysics*, vol. 5.

FIGURE 2.24

Structure of a star leaving the main sequence. In region A all of the hydrogen has been converted to helium, in region B no nuclear reactions have occurred, and in the hatched area hydrogen is burning into helium.

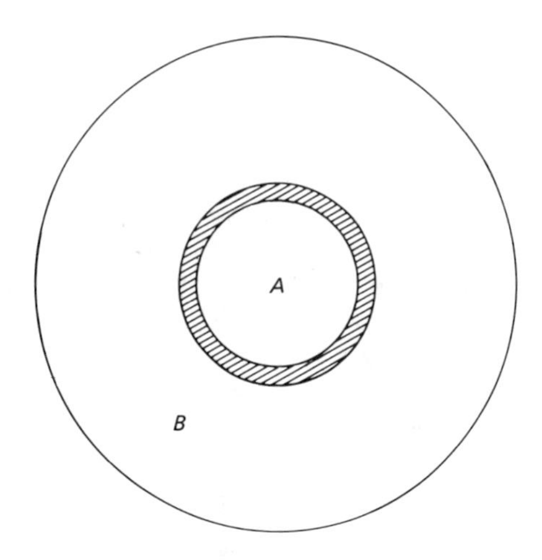

FIGURE 2.25

Theoretical HR diagram. Log ($L/L_\odot$) versus log T_{eff}, showing post-main-sequence evolution tracks for stars with masses in the range of $0.25-15M_\odot$. L, $L_\odot$—luminosities of star and sun; T_{eff}—effective surface temperature of star; $M_\odot$—mass of sun. The times to travel between numbered points are given in Table 2.8. For $M = 0.25$ and $0.5M_\odot$ the evolution is shown schematically only by broken lines. For $M = 1-2.25M_\odot$ the tracks are shown up to the point of core helium ignition. For $M = 3-15M_\odot$ the tracks are shown up to a point just before helium core exhaustion. The structure of a $5M_\odot$ star at the different numbered points is shown in Fig. 2.29. (From [I1].)

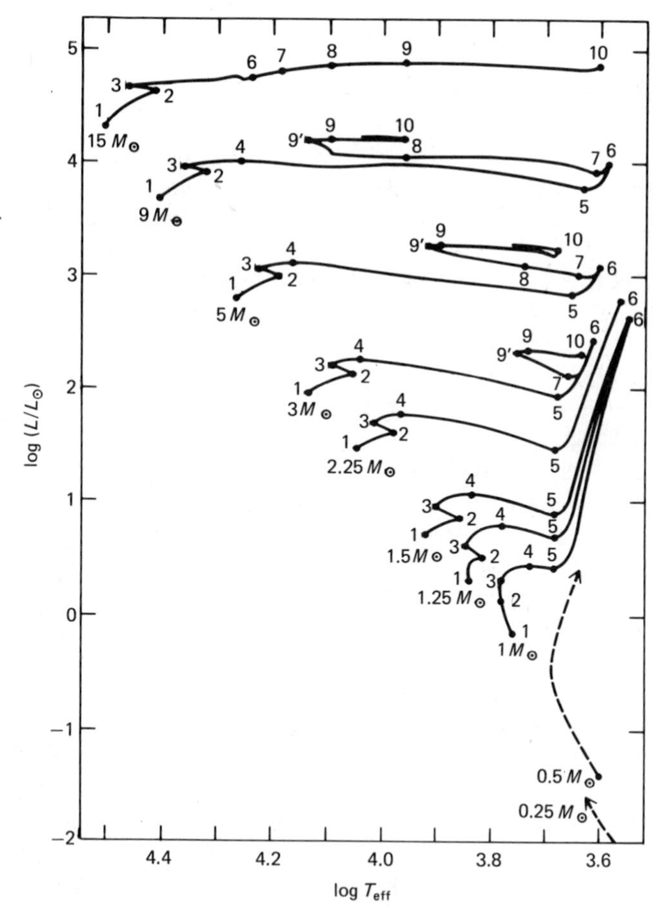

the time in units of 10^9 years to reach the numbered points on the evolutionary tracks. Now if we assume that all stars in a cluster formed at the same time, then the HR diagram of a cluster should join up the evolutionary positions of stars of different masses but of the same age. Such loci are called *isochrones* and are illustrated in Fig. 2.26. As a cluster ages, the upper part of the main sequence evolves away to the right, and the cluster turn-off point is a sensitive measure of the age of the cluster. Comparison with old open clusters like M67 and NGC 188 or with globular clusters gives estimates for the ages of these clusters, as shown in Table 2.9. The globular clusters have to be compared with evolutionary tracks for stars with a low heavy-element abundance (one tenth to one-hundredth of that in the sun). Fig. 2.27(a) shows a comparison of the HR diagram for the globular cluster M92 with theoretical isochrones at the ages of $10-18 \times 10^9$ years. Fig. 2.27(b) shows color–magnitude diagrams for three globular clusters and the old open cluster NGC 188, illustrating the effects of increasing heavy-element abundance. Here [Fe/H] denotes

$$\log \left\{ \frac{[n(\mathrm{Fe})/n(\mathrm{H})]_{\mathrm{cluster}}}{[n(\mathrm{Fe})/n(\mathrm{H})]_{\mathrm{sun}}} \right\}$$

which is related to the heavy-element abundance Z by

$$\log Z \simeq [\mathrm{Fe/H}] - 1.7. \tag{2.22}$$

The main effect is that stars on the red giant branch and the horizontal branch tend to be more luminous for lower heavy-element abundance.

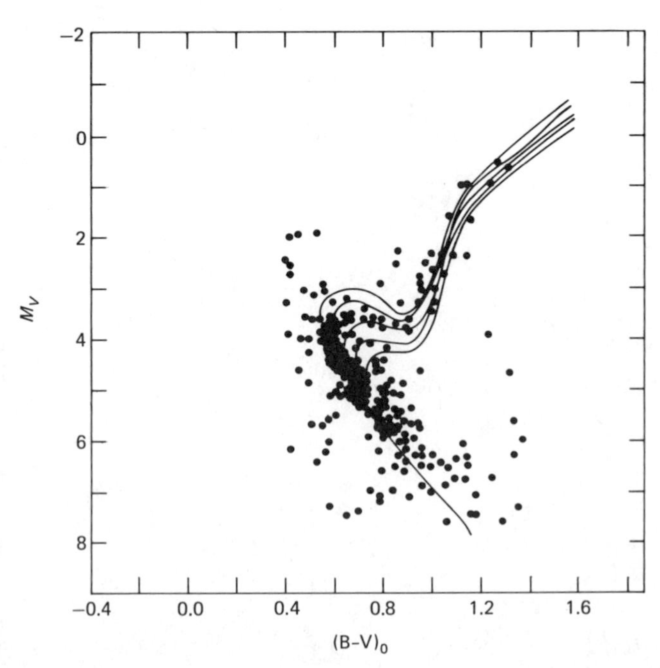

FIGURE 2.26

HR diagram, absolute magnitude M_V *versus* $(B-V)_0$*, the B-V color corrected for the effects of reddening by dust, for old disk cluster NGC 188. The absolute magnitude scale is determined by fitting to the Hyades main sequence (assuming* $\mu_{Hya} = 3.30$*). The theoretical isochrones are for* Y = 0.30, Z = 0.02, *and ages (from top to bottom) of 4, 5, 7, 10, and* 13×10^9 *years. The main-sequence turn-off point indicates an age of* $5-6 \times 10^9$ *years. (From [T3, p. 199].)*

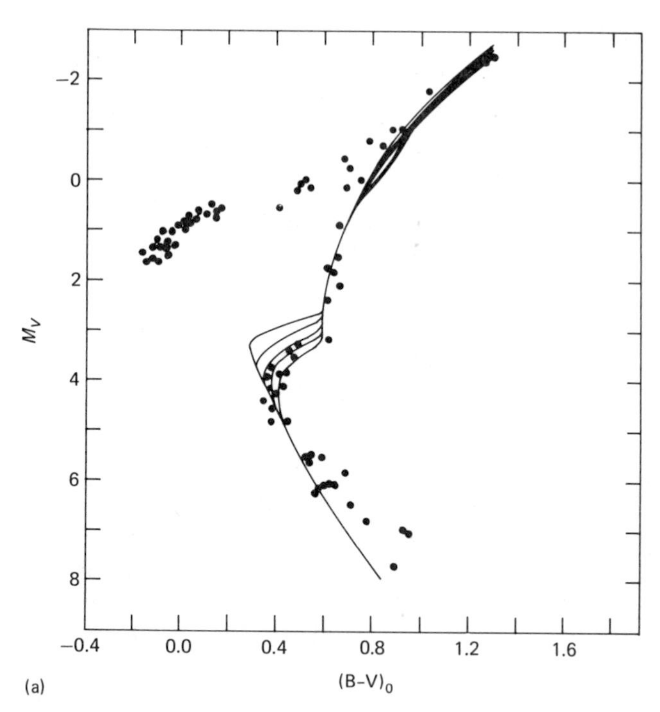

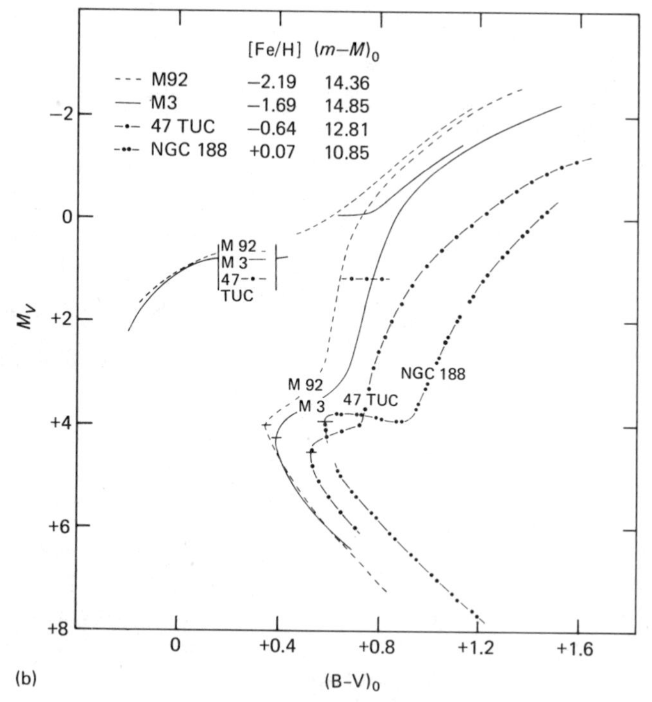

FIGURE 2.27

(a) HR diagram, absolute visual magnitude M_V *versus* $(B\text{-}V)_0$*, the B-V color corrected for the effects of reddening by dust, for globular cluster M92. The absolute magnitude scale is fixed by setting* $M_V = 0.6$ *for the RR Lyrae variable stars. The theoretical isochrones are for* $Y = 0.20$, $Z = 0.0001$, *at ages (from top to bottom) of 10, 12, 14, 16, and* 18×10^9 *years. The turn-off point indicates an age of* $14–16 \times 10^9$ *years. (From [T3, p. 199].)*
(b) HR diagram. Absolute visual magnitude M_V *versus* $(B\text{-}V)_0$*, the B-V color corrected for the effects of reddening by dust, for three globular clusters and old open cluster NGC 188, illustrating the effect of increasing heavy-element abundance. The values of [Fe/H] and the distance moduli to the clusters* $(m - M)_0$ *are shown inset. The location of the RR Lyrae variable stars is indicated by the pair of vertical lines. (From [S5].)*

TABLE 2.9

Ages and heavy-element abundances of open and globular clusters.

Open clusters	*Age* (10^9 yr)	[Fe/H]*	Z†	*Reference*
N6067	0.016			[H3]
M11	0.063			[H3]
Pleiades	0.063			[H3]
Praesepe	0.63			[H3]
Hyades	0.7	0.20	0.032	[H7]
N2477	0.7	0.10	0.020	[H7]
N5822	0.9	−0.10	0.016	[H7]
N2360	1.3	−0.07	0.017	[H7]
N7789	1.6	−0.25	0.011	[H7]
N752	1.7	−0.10	0.016	[H7]
N3680	1.8	0.00	0.020	[H7]
M67	3.2	0.00	0.020	[H7]
N6819	3.5	−0.16	0.014	[H7]
N2243	3.9	−0.51	0.0061	[H7]
N2506	4.0	−0.54	0.0058	[H7]
N2420	4.0	−0.39	0.080	[H7]
N188	5.0	0.00	0.020	[H7]
Mel 66	6.5	−0.64	0.0046	[D2]
Globular clusters				
M71	14.3 ± 2.9	−0.4	0.008	[S5]‡
47 Tuc	17.2 ± 3.4	−0.64	0.0046	[S5]
M5	19.0 ± 3.8	−1.58	0.0005	[S5]
M13	20.2 ± 4.0	−1.73	0.0004	[S5]
N6752	14.8 ± 3.0	−1.52	0.0006	[S5]
M3	17.2 ± 3.4	−1.69	0.0004	[S5]
M15	15.1 ± 3.0	−2.15	0.00015	[S5]
M92	15.6 ± 3.1	−2.19	0.0001	[S5]

* [Fe/H] is defined on p. 72.

† The heavy-element abundance Z is calculated from [Fe/H] using Eq. (2.22).

‡ The ages and Z values taken from [S5] are calculated assuming a helium abundance $Y = 0.23$.

We see in Fig. 2.25 that the later stages of the evolutionary tracks of stars are beginning to define an almost vertical locus in the HR diagram. This is the *Hayashi locus* for stars that are convective throughout, first derived by the Japanese astronomer Chushiro Hayashi in 1961. No stable stars can exist to the right of this curve, which is the *Hayashi forbidden region.* The stars evolve up this locus, the central

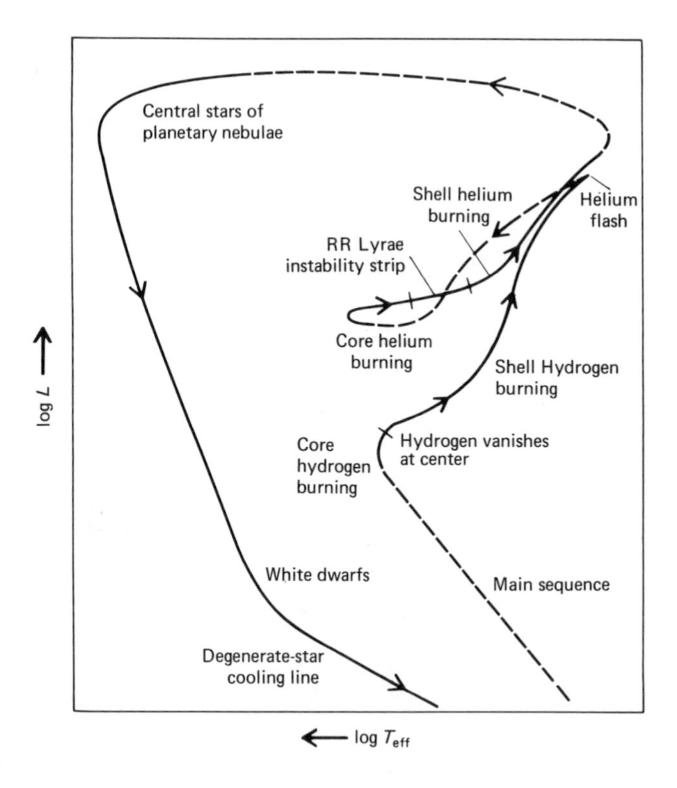

FIGURE 2.28

Schematic evolution track for a representative low-mass globular cluster star from the main sequence to its ultimate demise as a white dwarf. The major energy sources at several key phases are indicated. Dashed lines correspond to episodes of very rapid evolution, during which details of the structure of the star are, at present, not too well known. (From [M1].)

temperature steadily increasing until it is high enough for helium burning to commence. When it does so, the sudden increase of temperature in the degenerate material of the stellar core results in an explosive change in the central regions, and the star moves off rapidly to the left in the HR diagram. The details of this *helium flash* cannot be followed numerically in the models, and it is indicated by the dashed curve in Fig. 2.28. The star then settles to a stage of core helium burning together with hydrogen burning in a shell surrounding this core, and evolves horizontally to the right, defining the horizontal branch in globular clusters. When all the helium in the central regions has fused to carbon, a degenerate central core forms in which the temperature increases until eventually, if the star is massive enough, carbon is ignited in a second explosive event. Whether or not this happens, these final stages are accompanied by substantial loss of material from the surface of the star, possibly in a single event in which the outer layers of the star are ejected. The remaining core of the star has a very high surface temperature (100,000 K), and the star moves over to the far left of the HR diagram, illuminating the expelled gas as a planetary nebula before cooling off to become a white dwarf.

POST-MAIN-SEQUENCE EVOLUTION OF HIGH-MASS STARS ($M > 5M_\odot$)

The main difference between high- and low-mass stars when they start to exhaust their hydrogen and leave the main sequence is that high-mass stars have a convective core. The convective motion mixes fresh hydrogen down to the center to replace that fused to helium, and the whole convective core maintains a homogeneous composition. When hydrogen is finally exhausted, this happens uniformly throughout the convective core, and the mass of this core may then exceed a critical value, the *Schönberg–Chandrasekhar* limit. This is the maximum mass for a slowly contracting isothermal helium core to be able to support the pressure of the rest of the star. If the core mass exceeds this value, the core collapses rapidly and the outer layers readjust drastically. The star moves very quickly from the main sequence across to the red supergiant branch, so quickly that very few stars are found in the intervening region of the HR diagram (a zone known as the Hertzsprung gap). Stars whose helium cores have masses below the Schönberg–Chandrasekhar limit proceed to the red supergiant branch at a more leisurely rate.

The details of the post-main-sequence evolution of more massive stars are extremely complex. Fig. 2.25 illustrates models by Iben [I1] for stars of mass 5, 9, and $15M_\odot$, with helium abundance $Y = 0.27$ and heavy-element abundance $Z = 0.02$. The internal structure of a $5M_\odot$ star as a function of age is illustrated in Fig. 2.29. When hydrogen is exhausted in the core, hydrogen burning commences

FIGURE 2.29

Internal structure of a star of 5 $M_\odot$ as a function of age. M—mass inside a particular radius; M_s—total mass of star. In the heavily hatched areas the named nuclear reactions occur. Convection zones are shown with grey shading . At point 3, hydrogen burning in the core ceases and begins in a shell, continuing to point 11. At point 5, helium core burning begins, ending at point 10, where helium shell burning begins. At point 11, the star is approaching the Hayashi convective locus and has a deep outer convective zone. (From [K2].)

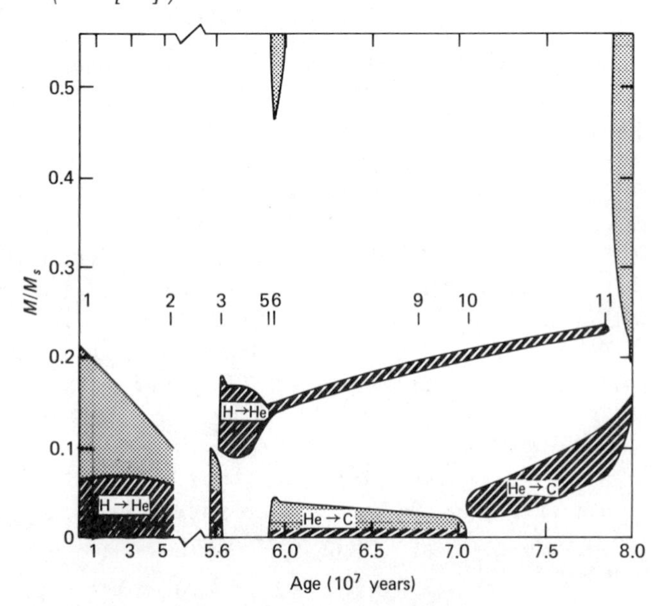

in a shell surrounding the helium core. Eventually the core becomes hot enough for helium to burn to carbon there. After a while the helium, too, becomes exhausted in the core, and helium burning starts in a shell. At this stage there are two shells in which nuclear reactions are occurring, an inner one in which helium is burning to carbon, and an outer one in which hydrogen is burning to helium.

A few stars are found in the Hertzsprung gap, and a very important star of this type is the Cepheid variable star. The outer layers of these stars pulsate in and out regularly on a time scale of 2–100 days, in a way analagous to the resonance of a mechanical system. This causes a cyclic variation in the brightness of the star (see Fig. 1.3). Cepheid variables occupy a rather narrow strip of the HR diagram (Fig. 2.30) and satisfy a very important relationship between period and luminosity (see

FIGURE 2.30

Composite HR diagram, absolute visual magnitude M_V *versus* $(B\text{-}V)_0$*, the B-V color corrected for the effects of reddening by dust, for 10 open clusters, showing the Cepheid instability strip* C*, with the loci occupied by Cepheids with a particular period* P*. The age of each cluster, whose main-sequence turn-off point is at a given* M_V*, is shown on the right. (From [S1].)*

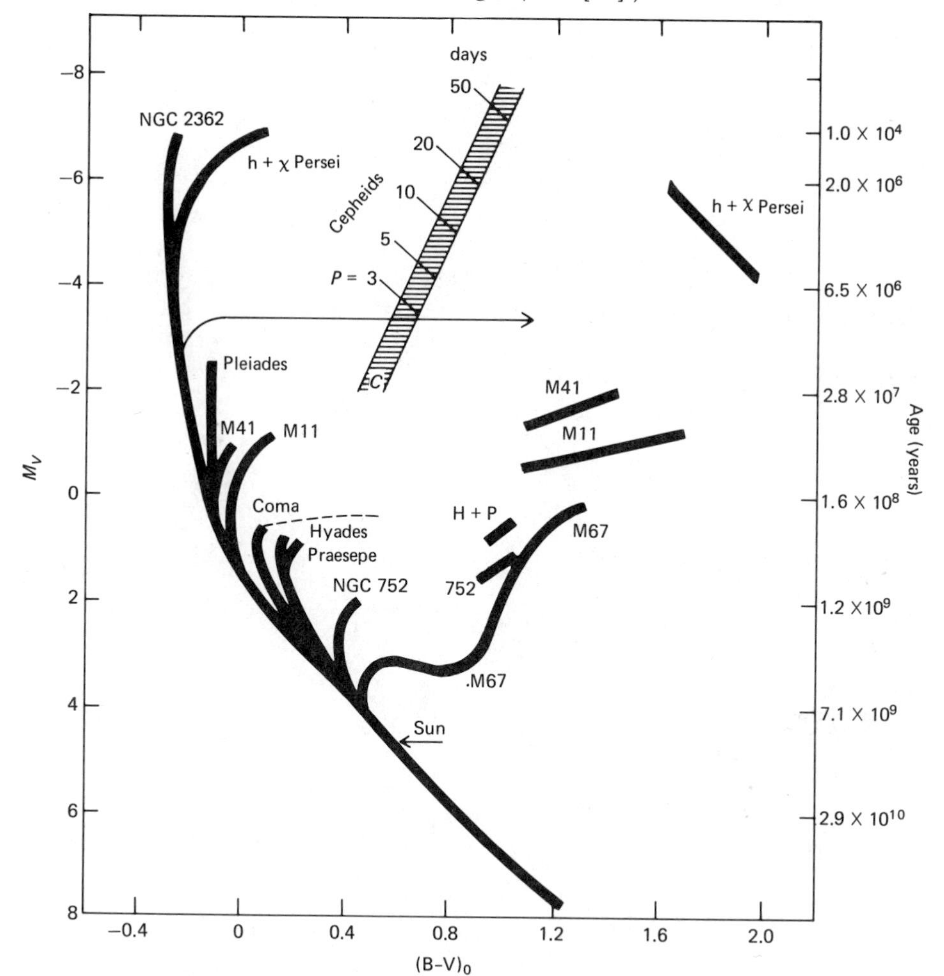

Fig. 3.1). As we shall see in Chap. 3, this relationship provides one of the most reliable of extragalactic distance indicators.

LATE STAGES OF EVOLUTION

The final stages of evolution of a star depend very sensitively on the mass of the star. A star of mass less than $0.1M_\odot$ never gets hot enough to burn hydrogen, so simply contracts and cools off to become a black dwarf (Fig. 2.31). Stars of mass between 0.1 and $0.4M_\odot$ evolve only a little way to the right of the main sequence and never achieve helium core burning before they evolve to the left of the HR diagram and cool to become white dwarfs. We have already shown in Fig. 2.28 the late evolution of stars of mass $1-2M_\odot$. It is believed that most stars in this mass range become planetary nebulae and that a star throws off a substantial fraction of its original mass before reaching this stage. Strong mass loss is indeed observed in stars on the upper part of the red giant branch, the *asymptotic giant branch,* where stars return to the red giant branch for the second time and are burning helium in one shell and hydrogen in another. Such stars often show long-period variations, with a period of 100 – 600 days, and are called *Mira variables* after the star Mira first studied by Fabricius in 1596. Also in this region of the HR diagram are found *carbon stars,* stars with carbon-rich atmospheres, believed to be the result of the convection of nucleosynthesis products to the star's surface. The exact mechanism of the mass loss is still not completely understood.

The late stages of a star of mass greater than $5M_\odot$ are more dramatic. Such a star will, when helium is exhausted at the center, start to fuse helium to carbon and oxygen, and so on through a complex sequence of reactions up to iron. It will develop an onionlike structure, with a different composition in each layer and nuclear burning proceeding in several different shells simultaneously (Fig. 2.32). However, the nuclear reactions by which iron fuses to make heavier elements are endothermic; they absorb energy rather than giving it out. The iron core therefore

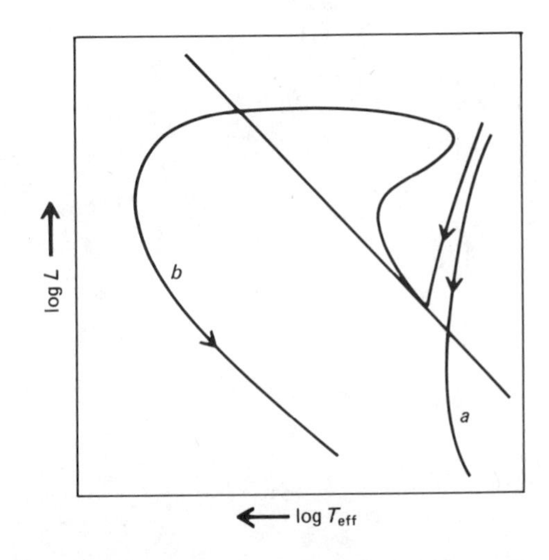

FIGURE 2.31
Schematic evolution of stars of very low mass in the HR diagram. Log L *versus log* T_{eff}. *Curve* a *— star of mass lower than* $0.1M_\odot$, *which cools off to become a black dwarf without ever commencing hydrogen burning; curve* b *— star of mass between 0.1 and* $0.4\ M_\odot$, *which after a long life on the main sequence becomes a red giant before moving to the left of the diagram and then cooling off to become a white dwarf. (From [T1].)*

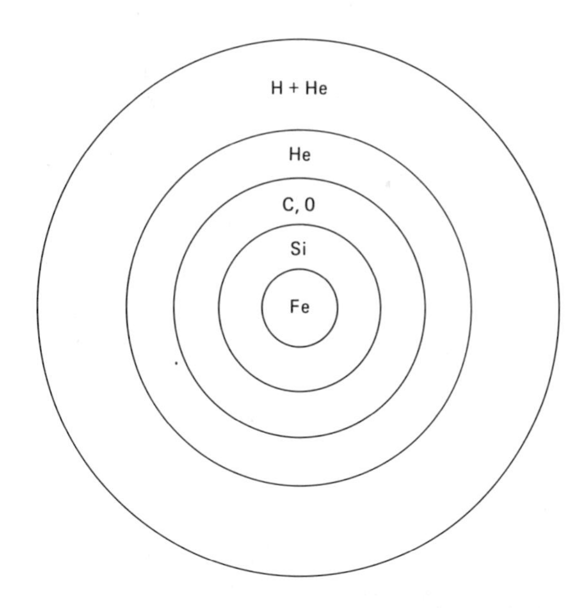

FIGURE 2.32
Chemical composition of highly evolved massive star with iron core. (From [T1].)

contracts and heats up, but no new energy is generated. Soon its mass exceeds the Schönberg–Chandrasekhar limit and so the core collapses. Eventually the temperature becomes so high that the heavy elements are dissociated into a mixture of protons and neutrons. Meanwhile the layers outside the iron core start to collapse too. There is now a powerful thermonuclear explosion which reverses the collapse of the outer parts of the star and blows them off. The core is compressed down to form a *neutron star.* The sudden energy generation brightens up the star by a huge factor, a thousand million or more. The star has become a *supernova.* The enormous luminosity of supernovae makes them detectable to huge distances, 100 Mpc or more. We shall study the appearance of supernovae in more detail in Sec. 3.2.

PRE-MAIN-SEQUENCE EVOLUTION

We know that new stars are being formed within clouds of interstellar gas in the disk of our Galaxy, but the details of the process are not yet understood. A cloud is compressed together by one of several possible mechanisms. Spiral galaxies appear to be unstable to the formation of a spiral-shaped compression wave which rotates through the disk, triggering star formation as it passes each locality. Another mechanism which can compress a cloud is the blast associated with a nearby supernova explosion. And a third possibility is that the large increase in pressure associated with the development of an HII region results in compression of any gas engulfed by the growing HII region. Regions of above average density then undergo gravitational collapse, increasing their density until they are opaque to their own radiation. As contraction continues, the central temperature steadily increases and the resulting "protostar" becomes a source of infrared radiation. When the central temperature becomes high enough, hydrogen burning starts, the contraction ceases, and the star has arrived on the main sequence.

During the pre-main-sequence phase we may expect to see the star completing its accretion of gas from the cloud in which the star is forming. Evidence is indeed

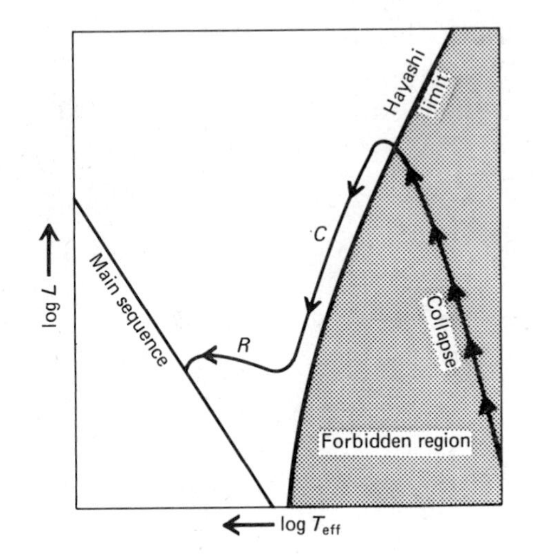

FIGURE 2.33

Schematic evolution track in HR diagram, log L *versus log* T_{eff} *for a pre-main-sequence star. On the Hayashi track* C *the star is convective throughout. On the portion of the track labeled* R *the core is radiative. Nuclear reactions begin when the star reaches the main sequence (From [M1].)*

seen for accretion in T Tauri stars, which are believed to be stars in the pre-main-sequence phase with masses of around $1M_\odot$ or less. The name arises from the prototype of this class, the star T Tauri in the Taurus Dark Cloud, a nearby cloud of gas and dust. There is also evidence for strong stellar winds in some T Tauri stars, and these winds probably sweep away the gas and dust which have not formed into the star or, if it has one, its planetary system.

The last stages of this contraction to the main sequence are indicate schematically in Fig. 2.33. The protostar contracts and increases its luminosity until it reaches the Hayashi locus for stars that are convective throughout. It then descends the Hayashi locus for a while before moving across to the main sequence.

2.8 Interstellar dust and gas

Spread between the stars of the Milky Way is interstellar matter, composed of 70–75% hydrogen, 25–30% helium, and about 2% heavy elements by mass. The hydrogen and helium are in the gaseous state, while the heavy elements can be in either the gaseous or the solid state, the latter in the form of tiny grains of interstellar dust.

Neglect of the effect of interstellar dust on starlight was the major factor in delaying understanding of the scale of our Galaxy and of the enormous distances to other galaxies. As noted in Chap. 1, the problem was that although the possibility of obscuration by dust was considered, measurement techniques were insufficiently refined to demonstrate its existence. The neglect of dust obscuration led Kapteyn to a model of the universe which consisted of a disklike distribution of stars with the sun near the center. Only when Shapley came to study the distribution of globular clusters, many of which are at fairly high galactic latitudes and not too drastically affected by dust obscuration, did the asymmetry of the Milky Way as seen from the sun become apparent.

Fig. 1.4 illustrates the overall visual appearance of the Milky Way from earth. The dark patches are now known to be giant clouds of molecules and dust. The dust absorbs and scatters most of the visible and ultraviolet light from more distant stars. As we shift to infrared wavelengths (Fig. 2.34), the true appearance of the Milky Way becomes apparent, for the extinction efficiency of the small dust grains declines with increasing wavelength. The grains have a characteristic size of 0.1 μm and are believed to be composed of silicates and amorphous carbon. A substantial fraction of the heavy elements in the interstellar gas may be locked up in the form of grains.

The extinction by dust at wavelength λ is measured by the number of magnitudes by which the brightness of a source is dimmed, denoted by A_λ. Fig. 2.35 shows a recent summary by B.D. Mathis and J.S. Savage of how A_λ depends on wavelength (see also App. A.8). We see that A_λ falls sharply as λ increases. This variation of A_λ with λ means that the color of a source is changed by intervening dust—the source is reddened. This can be quantified by measuring, say, B-V for distant stars of known spectral type and comparing the result with that for nearby unreddened stars of the

FIGURE 2.34

Full-sky photo of our Galaxy taken on infrared film. Note how much more clearly the central regions of the Galaxy stand out, compared with an optical picture (Fig. 1.4). (From [C5].)

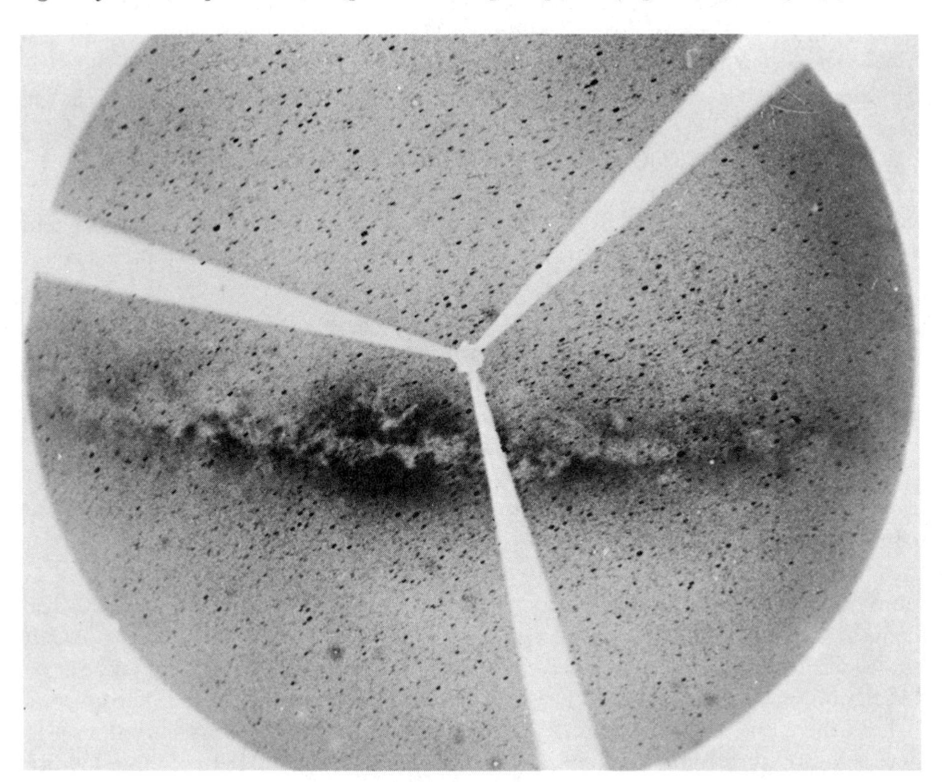

FIGURE 2.35

Average normalized interstellar extinction $E(m_\lambda\text{-}V)/E(B\text{-}V)$ *versus* $1/\lambda$, *where* λ *is the wavelength in micrometers.* $E(m_\lambda\text{-}V)$ *is the excess in the* $m_\lambda\text{-}V$ *color due to reddening by interstellar dust. The plotted locus can be converted to total normalized extinction* $A_\lambda/E(B\text{-}V)$, *where* A_λ *is the extinction at wavelength* λ *in magnitudes, by adding 3 to the quantity plotted. (Adapted from [S7].)*

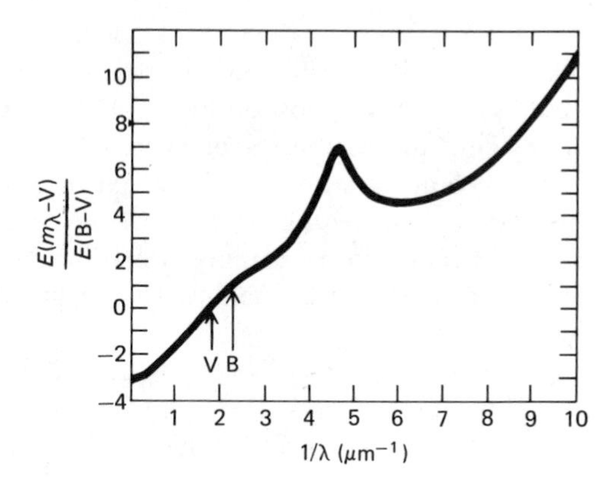

same type. The difference E(B-V) is called the *color excess* and is proportional to the visual extinction A_V,

$$A_V = R \cdot E(\text{B-V}) \tag{2.23}$$

where R is the coefficient of proportionality, called the reddening ratio (or the ratio of total to selective absorption), and has a value of about 3.0 (Box 2.4). Once R is known, a measurement of the color excess E(B-V) can be converted to the visual extinction A_V in magnitudes, using Eq. (2.23). Eq. (2.23) would hold exactly if B and V were measures of monochromatic fluxes of radiation. However, because the photometric bands have a finite width, R will depend slightly on the spectrum of the star and on E(B-V) (Box 2.4).

The interstellar gas in our Galaxy is found in four main phases [S11]:

1 GIANT MOLECULAR CLOUDS. Composed largely of molecular hydrogen, these vast clouds have a size of about 100 pc, and the total mass of a cloud is about 100,000 $M_\odot$, with a mean density of hydrogen molecules $\bar{n} \sim 500$ cm^{-3}, about 1000 times the average density of interstellar gas. However, each cloud is really a complex of smaller clouds, some of which are very much denser. In these dense molecular clouds, new O and B stars are often seen to be in the process of formation. Lower mass stars ($\lesssim 1 M_\odot$) may be forming steadily throughout the complexes, since many T Tauri stars, which as noted above are believed to be the precursors of solar-type stars, are found in the nearby well-studied molecular clouds. The correlation of dust with molecules can be studied for these nearby clouds ($d < 1$ kpc) by comparing the number of magnitudes of visual extinction by dust A_V, inferred from star counts toward the cloud and toward adjacent regions of the sky, with the total number of molecules in a cylindrical column of 1 cm^2 cross-sectional area (the column density). The latter may be estimated from ultraviolet studies of hydrogen molecules or from millimeter-wave studies of carbon monoxide. The amount of extinction by dust can also be measured by the reddening of the light from stars behind the cloud and using Eq. (2.23). Fig. 2.36 illustrates the distribution on the sky of dust clouds found by star counts. The dust plays a crucial role in allowing

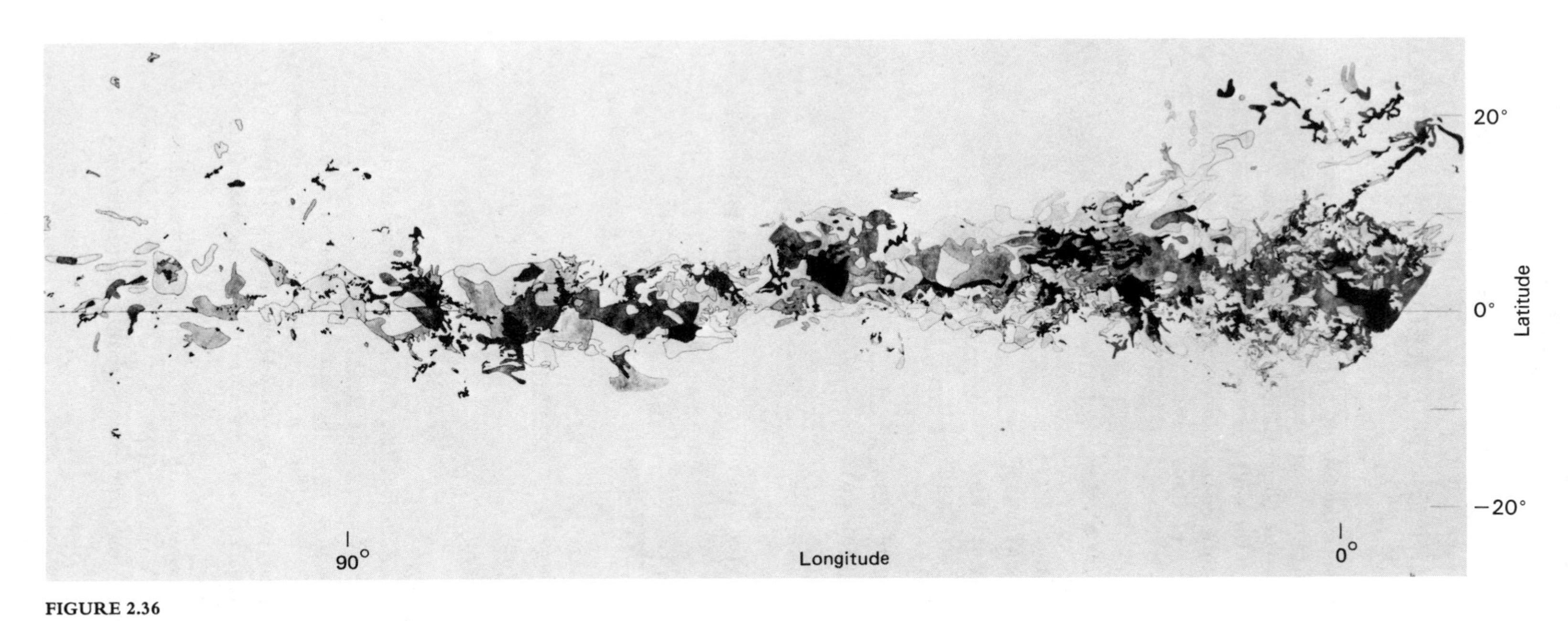

FIGURE 2.36

Distribution on the sky, in galactic coordinates (l, b), *of dust clouds found by star counts, for* $0° \leq l \leq 120°$. *Darker shading shows clouds with higher extinction. (From [L4].)*

BOX 2.4 THE VALUE OF THE REDDENING RATIO $R = A_V/E(\text{B-V})$

Most studies of interstellar dust in the Galaxy yield $R \sim 3.1$ [S7] with an uncertainty of ± 0.2 [M1], though certain regions (such as the ρ Ophiuci dust cloud) give higher values of R, probably due to larger than average dust grains.

Blanco [B8] pointed out that the width of the B and V photometric bands should result in a dependence of R on the type of star being observed and on the total dust extinction A_V. His calculations gave:

Assumed Stellar Spectrum	A_V	R
Blackbody, $T = 25{,}000$ K	1.02	3.0
	8.01	3.24
Vega (A0, $T \sim 10{,}000$ K)	1.01	3.07
	7.86	3.28
Blackbody, $T = 8000$ K	1.00	3.17
	7.79	3.38
β Pegasus (M2, $T \sim 3000$ K)	0.99	4.01
	7.69	4.12

Blanco summarized his results for stars of spectral type G5 and earlier by

$$A_V = [3.1 + 0.3(\text{B-V})_{\text{intrinsic}}] \frac{E(\text{B-V})}{1 - 0.03A_V}.$$

More recently Olsen [O2] derived the approximation

$$R = 3.25 + 0.25(\text{B-V})_{\text{intrinsic}} + 0.05E(\text{B-V})$$

valid for $(\text{B-V})_{\text{intrinsic}} < 1.4$ and $E(\text{B-V}) < 1.5$. This is the formula adopted by de Vaucouleurs for his distance scale papers [V2].

Lee [L1] deduced $R = 3.6$ from infrared and visual observations of nearby M supergiants. Morgan and Nandy [M3] found a value $R = 3.21 \pm 0.13$ for early type supergiants in the Large Magellanic Cloud, while Humphreys [H10] found a good fit to infrared observations of late type supergiants in the Large Magellanic Cloud using Lee's color ratios.

In this book I follow Sandage and Tammann in adopting $R = A_V/E(\text{B-V}) = 3.0$, independent of spectral type, so that $A_B/E(\text{B-V}) = 4.0$. The error associated with this assumption is likely to be small compared with other uncertainties in the distance scale.

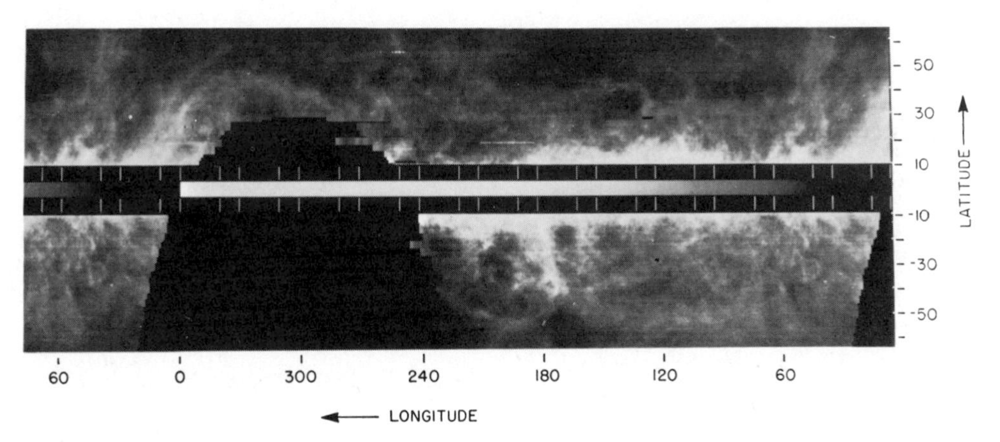

FIGURE 2.37

Distribution on the sky, in galactic coordinates (l, b), of neutral atomic hydrogen. Brighter shading denotes regions of higher column density of atomic hydrogen. (From [H6].)

hydrogen molecules to form in these clouds, shielding the molecules from the general ultraviolet radiation field in the Galaxy, which would tend to dissociate the molecules.

2 NEUTRAL HYDROGEN CLOUDS. These are gas clouds of lower column density in which $A_V < 1$. Such clouds do not provide sufficient shielding for molecules to form, and the hydrogen here is therefore found mainly in neutral atomic form (referred to as HI, in distinction to HII, ionized hydrogen, and H_2, molecular hydrogen). The comparatively cool neutral hydrogen atoms can be detected through their characteristic radio emission at a wavelength of 21 cm. (This effect was predicted by H.C. van de Hulst in 1944 and first detected by H.I. Ewen and E.M. Purcell in 1951.) In these neutral clouds the density of hydrogen atoms lies in the range of 0.1 – 100 cm^{-3}. The distribution on the sky of neutral hydrogen is illustrated in Fig. 2.37. Again there is a good correlation of visual extinction, characterized by the reddening E(B-V) and the column density of neutral hydrogen atoms N(HI) of the form

$$N(\mathrm{HI}) = 4.8 \times 10^{21}\, E(\mathrm{B\text{-}V}) \quad [\mathrm{cm^{-2}}]. \tag{2.24}$$

Fig. 2.39 shows the correlation of the total column density in hydrogen atoms, $N(\mathrm{HI}) + 2N(\mathrm{H_2})$, with the reddening, with mean line

$$N(\mathrm{HI} + \mathrm{H_2}) = 5.8 \times 10^{21}\, E(\mathrm{B\text{-}V}). \tag{2.25}$$

3 IONIZED GAS. Massive O and B stars are copious sources of ultraviolet radiation and ionize the gas in their vicinity out to considerable distance. App. A.13

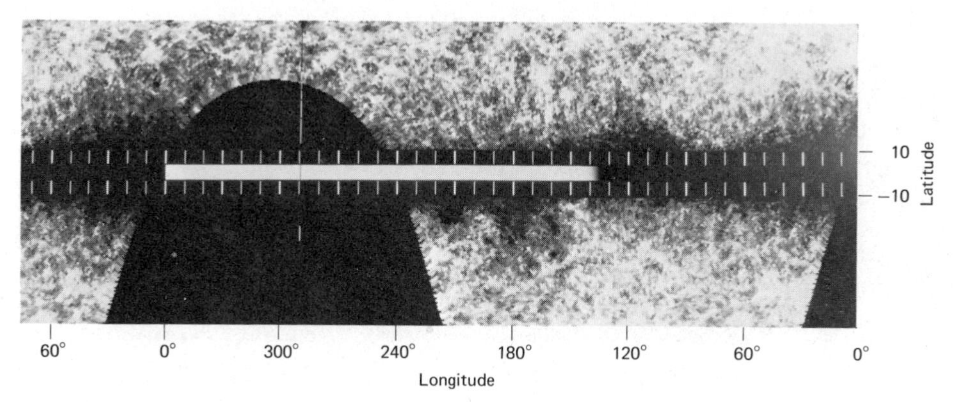

FIGURE 2.38

Distribution on the sky, in galactic coordinates (l, b), of the Lick Observatory counts of galaxies. Brighter shading denotes directions of higher galaxy numbers per square degree. (From [H6].)

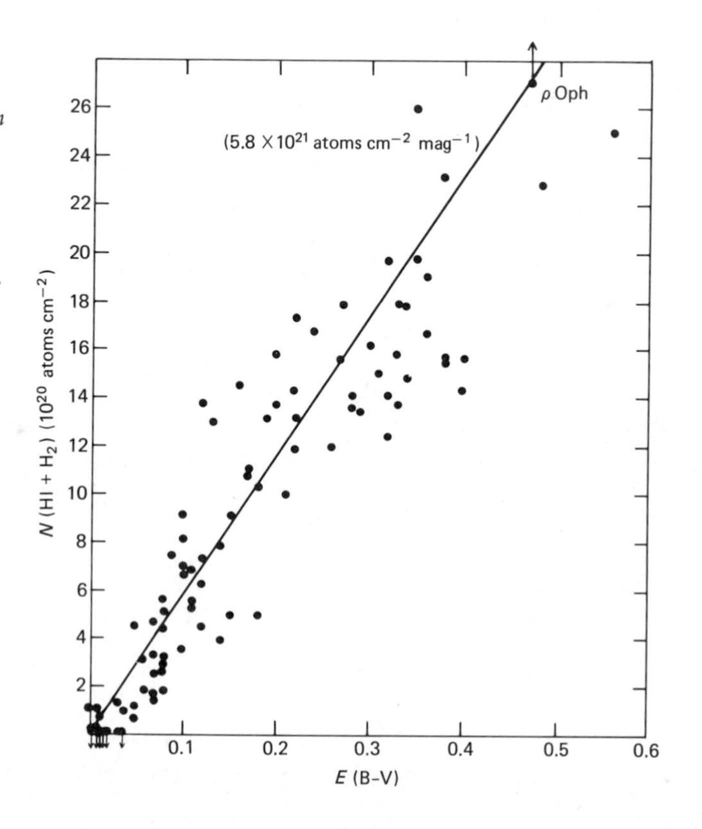

FIGURE 2.39

Correlation between total column density of hydrogen atoms N(HI + H_2), including those in molecules, and interstellar reddening E(B-V). The straight line is Eq. (2.25). (From [S7].)

shows the extent of the ionized zone, or HII region, for stars of different spectral types, neglecting the effect of dust. Since massive stars must be young objects, HII regions, which can be detected through their emission lines and continuous (free–free) radiation at optical, infrared, and radio wavelengths, are good indicators of where new massive stars are forming. Thus HII regions, young OB associations, and dense molecular clouds are often associated with each other. Much of the optical and ultraviolet radiation from young HII regions is absorbed by dust and reemitted at infrared wavelengths.

4 HOT DIFFUSE MEDIUM. Spread between the clouds of neutral and atomic hydrogen is a tenuous very hot gas ($n_H \sim 10^{-3}$ cm^{-3}, $T \sim 10^5$ K), believed to be the accumulated remnants of supernova explosions. This hot gas only contributes a few percent of the total mass of gas in the Galaxy.

Fig. 2.40 illustrates where these four different gas phases lie in a plot of hydrogen column density A_V against mass, with indications of how the transition is made from one phase to another.

The distribution on the sky of neutral atomic hydrogen can be compared with the number of galaxies per square degree from the Lick observatory galaxy counts, shown in Fig. 2.38. Although the true number of galaxies brighter than a given magnitude may vary from direction to direction due to galaxy clustering, there is no evidence that any large-scale anisotropy is present. The systematic decline in the number of galaxies per square degree as the plane of the Milky Way is approached therefore indicates the role of dust in dimming the light from the galaxies. Carl Heiles and E.B. Jenkins [H6] have shown that there is a good correlation between the two distributions shown in Figs. 2.37 and 2.38.

The ideal way to estimate the extinction to extragalactic objects is to estimate by how much stars far out in the halo of the Galaxy, near the desired line of sight, are reddened. This has been done for a number of galaxies (see Table 3.9). Alternatively the amount of extinction could be estimated from the average number of galaxies per square degree or from the column density in neutral hydrogen, for example from the maps of neutral hydrogen which Heiles has made of most of the sky [H5]. The latter is preferable since galaxy counts are affected by galaxy clustering. In 1978 David Burstein and Carl Heiles carried out the most elaborate study to date of the correlation between reddening, galaxy counts, and neutral hydrogen column density [B10]. From small discrepancies in the latter two methods they conclude that the gas-to-dust ratio [equivalent to the constant in Eq. (2.24)] varies with direction on the sky, and they give a prescription for deducing E(B-V) from the neutral hydrogen density and galaxy counts combined. They also give contour maps of the distribution of E(B-V) in galactic coordinates [B11].

In practice the interstellar extinction has usually been estimated from some simple analytical formula, of which the most widely used is

$$A_V = \alpha \csc b \tag{2.26}$$

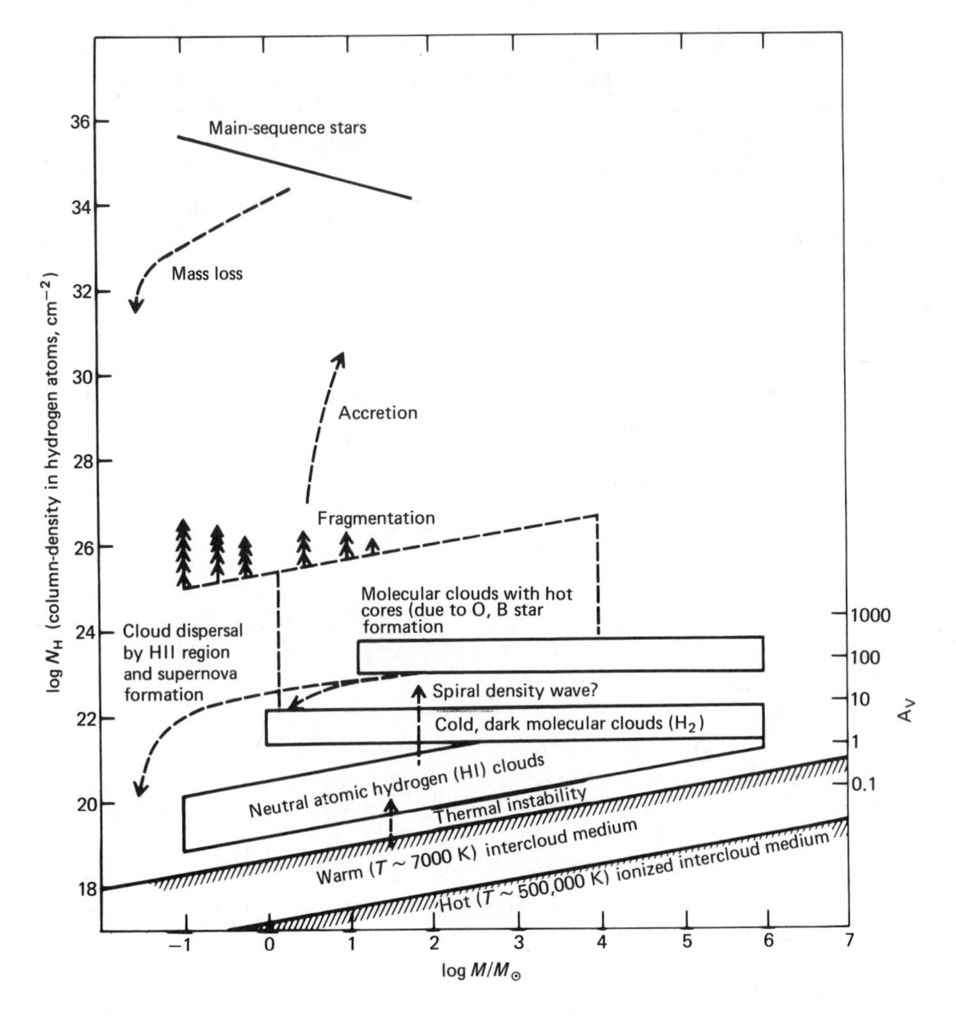

FIGURE 2.40

Cycle of gas and stars in the Galaxy in a plot of hydrogen column density against mass, in solar units. The shaded lines at the bottom denote the intercloud medium, the boxes indicate the locations of different types of interstellar gas clouds, and the solid curve at the top shows the location of main-sequence stars. Broken lines indicate mechanisms by which gas changes from one phase to another.

where b is the galactic latitude and α is a coefficient of proportionality. This formula corresponds to the assumption of a plane parallel dust distribution, with dust density depending only on distance from the galactic plane. This is certainly a gross oversimplification of the dust distribution in the Galaxy. A more sophisticated formula is used by the de Vaucouleurs and Corwin in the *Second Reference Catalogue of Bright Galaxies* [V3], in which α varies with galactic latitude and longitude according to a formula depending on 21 parameters (Box 2.5). This formula has been determined from galaxy counts, bright galaxy colors, and optical-to-radio-emission ratios, and is intended to give the average extinction with a resolution of about 40° on the sky. The most controversial aspect of their interstellar extinction formula is that it predicts about 0.2 mag of extinction in the polar caps, whereas there is a substantial body of evidence from reddening and neutral hydrogen studies for essentially zero extinction toward the north galactic pole and little toward the south galactic pole (see, for example, the studies by Burstein and Heiles [B10], [B11]). The latter authors argue that galaxy counts, on which the formula by de Vaucouleurs et al. is mainly based, show too much scatter to determine the slope of the csc b law near the poles. Extinction with no associated reddening could in principle be caused by large dust grains.

Allan Sandage proposed in 1973 an ad hoc modification of Eq. (2.26) to take account of the extinction-free polar caps (see Box 2.5), equivalent to a plane parallel distribution of dust with two conical dust-free regions centered on the sun [S3]. Apart from the oversimplified and unrealistic nature of this picture, the Sandage formula gives interstellar extinction values lower than those obtained directly from reddening studies (see Table 3.9). The formula of de Vaucouleurs et al., on the other hand, tends to give higher values than those found from reddening studies. Sandage's formula was based on a picture of the interstellar dust as distributed like cirrus clouds in the atmosphere [S4], a picture which has been amply confirmed by the Infrared Astronomy Satellite (IRAS). IRAS certainly sees patchy emission from interstellar dust near the galactic poles; a preliminary estimate for the value of A_V for this dust is ~ 0.1 mag.

The best compromise appears to be the formula proposed in 1981 by Richard Fisher and Brent Tully [F2], based on the Burstein and Heiles study. They give an approximate formula for the distribution of neutral hydrogen column density as a function of galactic latitude and longitude (Box 2.5) and combine this with Burstein

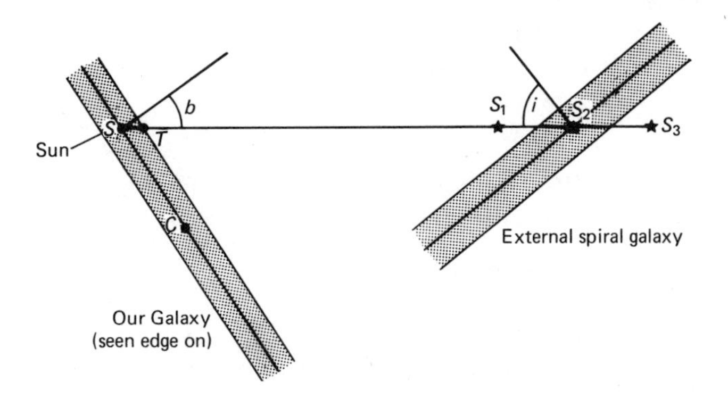

FIGURE 2.41

Geometry of dust in an external spiral galaxy. In addition to the obscuration due to dust in the Milky Way (along ST*), a star in an external spiral galaxy will suffer internal obscuration depending on the inclination of the galaxy* i *and on whether the star is above (S_1), in the middle of (S_2), or behind (S_3) the dust layer.*

BOX 2.5 INTERSTELLAR REDDENING CORRECTIONS

1 SANDAGE [S3] proposed the reddening correction

$$E(\text{B-V}) = \begin{cases} 0.03(\csc b - 1), & |b| < 50^\circ \\ 0, & |b| \geq 50^\circ \end{cases}$$

where b is the galactic latitude. Then $A_{pg} = A_B = 4E(\text{B-V})$ and $A_V = 3E(\text{B-V})$. This is based on evidence for essentially no reddening at the galactic poles and is meant to represent a cirruslike cloud distribution. The above formula leads to a discontinuity in the reddening correction at $|b| = 50^\circ$, so the formula is interpolated smoothly between $|b| = 40^\circ$ and 50°, that is, for $|b| = 40^\circ, 42^\circ, 44^\circ, 46^\circ, 48^\circ$, and 50°, $A_V = 3E(\text{B-V}) = 0.05, 0.04, 0.03, 0.02, 0.01$, and 0.0, respectively.

2 de VAUCOULEURS ET AL. [V3] adopt a formula* depending on both galactic latitude b and longitude l,

$$A_B = \begin{cases} 0.19(1 + S_N \cos b)\,|C|, & b > 0 \\ 0.21(1 + S_S \cos b)\,|C|, & b < 0 \end{cases}$$

where

$$\begin{aligned} S_N(l) &= 0.1948 \cos l + 0.0725 \sin l + 0.0953 \cos 2l \\ &\quad - 0.0751 \sin 2l + 0.0936 \cos 3l + 0.0639 \sin 3l \\ &\quad + 0.0391 \cos 4l + 0.0691 \sin 4l \\ S_S(l) &= 0.1749 \cos l - 0.0112 \sin l + 0.1438 \cos 2l \\ &\quad - 0.0180 \sin 2l - 0.0897 \cos 3l - 0.0013 \sin 3l \\ &\quad + 0.0568 \cos 4l + 0.0433 \sin 4l \\ C &= \csc(b + 0^\circ.25 - 1^\circ.7 \sin l - 1^\circ.0 \cos 3l). \end{aligned}$$

This formula predicts $A_B = 0.19$ at the north galactic pole and 0.21 at the south galactic pole.

---

and Heiles's version of Eq. (2.24), omitting a term describing the variation of the gas-to-dust ratio with direction. The Fisher and Tully formula gives good agreement with reddening studies toward nearby galaxies (see Table 3.9), and with the Burstein and Heiles contour maps of E(B-V) [B11]. In this book I therefore adopt the Fisher–Tully formula where direct reddening estimates are not available.

However, it is important to bear in mind that the interstellar extinction has a very patchy distribution, so that the expressions for the variation of extinction with direction summarized in Box 2.5 have only an average validity. Ideally the extinc-

To calculate $E(\text{B-V})$ and A_V, the following relation is used:

$$A_B = (R + 1)E(\text{B-V})$$

where

$$R = \frac{A_V}{E(\text{B-V})} = \frac{3.1 + 0.3(\text{B-V})}{1 - 0.02A_B} \quad \text{(See Box 2.4)}$$

3 FISHER AND TULLY [F2] use the correlation between extinction and neutral hydrogen column density of Burstein and Heiles [B10],

$$A_B = -0.149 + 6.41 \times 10^{-4} N_H$$

where N_H is in units of 10^{18} atoms per square centimeter, in which a term representing the variation of the gas-to-dust ratio with direction has been dropped. A_B is taken as 0 if $N_H \leq 232$ (this can be interpreted as being due to an isotropic distribution of gas which is free of dust). They model the variation of N_H with direction as

$$N_H = \begin{cases} 323 \csc |b|, & b < 0 \\ 323 \csc |b| + [105 - 3.8b - 89 \cos(l - 140°)], & b > 0. \end{cases}$$

This gives $A_V = 0.06$ at the south galactic pole and 0.00 at the north galactic pole. They note that the above formula fails to predict the larger values of N_H at $150° < l < 195°$ and $b > -45°$ in the south, and at $330° < l < 20°$ and $b < 40°$ in the north.

Alternatively, $E(\text{B-V})$ can be read from the contour maps of Burstein and Heiles [B11] and then combined with R from Box 2.4.

*** The constants in this formula have been provided by G. de Vaucouleurs (private communication) and differ from those given in the introduction to the *Second Reference Catalogue of Bright Galaxies* [V3], though they were the constants actually used for the entries in the catalog.**

---

tion should be measured toward each individual object under study.

In correcting the magnitude of a galaxy for the effects of dust, we have to correct not only for the extinction in our own Galaxy, but also for the effects of extinction in the galaxy we are observing. A star in the disk of a distant galaxy may be expected to suffer some extinction due to the dust in the disk of this galaxy, unless the star lies by chance on the near side of the disk (Fig. 2.41).

Box 2.6 summarizes the forms of the corrections for internal extinction proposed by E. Holmberg, de Vaucouleurs, and Fisher and Tully, and a comparison of

BOX 2.6 CORRECTION FOR INTERNAL EXTINCTION IN GALAXIES

The inclination angle is defined as $i = 0°$ for a face-on galaxy and $i = 90°$ for an edge-on galaxy. We have to distinguish two types of correction:

1 The correction of the total magnitude of the galaxy for the effects of internal extinction as a result of its inclination to our line of sight. This *inclination correction* restores the galaxy's integrated magnitude to the value it would have if the galaxy were face on. A further correction for the effect of the dust between us and the midplane of the galaxy yields the *absorption-free magnitude,* the magnitude the galaxy would have if it had no interstellar dust.
2 The *total internal extinction correction* is the internal extinction suffered by a star situated in the plane of symmetry of the galaxy, for example, a Population I object with no additional extinction due to local (circumstellar) dust. This might be expected to depend on the position of the star in the galaxy (see, for example, the observations by Kumar [K3] in M31), but in the absence of detailed models of the dust distribution in external galaxies we shall assume that this correction is the same as that required to give an absorption-free magnitude.

Three forms for these corrections have been proposed:

1 HOLMBERG [H8], by analogy with the csc b law for interstellar extinction in our Galaxy (eqn. 2.26), proposed an inclination correction

$$A^{i}_{pg} = \alpha(\sec i - 1)$$

where

$$\sin^2 i = \frac{1 - q^2}{1 - q_0^2}$$

with $q = b/a$, the ratio of the observed minor and major axes of the galaxy, and $q_0 = 0.2$, and where

$$\begin{aligned} \alpha &= 0.43 \text{ for galaxies of type Sa-b} \\ &= 0.28 \text{ for galaxies of type Sc} \\ &= 0 \text{ for ellipticals and lenticulars.} \end{aligned}$$

The total internal extinction correction (that is, the correction to absorption-free magnitude) is obtained by adding α to A^{i}_{pg}. Holmberg notes that as $i \to 90°$, the observed correction in fact tends to a finite value ($A^{i}_{pg} = 1.33$ for Sa-b and 1.03 for Sc), so his formula breaks down at large inclinations.

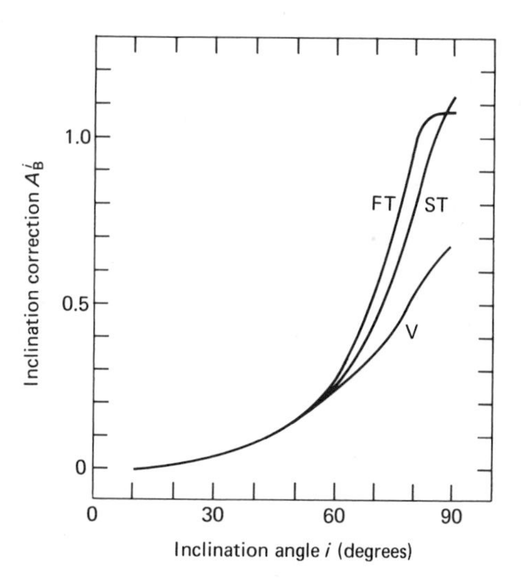

Comparison of the inclination correction predicted by the three formulas discussed. ST—Sandage and Tammann's version of the Holmberg correction; V—correction proposed by de Vaucouleurs et al.; FT—correction proposed by Fisher and Tully. In each case the formula for Sc galaxies is shown. The agreement is excellent for inclination angle $i < 55°$.

Sandage and Tammann [S6] adopt Holmberg's formula, except that they use $\sec i = a/b$ and adopt $\alpha = 0.43$ for types Sa-b, 0.28 for Sbc-Sdm, and 0.15 for Sm-Im.

2 de VAUCOULEURS ET AL. [V3] adopt

$$A_B^i = \alpha(T) \log\left(\frac{a}{b}\right)$$

where T is the numerical index of stage along the revised Hubble sequence (see Table 2.10) and

$$\alpha(T) \begin{cases} 0, & T \leq -4 \\ 0.2(T+4), & -3 \leq T \leq -1 \\ 0.7, & T = 0 \\ 0.8, & 1 \leq T \leq 8 \\ 0.1T, & T > 8. \end{cases}$$

To get the total internal extinction correction, a further 0.17 is added for all $T \geq -3$.

3 FISHER AND TULLY [F2] propose for the inclination correction

$$A_B^i = -\alpha[\ln(e^{-\sec i} + \beta) + 1]$$

where $\alpha = 0.3$ and $\beta = 0.01$ for $T \geq 1$ and $\alpha = 0$ otherwise.

This is chosen to reduce to $\alpha(\sec i - 1)$ for small i, and to 3.6α for $i \sim 90°$. They do not give an estimate for the total extinction correction.

these corrections for galaxies of spiral type Sc is shown. It can be seen that there is good agreement for inclination angle $i < 55°$. But the adoption of corrections as high as those predicted by the Holmberg and Fisher–Tully laws for large inclination angles leads to excessively high corrected luminosities for Virgo cluster galaxies studied by Sandage and Tammann (see Sec. 4.4). I therefore adopt the de Vaucouleurs correction in this book.

2.9 Kinematic distance methods

We saw in Sec. 2.3 that the bulk of the material of the disk of our Galaxy moves in a basically circular motion, with a circular velocity $\Theta_0(\varpi)$, which will vary with radial distance from the galactic center ϖ. If we can establish the form of $\Theta_0(\varpi)$, then it is possible to obtain an estimate of the distance of Population I material from its radial velocity, though this distance is usually ambiguous.

Fig. 2.42 summarizes our present knowledge of $\Theta_0(\varpi)$ for the regions of the Galaxy interior to the sun, derived from observations with the 21-cm line of neutral hydrogen. For directions interior to the sun [$0° \leq l < 90°$ and $270° < l \leq 360°$, see Fig. 2.43(a)] the highest radial velocity seen along a particular direction in the galactic plane can be presumed to be the velocity of material streaming tangential to the line of sight. (This is valid provided the angular velocity $\Theta_0(\varpi)/\varpi$ decreases with ϖ, which is satisfied for all but the innermost regions of the Galaxy.) The solid line in Fig. 2.42 shows the prediction of a detailed model for the mass distribution in our Galaxy.

For regions beyond the sun, Leo Blitz [B9] has investigated the rotation law using millimeter-wavelength observations of carbon monoxide in molecular clouds, and obtained distances from OB associations associated with the clouds. The rotation curve shows no sign of falling off with distance as predicted by the model shown in Fig. 2.42 and, as we shall see, this implies the existence of a massive halo of dark material surrounding our Galaxy.

Now if we observe the radial velocity V_r of an individual Population I object, for

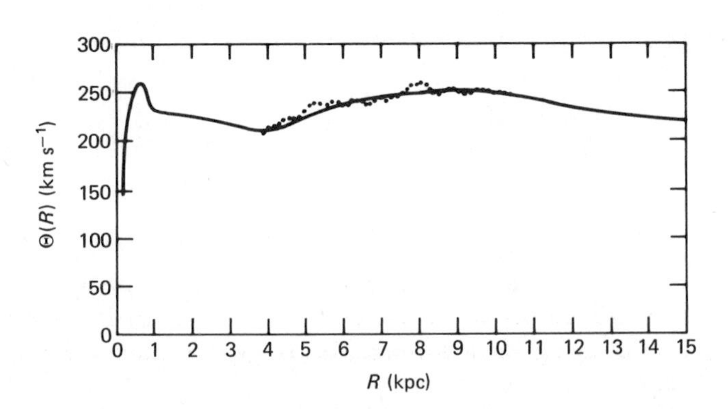

FIGURE 2.42

Rotation curve Θ (R) for inner parts of our Galaxy as deduced from 21-cm radio observations. Individual data points are plotted as dots; the smooth curve is from dynamical models. (From [B12].)

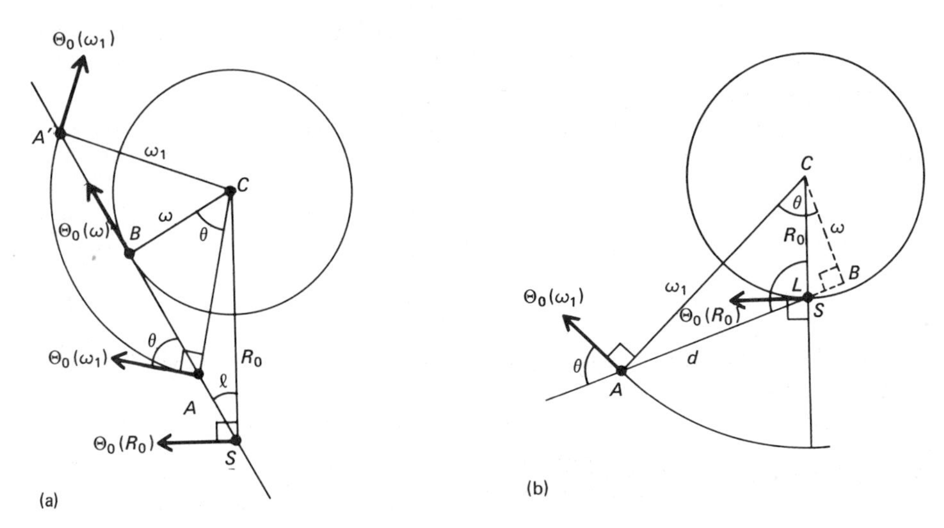

FIGURE 2.43

Geometry of galactic rotation for sources in galactic plane. (a) $0° < \mathrm{l} < 90°$; $270° < \mathrm{l} < 360°$. *(b)* $90° < \mathrm{l} < 270°$. C—*center of the Galaxy;* S—*sun;* A, A′—*source being observed;* B—*foot of perpendicular drawn from* C *to* AS; CS = R_0—*distance from sun to galactic center;* AC(= A′c) = ϖ_1—*distance of source from galactic center;* BC = ϖ; *angle* $\mathrm{B\hat{C}A} = \theta$; *angle* ACS = l—*galactic longitude of source. Note that* $\varpi = \varpi_1 \cos\theta = R_0 \sin \mathrm{l}$.

example, an HII region or a molecular cloud, we can obtain a *kinematic distance* by assuming that it is moving in a circular orbit. The radial velocity is then simply the component of the galactic rotation velocity relative to the sun, along the line of sight. From Fig. 2.43,

$$V_r = \left[\frac{\Theta_0(\varpi_1)}{\varpi_1} - \frac{\Theta_0(R_0)}{R_0}\right] R_0 \sin l \tag{2.27}$$

where ϖ_1 and R_0 are the distances of the source and the sun from the galactic center, and l is the galactic longitude. From measurement of V_r, ϖ_1 can be deduced from the known form of $\Theta_0(\varpi)$.

For $0° \leq l < 90°$ and $270° < l \leq 360°$, the distance of the source can then be calculated as

$$d = R_0 \cos l \pm R_0|\sin l|\left[1 - \left(\frac{R_0 \sin l}{\varpi_1}\right)^2\right]^{1/2} \tag{2.28}$$

the $\pm$ sign depending on whether the source is at point A' or at point A in Fig. 2.43. Thus there are two locations along the line of sight in a particular direction (A', A) at which material partaking of the galactic rotation would have the same radial velocity V_r. These are referred to as the far and near kinematic distances, and other arguments must be used to decide which is correct (such as evidence for absorption by intervening material, or that assumption of the far kinematic distance for a source leads to a size or luminosity that is impossibly high).

If $90° \leq l \leq 270°$, on the other hand, the distance can be obtained unambiguously from [see Fig. 2.43(b)]

$$d = R_0|\sin l|\left[1 - \left(\frac{R_0 \sin l}{l}\right)^2\right]^{1/2} - R_0|\cos l| \tag{2.29}$$

with ϖ_1 given by Eq. (2.27).

For stars near the sun, Eq. (2.27) can be approximated by expanding $\Theta_0(\varpi)$ in a Taylor series as

$$\Theta_0(\varpi) = \Theta_0(R_0) + (\varpi - R_0)\left(\frac{d\Theta_0}{d\varpi}\right)_{\varpi=R_0} + \cdots. \tag{2.30}$$

Then

$$\begin{aligned} V_r &\simeq \left\{\left(\frac{d\Theta_0}{d\varpi}\right)_{\varpi=R_0} - \frac{\Theta_0(R_0)}{R_0}\right\}(\varpi_1 - R_0)\sin l \\ &\simeq Ad \sin 2l \end{aligned} \tag{2.31}$$

where

$$2A = \left(-\frac{d\Theta_0}{d\varpi} + \frac{\Theta_0}{\varpi}\right)_{\varpi=R_0} \tag{2.32}$$

since $d \simeq -(\varpi_1 - R_0)\cos l$.

Similarly the tangential velocity of the star can be written

$$\begin{aligned} V_t &= \Theta_0(\varpi_1)\sin\theta - \Theta_0(R_0)\cos l \\ &\simeq d(B + A\cos 2l) \end{aligned} \tag{2.33}$$

where

$$2B = \left(-\frac{d\Theta_0}{d\varpi} - \frac{\Theta_0}{\varpi}\right)_{\varpi=R_0}. \tag{2.34}$$

A and B are called Oort's constants, after the Dutch astronomer Jan Oort, and have been determined for stars near the sun as [M1]

$$A = 16 \pm 2 \text{ km s}^{-1}\text{ kpc}^{-1}$$

$$B = -11 \pm 3 \text{ km s}^{-1}\text{ kpc}^{-1}.$$

Oort's constants relate the radial and tangential velocities for a Population I object, as observed from the sun, to the distance and galactic longitude of the object. In 1964 the International Astronomical Union adopted the values $A = 15$ and $B = -10$. These values, along with the value of R_0, the distance of the galactic center from the sun, are now under reconsideration.

The rotation curve can be used for another important purpose, to determine the mass distribution in our Galaxy. To illustrate how this is done, suppose the mass in the central regions were very much greater than the mass in the disk and the halo. Then from the balance of gravitation and centrifugal force we would expect

$$\frac{GM}{\varpi^2} = \frac{\Theta_0^2(\varpi)}{\varpi}$$

or

$$\Theta_0(\varpi) \propto \varpi^{-1/2}$$

that is, the rotation velocity falls off as the inverse square root of the distance from the center of the Galaxy. This is referred to as the Keplerian rotation law, and the mass inferred from $\Theta_0(\varpi)$,

$$M = \frac{\varpi\Theta_0^2(\varpi)}{G}$$

is called the Keplerian mass. More generally, it is clear that the actual form of $\Theta_0(\varpi)$ gives us information about the mass distribution in the Galaxy. The fact that the rotation velocity remains approximately constant beyond the solar circle, and shows no sign of tending toward the Keplerian form expected at large distances from the main mass concentration, demonstrates the existence of a massive halo around our Galaxy. This is believed to extend to a distance of about 50 kpc from the center of the Galaxy and to comprise about 90% of the mass of the Galaxy. Since no stars or light of any form are seen associated with this halo material, it is referred to as the *dark halo.* This phenomenon was first noticed in external spiral galaxies, some rotation curves of which are shown in Fig. 2.44. Again, at large radial distances from the centers of the galaxies, we see the rotation velocity remaining approximately constant in several instances.

What this dark matter could be remains a subject for speculation. The possibility that it could be normal stars ($M \gtrsim 0.1M_\odot$) has probably been eliminated by arguments based on star counts [S8], [R1]. The remaining possibilities are objects of even lower mass (0.01 or $0.001M_\odot$), that is, black dwarfs or "Jupiters"; black holes, that is, objects which have collapsed so far that light can no longer escape from them; and neutrinos with nonzero rest-mass energy in the range 3–30 eV.* Normally neutrinos are assumed to have zero rest mass like photons, but in grand unified theories

* 1 eV = 1.6×10^{-12} erg.

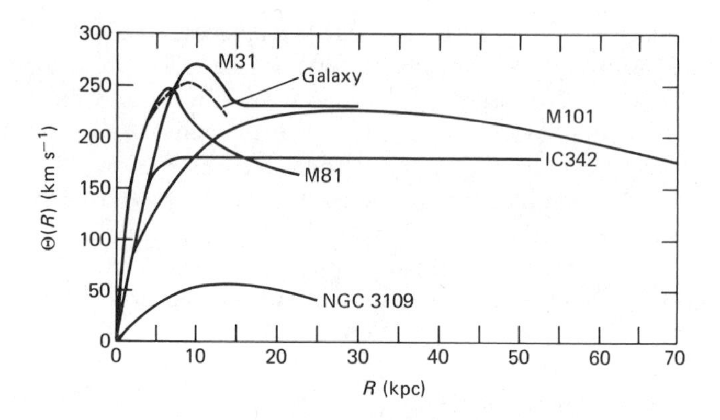

FIGURE 2.44

Rotation curves of spiral galaxies, including our Galaxy, from 21-cm line radio measurements. The vertical axis is $\Theta(R)$ *and the horizontal axis is the radial distance* R *from the center of the galaxy. Note that the rotation curves of M31 and IC342 show no sign of a decline of* $\Theta(R)$ *with* R *at large values of* R*, indicating the presence of a massive dark halo. (From [M1].)*

(see p. 21) it is expected that neutrinos would have a nonzero rest mass, and recently some observations have seemed to support this hypothesis.

Another method of estimating the distances of molecular clouds, which has been developed recently by R. Genzel and co-workers, involves very long baseline interferometric (VLBI) studies of water maser sources associated with regions of star formation [G1], [G2], [S9]. It has been found that in the hot high-density molecular gas near a newly formed star, the water molecules can produce maser emission at radio wavelengths. (A maser occurs when favorable alignments of atoms lead to amplification of stimulated emission from atom to atom.) The emission tends to come from a cluster of small spots in the star-forming cloud. The high positional accuracy of VLBI measurements, often as good as $\pm 0''.0001$, allows the proper motions of the masering spots in each region to be estimated. Combined with measurements of the radial velocities of the spots and a simple model (for example radial expansion or random motions), distances to molecular clouds have been obtained which are in good agreement with other estimates. This method is analogous to the method of statistical parallaxes described in Sec. 2.3.

2.10 Spiral arms, galaxy types, and luminosity classes

The fact that our Galaxy contains a disk in which gas and young stars are found strongly suggests that it is a spiral galaxy like our neighbor, M31. Direct evidence for spiral arms can be found by studying the distribution on the galactic plane of young objects, like HII regions, open clusters, and OB associations, for which we know the distance. Fig. 2.45 shows the distributions of

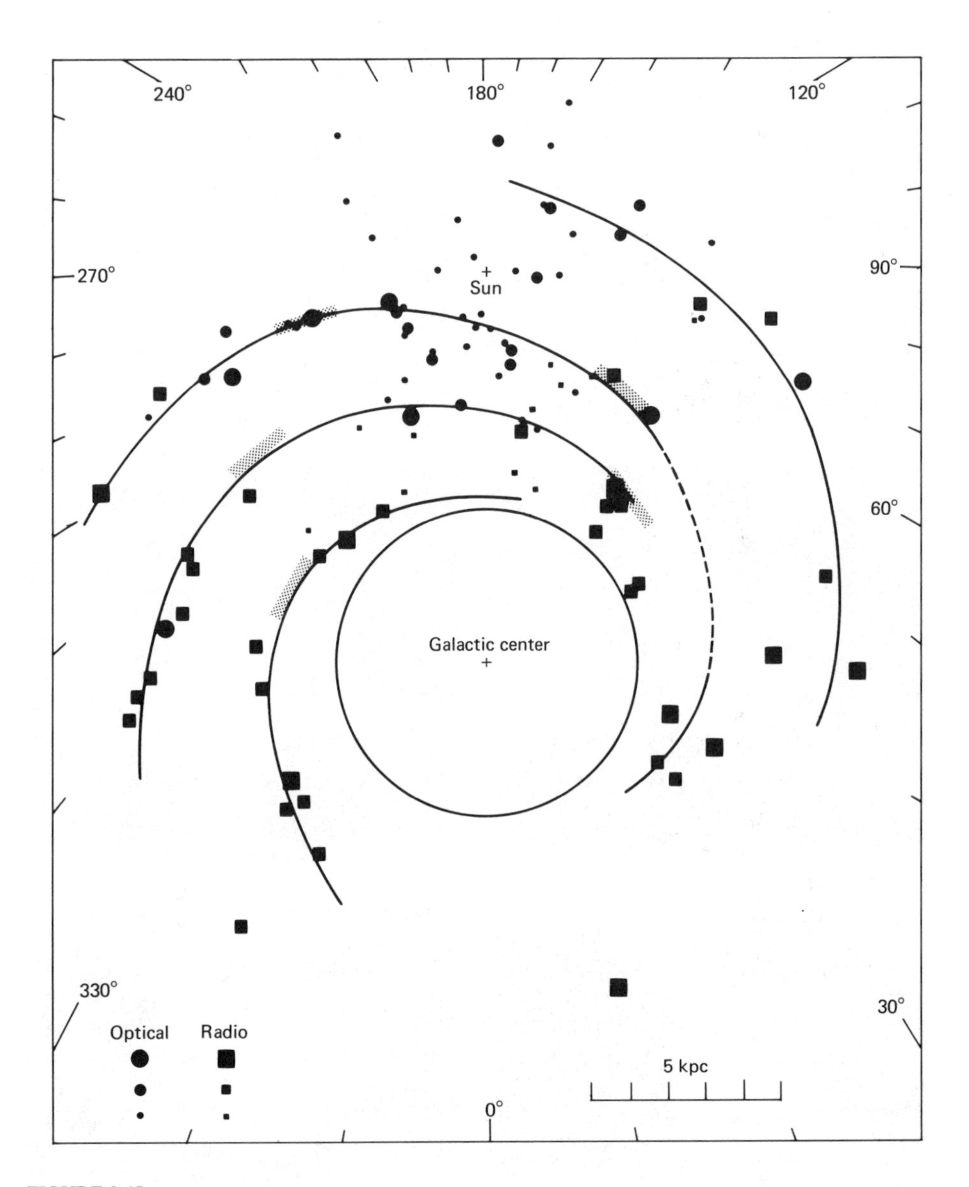

FIGURE 2.45

Spiral model of our galaxy obtained from young HII regions. ■—optical data; ■—radio data. Hatched areas denote intensity maxima in the radio continuum and in neutral hydrogen. At the edge the galactic longitude l *viewed from the sun is indicated. The scale is shown at the lower right. (From [G3].)*

FIGURE 2.46

Photographs of some normal galaxies. (a) M104, type Sa, seen almost edge on. Note the band of dust lying in the disk of the galaxy, and the prominent nuclear bulge. (b) M81, type Sab. Note the tightly wound spiral arms. (c) NGC 4565, type Sb, seen edge on. (d) M51, type Sbc. (e) M101, type Sc. Note the loosely wound spiral arms and the insignificant nucleus. (f) M87, type E1. Note the smooth, symmetrical appearance. (Courtesy of Mt. Palomar Observatory.)

young HII regions on the galactic plane. The evidence that these and other young Population I objects like Cepheids or neutral hydrogen clouds are concentrated into spiral arms, and hence that the formation of massive O and B stars takes place in the arms, is overwhelming. However, the overall pattern of spiral arms in our Galaxy is very hard to trace out. Fig. 2.46 shows the visible appearance of some nearby spiral galaxies, where the spiral structure is much clearer.

There are a number of different classification schemes for spiral galaxies, of which the first and simplest is that of Hubble. He divided spirals into two main groups: (1) barred spirals, in which a luminous bar is present across the nucleus of the galaxy and where the spiral arms start from the two ends of the bar, and (2) normal spirals, which do not have a bar and in which the spiral arms start from the nucleus. Each group is subdivided into a sequence of types a, b, and c, in which the spiral arms progressively become more loosely wound and the contrast in brightness between the nucleus and the disk declines. Examples of these types can be seen in Figs. 2.46 and 2.47. Fig. 2.48 illustrates Hubble's original classification of the main spiral galaxy types. Because of the shape of this diagram, it is known as the tuning-fork classification. The three other main types of galaxy are also placed in this figure: (3) elliptical galaxies, which have an elliptical shape and usually contain little or no Population I material (they are subdivided according to their degree of ellipticity, characterized by $n = 10(1 - b/a)$, where a and b are the angular radii of the major and minor axes); (4) lenticular or S0 galaxies, which are similar to ellipticals in having little or no Population I material, but whose lens shape indicates the presence of a disk component, though one that contains very little gas or young stars; and (5) irregular galaxies, which have an amorphous structure and a poorly defined nucleus and are rich in gas and dust.

Modifications of this scheme have been proposed by several astronomers. Gerard de Vaucouleurs has introduced a further dimension into the scheme for spirals depending on whether the galaxies show a ring structure or are S-shaped (in addition to whether they are barred or normal). He also defines a useful parameter T, which ranges from -5 for ellipticals to $+10$ for irregulars (Table 2.10). W.W. Morgan proposed a classification based mainly on the degree of central concentration of the light of galaxies (the Yerkes system). Sidney van den Bergh has placed the S0 galaxies in a sequence a, b, c, parallel to the spirals, and proposes an intermediate group, the anemic spirals, in which only a weak Population I component is present.

About 90% of galaxies fit into the above classification schemes, though many earn the additional label p for peculiar. A minority of galaxies have truly bizarre appearances (two are shown in Fig. 2.49), and in many cases this is due to tidal interaction between two nearby galaxies. A few percent of galaxies show signs of activity in their nuclei, in the form of compact nonthermal sources of visible and ultraviolet radiation, and show evidence of very rapid large-scale motions with velocities of tens of thousands of kilometers per second. These are the Seyfert galaxies, discovered by Carl Seyfert in 1943, in which the nuclear source contributes from a few percent to 50% of the total light. A more extreme type of activity is seen in quasars, in which the nuclear source may be 100 or 1000 times more luminous than the stellar component and the galaxy looks indistinguishable from a star on a

FIGURE 2.47

Photographs of barred spirals. (Left) NGC 1300, type SBb. (Right) NGC 7741, type SBc. (Courtesy of Mt. Palomar Observatory.)

FIGURE 2.48

Hubble's tuning-fork classification of galaxies. (From [H9].)

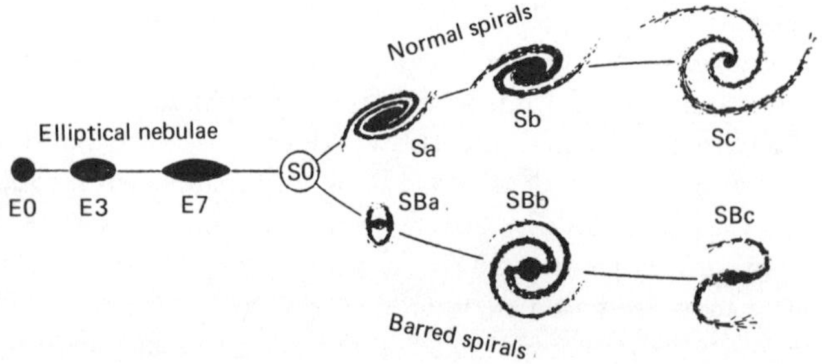

TABLE 2.10

De Vaucouleurs galaxy classification parameter T.

T	*Hubble type*	*Revised Hubble type*
−6		cE
−5	E	E
−4		E^+
−3	E-SO	L^-
−2	SO	L
−1		L^+
0	SO/a	SO/a
1	Sa	Sa
2	Sab	Sab
3	Sb	Sb
4	Sbc	Sbc
5		Sc
6	Sc	Scd
7		Sd
8	Sc-Irr	Sdm
9		Sm
10	Irr I	Im
11		cI
12*	Irr II	IO

* Non-Magellanic irregulars (Hubble type Irr II) are given the index $T = 0$ by de Vaucouleurs, $T = 12$ by Fisher and Tully [F2].

SOURCE: Compiled from [V3].

FIGURE 2.49

Photographs of two peculiar galaxies. (Left) NGC 2623, probably two closely interacting galaxies. (Right) M82, probably a galaxy that has run into a large cloud of gas and dust. (Courtesy of Mt. Palomar Observatory.)

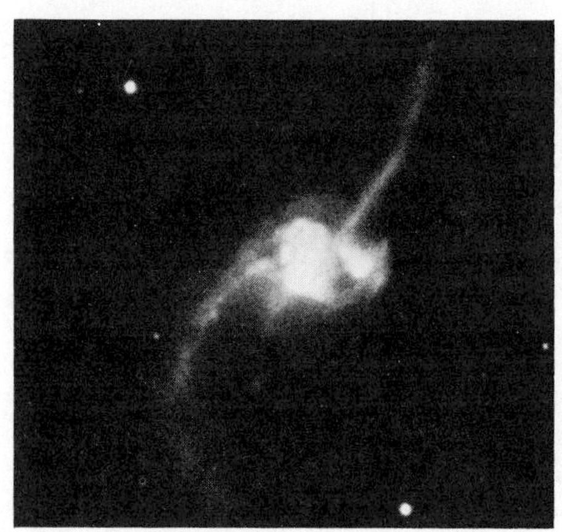

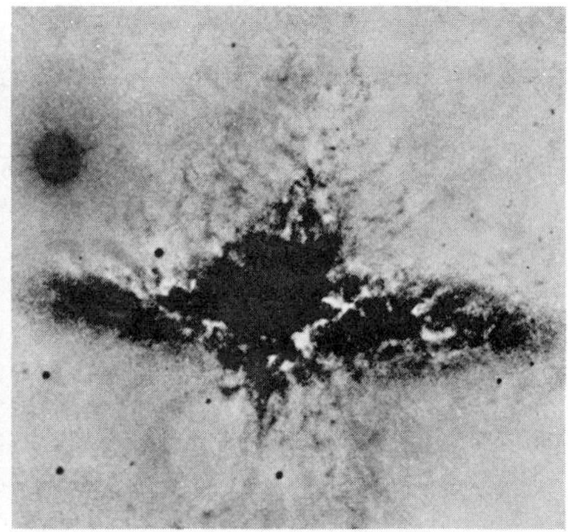

photographic plate. Intermediate cases, in which the galaxy has a very strong central starlike component (N galaxies in Morgan's classification) have been found in radio surveys and in lists of galaxies with strong ultraviolet excess published by B.E. Markarian. Fritz Zwicky listed many compact galaxies in which the galaxy looks stellar on a Palomar Sky Survey plate (taken with the 48-in Schmidt telescope, that is, galaxy diameter $\leq 2''$, but not on a plate taken with a large telescope. The IRAS satellite has found many examples of 'starburst' galaxies, in which the galaxy appears to be undergoing an exceptional burst of star formation and the far-infrared luminosity of the galaxy exceeds its visible luminosity by a factor ranging from 2 to 100.

Van den Bergh has introduced a very important additional parameter into the classification of spirals, the luminosity class, based on the clarity and contrast of the spiral arms (Fig. 2.50). The luminosity of the galaxy, as well as other properties, seem to be correlated with this parameter, which makes it a useful distance indicator (see Sec. 4.7).

At this point it is worth saying something about current theories of how these beautiful spiral patterns came to be produced. There are three main theories:

1 DENSITY-WAVE THEORY. In this theory the combined gravitational attraction of the stars of the galaxy as they weave in and out of the disk generates a spiral-shaped instability which rotates around the galaxy at constant angular velocity (the pattern speed). The gas of the disk experiences a compression as the density wave passes, and this leads to the collapse of some clouds of gas and their fragmentation into stars.

2 STOCHASTIC STAR FORMATION. Here the emphasis is on the way that OB star formation in one cloud can trigger star formation in another nearby cloud via supernova explosions or expanding HII regions. When this effect is combined with the differential rotation of the galaxy, long streamers of star-forming regions are drawn out tangential to the radius vector. The net effect can yield results very similar to the more ragged spiral galaxies. However, the more symmetrical spiral patterns, the "grand-design" spirals as theoreticians call them, can hardly be formed in this way.

3 GALAXY–GALAXY INTERACTION. Recent work on stellar dynamics has shown the importance of galaxy–galaxy interaction on spiral formation and suggested that the existence of a small companion galaxy may play a big part in the generation of grand-design spirals. Galaxy–galaxy interaction may also cause enhancement of star formation in galaxies and may be one of the causes of activity in galactic nuclei.

Whichever of these theories is correct (all three may contribute), the rate of star formation in a galaxy is strongly correlated with the morphology. The fraction of a galaxy's mass in the form of gas increases steadily as we go along the Hubble sequence from ellipticals to irregulars. This can be interpreted as implying that star formation has proceeded more slowly in irregulars than in spirals, and more slowly

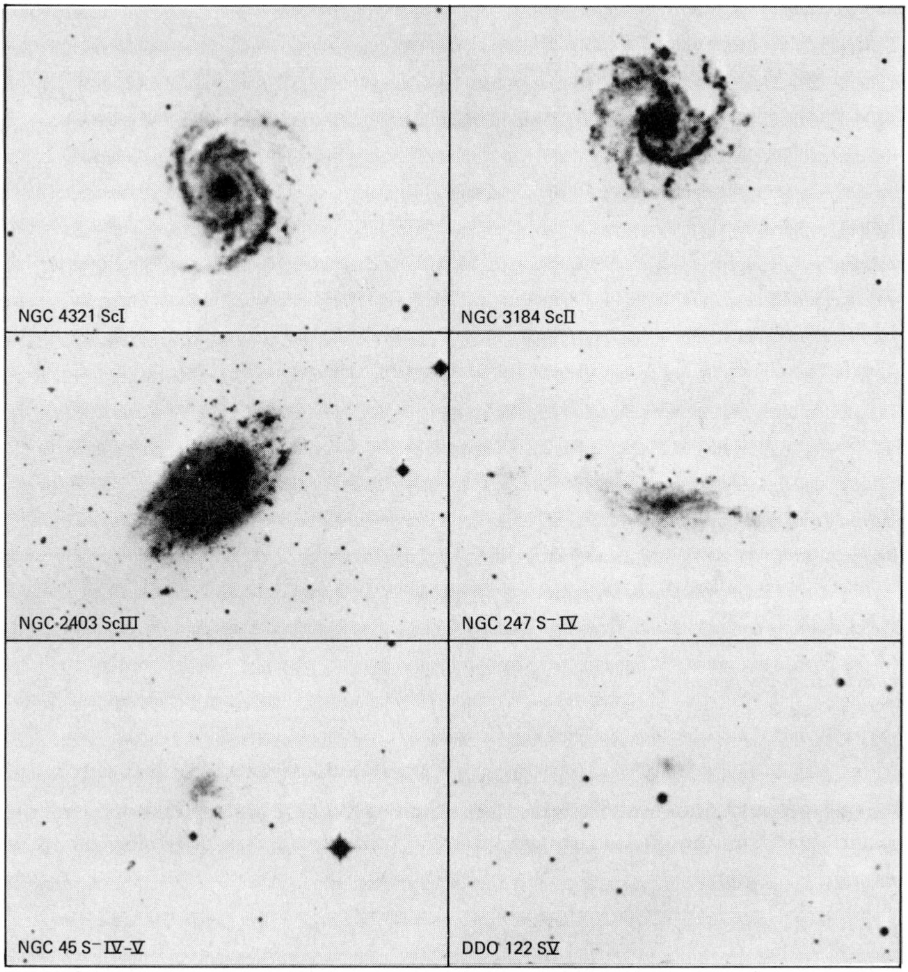

FIGURE 2.50

Examples of van den Bergh's luminosity classification of spirals. The lower the luminosity class, the thicker and more prominent are the spiral arms. (From [B4].)

in spirals than in ellipticals. Indeed most star formation in the majority of ellipticals was probably complete soon after formation of the galaxy.

2.11 *The age of the Galaxy*

The oldest stars in our Galaxy are those in globular clusters. We saw in Sec. 2.7 that the age of globular clusters can be determined from their main-sequence turn-off point and, more generally, by comparing theoretically predicted isochrones with the observed HR diagrams.

FIGURE 2.51

Schematic representation of the variation of heavy-element abundance, measured by [Fe/H], with age for selected open (+) and globular clusters (■), the solar neighborhood (×), and the sun (■). The globular clusters are consistent with lying in the hatched box, the width of which gives the time scale for the collapse of the halo during formation of the galaxy. The solid line indicates the evolution of heavy-element abundance with time in the Galaxy. (From [S5].)*

** See eqn (2.22) (p. 72) for relationship between [Fe/H] and Z.*

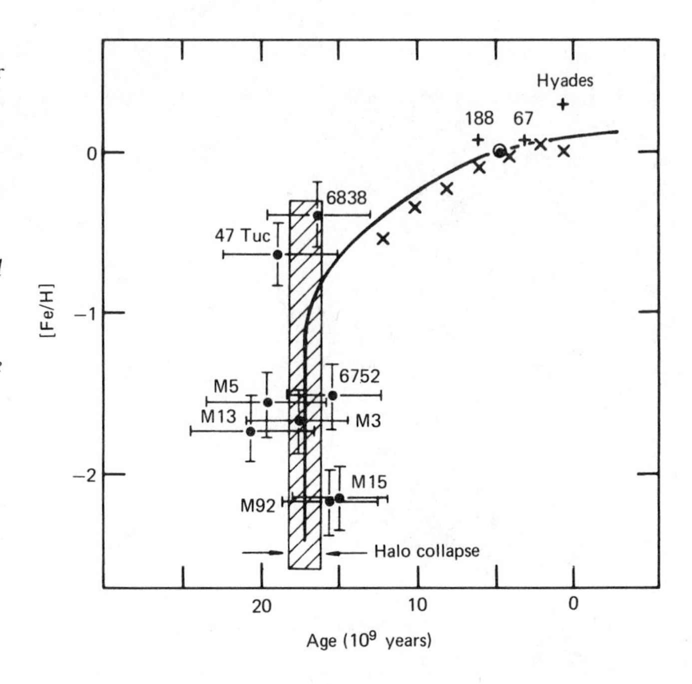

Figure 2.51 shows a plot due to Sandage of the heavy-element abundance Z in selected globular and open clusters against their age. There are several points to note about this diagram. First, the age of the globular clusters is in the range of $14-20 \times 10^9$ years (see Table 2.9). A similar conclusion has been reached in a recent (1982) study by Don VandenBergh, who found an age range of $15-18 \times 10^9$ years [VI]. This removes a discrepancy between earlier studies, which found lower ages for the metal-rich globular clusters like 47 Tucanae, and the widely held theory that globular clusters formed during a phase of rapid collapse of the protogalaxy.

Secondly, the old open clusters show a trend of increasing heavy-element abundance Z with time. This arises because the more massive stars in an old cluster rapidly complete their evolution and contaminate the gas, from which the later clusters form, with heavy elements. Third, the metal abundance in the oldest globular clusters, though low ($Z \sim 10^{-4}$), is not zero, and so some explanation has to be found for how these first heavy elements were made.

Some of the sources of uncertainty in the globular cluster ages have been reviewed by Russell Cannon [C1]. These include observational errors, uncertainties in stellar interior opacities, deficiencies in the simple convective theory usually employed, doubts about neutrino production, errors in the transformation from observed to theoretically calculated quantities, and uncertainties in the chemical-composition parameters. An independent estimate of globular cluster ages by T.S. van Albada and co-workers [A1] gave values in the range of $11-14 \times 10^9$ years, though the method may not yet be completely reliable.

To summarize, there is now a considerable consensus among theorists and observers favoring globular clusters ages in the range of $14-20 \times 10^9$ years. The greatest uncertainty is probably the efficiency with which the material in stellar interiors is mixed. If there is more mixing than generally supposed (for example, due to diffusion or to convection more efficient than assumed in present calculations), then this would tend to accelerate evolution, and age estimates would be lowered. On the other hand, as Bruce Carney points out [C2], strong magnetic fields or rapid rotation could inhibit convective mixing, and age estimates would then have to be raised.

A second, very important method of estimating the age of the Galaxy is through *nucleocosmochronology,* the dating of material by the abundance of radioactive isotopes in it. We saw in Sec. 1.8 that this technique leads to an estimate for the age of the solar system of 4.55×10^9 years. The most important isotopes for estimating the age of the Galaxy are the isotopes of thorium and uranium ^{232}Th, ^{235}U, and ^{238}U, which decay, with half-lives of 13.9, 0.7, and 4.5×10^9 years, to different isotopes of lead (^{208}Pb, ^{207}Pb, and ^{206}Pb). The decay chain of these isotopes is such that at any time the material is almost entirely composed of the original isotope and the decay product, the intermediate products being relatively short-lived. Another important process is the β-decay of the isotope of rhenium ^{187}Re to the isotope of osmium ^{187}Os, with a half-life of 4.3×10^{10} years.

Now these radioactive elements are believed to be made by rapid neutron capture (the r-process) during supernova explosions, and theory predicts the initial ratios of the numbers of atoms of the different isotopes. For example, the initial ratio of ^{232}Th to ^{238}U is predicted to be $^{232}\mathrm{Th}/^{238}\mathrm{U} = 1.6$. To get the ratio observed in terrestrial rocks today, $^{232}\mathrm{Th}/^{238}\mathrm{U} = 4$, the radioactive isotopes must have been decaying for 9.8×10^9 years if they were all made at the same time (Fig. 2.52). If, on the other hand, and this is believed to be nearer to the true situation in our Galaxy, the radioactive elements were made at a uniform rate right up to the formation of the solar system, then an age for the Galaxy of 15×10^9 years is found. E.M.B. Symbalisty and Dave Schramm [S12] reviewed in 1981 all the available nucleocosmochronological evidence and concluded that the age of our Galaxy lies in the range of $8.7-15.8 \times 10^9$ years. The region of overlap of the two methods, globular clusters and nucleocosmochronology, is then $14-16 \times 10^9$ years, though the uncertainties in both methods mean that a wider range of ages for the Galaxy than this is credible. The importance of this estimate for cosmology will become clear in Sec. 6.3.

We can now give a rough history of our Galaxy. About 15 billion years ago the oldest Population II stars formed, mainly in globular clusters in the halo. We observe these old stars to have high velocities relative to our local frame of reference, implying rapid motion in and out of the Galaxy. It is natural to suppose that this motion is a relic of the state of motion of the gas from which the stars formed and hence that the protogalactic gas was in a state of collapse. The massive stars in this first generation evolved rapidly and sprayed the halo with heavy elements; and the conservation of angular momentum implies that the angular velocity of the gas steadily increased as the protogalaxy collapsed. Eventually further collapse in the radial direction was prevented by centrifugal force. The gas then continued collaps-

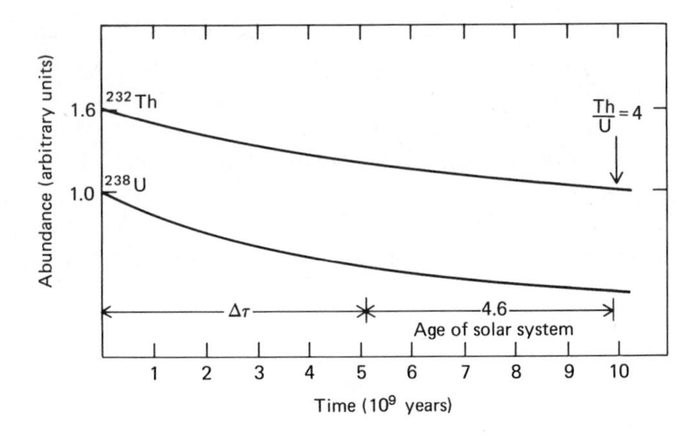

FIGURE 2.52

Determining the age of our Galaxy from relative abundances of radioactive isotopes ^{232}Th and ^{238}U on the earth. In the simple model illustrated here it is assumed that at t = *0 these isotopes are made in supernova explosions in the predicted ratio $^{232}Th/^{238}U = 1.6$. To obtain the ratio observed today, $^{232}Th/^{238}U = 4$, radioactive decay must have continued for about 9.8 billion years. If instead we assume that these isotopes were made steadily throughout the lifetime of the Galaxy up to the time the solar system formed, then the time from the formation of the Galaxy to the formation of the solar system $\Delta\tau$ is doubled, giving a total age of 15 billion years for the Galaxy. (From [S10].)*

ing in the direction perpendicular to the galactic plane, forming a disk. The Population II stars continued with the rapid in and out motions with which they were born, and they define the halo of the Galaxy. The compression of the residual gas into a disk with a strong concentration at the center resulted in a new generation of stars, the first stars of Population I. The gas gradually settled into a thin layer, and star formation has continued steadily there to the present day, the formation of the more massive O and B stars being concentrated to the spiral arms.

There are two problems with this scenario. First, the number of old Population I stars with low metal abundance is much lower than expected. This could either be because the first generation of disk stars consisted only of stars of mass $> 10\ M_\odot$, say, none of which survive today. This is called the prompt initial enrichment (PIE) scenario. Alternatively the disk may grow steadily by accretion of gas from the halo so that very few stars were formed from the initially metal-poor gas. The second problem is that the oldest halo (Population II) stars have a nonzero heavy-element abundance ($\sim 10^{-4}$ by mass). These heavy elements could either have been made in a generation of massive stars which formed very early in the history of the Galaxy and which have left no observable remnants today, or have been produced before galaxies formed, in a pregalactic generation of massive objects (see Sec. 5.1).

As star formation and evolution proceed, hydrogen is steadily converted to helium in the Galaxy. However, it can easily be shown that most of the helium in stars today was made before our Galaxy formed. If stars are always formed with the same distribution of masses [the initial mass function (IMF)], then we expect the

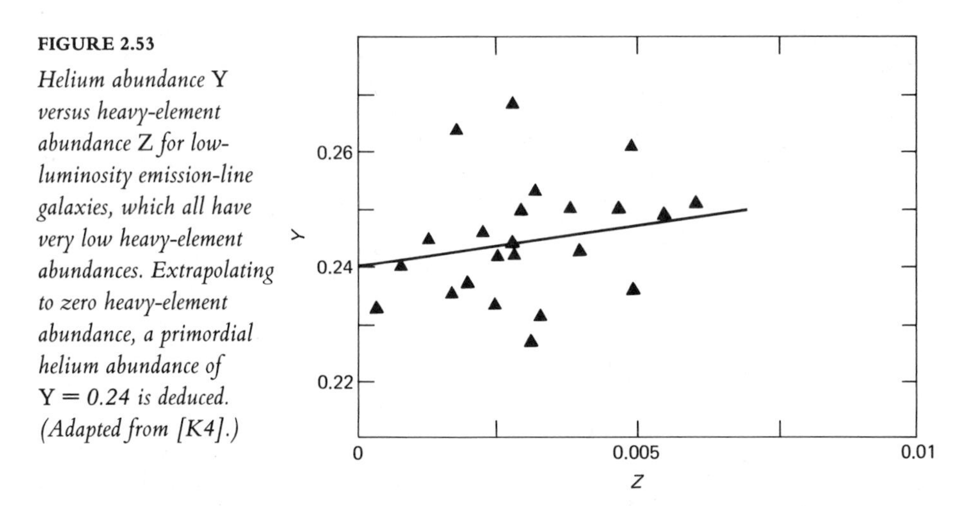

FIGURE 2.53

Helium abundance Y versus heavy-element abundance Z for low-luminosity emission-line galaxies, which all have very low heavy-element abundances. Extrapolating to zero heavy-element abundance, a primordial helium abundance of Y = 0.24 is deduced. (Adapted from [K4].)

amount of helium produced per unit mass of heavy elements produced, dY/dZ, to be a constant. For example, if stars are born with an initial mass function similar to that for stars in the solar neighborhood, then it is found that $dY/dZ \sim 2$.

If we now study the trend of helium abundance Y with heavy-element abundance Z in external galaxies, and extrapolate this trend back to zero heavy-element abundance, we find that the helium abundance does not tend to zero, but to a finite

TABLE 2.11

Helium abundance in our Galaxy and other galaxies.

Location	*Y*		*Reference*
Helium abundance determinations			
Galactic HII regions with $Z \simeq 0.02$	0.32 ± 0.01		[O1]
Orion nebula	0.28 ± 0.01		[O1]
Sun	0.22–0.24		[O1]
HII regions more than 10 kpc from galactic center	0.23–0.25		[O1]
Globular clusters	0.22 ± 0.04		[O1]
Large and small Magellanic clouds	0.25 ± 0.01		[O1]
Young galaxies	0.23 ± 0.02		[O1]
Low-luminosity galaxies	0.24 ± 0.01		[O1]
Emission-line dwarf galaxies	0.25 (+0.01, −0.02)		[O1]
Recent estimates of primordial helium abundance Y_p	Y_p	$\frac{dY}{dZ}$	
Irregular and compact galaxies	0.228 ± 0.014	2.8 ± 0.6	[L2]
Dwarf galaxies with emission lines	0.243 ± 0.010	0.7 ± 3.5	[K4]
Both combined (Fig. 2.53)	0.240 ± 0.007	1.25 ± 2.0	[K4]

value. Fig. 2.53 shows a plot of Y against Z for some nearby irregular, compact, and dwarf galaxies. From studies like these, the initial primordial helium abundance in galaxies is deduced to be in the range 0.23–0.25 by mass (Table 2.11). This helium is believed to have been made during the fireball phase of a hot big-bang universe and is in fact one of the best pieces of evidence that we live in this type of universe. This story is the subject of Chap. 5. Meanwhile, having surveyed and mapped the main features of our own Galaxy and having, we hope, put the distance scale on a secure basis there, we are ready to take the next steps up the cosmological distance ladder and try to estimate the distances to the nearest galaxies.

BIBLIOGRAPHY

[A1] Albada, T.S. van, Dickens, R.J., and Weavers B.M.H.R., "Far-Ultraviolet Photometry of Globular Clusters with ANS - III. Globular Cluster Ages," *Month. Not. R. Astron. Soc.,* 1981, **196,** pp. 823–833.

[A2] Alter, G., Ruprecht, J., and Vanyset, J., *Catalogue of Star Clusters and Associations,* 2nd ed., Akadamia Kiado, Budapest, 1970.

[B1] Bahcall, J.N., "Solar Neutrinos: Theory versus Observation," *Space Sci. Rev.,* 1979, **24,** pp. 227–251.

[B2] Balona, L.A., "Applications of the Method of Maximum Likelihood to the Determination of Cepheid Radii," *Month. Not. R. Astron. Soc.,* 1977, **178,** pp. 231–243.

[B3] Barnes, T.G., Dominy, J.F., Evans, D.S., Kelton, P.W., Parsons, S.B., and Stover, R.J., "The Distances of the Cepheid Variables," *Month. Not. R. Astron. Soc.,* 1977, **178,** pp. 661–674.

[B4] van den Bergh, S., "A Preliminary Luminosity Classification of the Late-Type Galaxies," *Astrophys. J.,* 1960, **131,** pp. 215–223.

[B5] van den Bergh, S., "The Distance Scale within the Local Group," in *Déclages vers le rouge et expansion de l'univers—l'évolution des galaxies et ses implications cosmologiques* (IAU Colloq. 37), Balkowski, C., and Westerlund, B.E., Eds., Editions CNRS, Paris, 1977, pp. 13–22.

[B6] Blaauw, A., "The Concept of Stellar Populations," in *Stars and Stellar Systems,* Vol. 5: *Galactic Structure,* Blaauw, A., and Schmidt, M., Eds., University of Chicago Press, 1965, pp. 435–454.

[B7] Blaauw, A., "The Calibration of Luminosity Criteria," in *Problems of Calibration of Absolute Magnitudes and Temperatures of Stars* (IAU Symp. 54), Hauck, B., and Westerlund, B.E., Eds., Reidel, 1973, pp. 47–56.

[B8] Blanco, V.M., "Some Remarks on the (U.B.V.) System," *Astrophys. J.,* 1956, **123,** pp. 64–67.

[B9] Blitz, L., "The Rotation Curve of the Galaxy to $R = 16$ Kiloparsecs," *Astrophys. J.,* 1980, **231,** pp. L115–119.

[B10] Burstein, D., and Heiles, C., "HI, Galaxy Counts, and Reddening: Variations in the Gas-to-Dust Ratio, the Extinction at High Galactic Latitudes, and a New Method for Determining Galactic Reddening," *Astrophys. J.,* 1978, **225,** pp. 40–55.

[B11] Burstein, D., and Heiles, C., "Reddening Derived from HII and Galaxy Counts: Accuracy and Maps," *Astr. J.,* 1982, **87,** pp. 1165–1189.

[B12] Burton, W.B., "The Morphology of Hydrogen and of Other Tracers in the Galaxy," *Ann. Rev. Astron. Astrophy.,* 1976, **14,** pp. 275–306.

[C1] Cannon, R.Q., "Ages of Galactic and Extragalactic Star Clusters of Various Abundances," in *Highlights of Astronomy,* vol. 6, West, R.M., Ed., Reidel, 1983, pp. 109–117.

[C2] Carney, B.W., "Globular Cluster Ages," in *Highlights in Astronomy,* vol. 6, West, R.M., Ed., Reidel, 1983, pp. 255–266.

[C3] Cassinelli, J.P., Mathis, J.S., and Savage, B.D., "Central Object of the 30 Doradus Nebula, a Supermassive Star, *Science,* 1981, **212,** pp. 1497–1501.

[C4] Clube, S.V.M., and Dawe, J.A., "Statistical Parallaxes and the Fundamental Distance Scale. II—Applications of the Maximum Likelihood Technique to RR Lyrae and Cepheid Variables," *Month. Not. R. Astron. Soc.,* 1980, **190,** pp. 591–610.

[C5] Code, A.D., and Houck, T.C., "Wide-Angle Infrared Photographs of the Southern Milky Way," *Astrophys. J.,* 1956, **121,** pp. 553–554.

[C6] Counselman, C.C., III, "Radio Astrometry," *Ann. Rev. Astron. Astrophys.,* 1976, **14,** pp. 197–214.

[D1] Delhaye J., "Solar Motion and Velocity Distribution of Common Stars," in *Stars and Stellar Systems,* vol. 5: *Galactic Structure,* Blaauw, A., and Schmidt, M., Eds., University of Chicago Press, 1965, pp. 61–84.

[D2] Demarque, P., "Ages and Abundances of Globular Clusters and the Oldest Open Clusters," in *Star Clusters* (IAU Symp. 85), Hesser, J.E., Ed., Reidel, 1980, pp. 281–303.

[F1] Fenkart, R.F., and Binggeli, B., "A Catalogue of Galactic Clusters Observed in Three Colours," *Astr. Astrophys.,* suppl. ser., 1979, **35,** pp. 271–275.

[F2] Fisher, J.R., and Tully, R.B., "Neutral Hydrogen Observations of a Large Sample of Galaxies," *Astrophys. J.,* suppl. ser., 1981, **47,** pp. 139–200.

[F3] Fowler, W.A., "Nucleosynthesis in Stars," *Mem. Soc. R. Sci. Liège,* ser. 4, 1954, **14,** pp. 88–111.

[G1] Genzel, R., Reid, M.J., Moran, J.M., and Downes D., "Proper Motions and Distances of H_2O Maser Sources. I—The Outflow in Orion-KL," *Astrophys. J.,* 1981, **244,** pp. 884–902.

[G2] Genzel, R., Downes, D., Schneps, M.H., Reid, M.J., Moran, J.M., Kogan, L.R., Kostenko, V.I., Matveyenko, L.I., and Ronnang, B., "Proper Motions and Distances of H_2O Maser Sources. II—W5I," *Astrophys. J.,* 1981, **247,** pp. 1039–1051.

[G3] Georgelin, Y.M., and Georgelin, Y.P., "The Spiral Structure of our Galaxy Determined from HII Regions," *Astr. Astrophys.,* 1976, **49,** pp. 57–79.

[G4] Gleise, W., "Hertzsprung-Russell Diagrams and Colour-Luminosity Diagrams for the Stars Nearer than Twenty-Two Parsecs," in *The HR Diagram* (IAU Symp. 80), Davis Phillip, A.G., and Hayes, D.S., Eds., Reidel, 1978, pp. 79–88.

[G5] Graham, J.A., "The Galactic Distance Scale," in *Large-Scale Characteristics of the Galaxy* (IAU Symp. 84), Burton, W.B., Ed., Reidel, 1979, pp. 195–200.

[G6] Greenstein, J.L., "The Evidence for a Universal Helium and Deuterium Abundance," *Phys. Scripta,* 1980, **21,** pp. 759–768.

[H1] Hanson, R.B., "The Hyades Cluster Distance," in *Star Clusters* (IAU Symp. 85), Hesser, J.E., Ed., Reidel, 1980, pp. 71–80.

[H2] Harris, W.E., "Spatial Structure of the Globular Cluster System and the Distance to the Galactic Center," *Astr. J.,* 1976, **81,** pp. 1095–1116.

[H3] Harris, W.E., and Racine, R., "Globular Clusters in Galaxies," *Ann. Rev. Astron. Astrophys.,* 1979, **17,** pp. 241–274.

[H4] Heck, A., "Some Methods of Determining the Stellar Absolute Magnitudes," *Vistas in Astron.,* 1978, **22,** pp. 221–264.

[H5] Heiles, C., "An Almost Complete Survey of 21-cm Line Radiation for $b > \simeq 10^0$. IV—The HI Column Density as a Function of Position of the Sky, *Astr. Astrophys.,* suppl. ser., 1975, **20,** pp. 37–55.

[H6] Heiles, C., and Jenkins, E.B., "An Almost Complete Survey of 21-cm Line Radiation for $b > \simeq 10^0$. V—Photographic Presentation and Qualitative Comparison with other Data," *Astr. Astrophys.,* 1976, **21,** pp. 333–360.

[H7] Hirschfeld, A., McClure, R.D., and Twarog, B.A., "Measurements of Abundances and Ages of Old Disk Stars," in *The HR Diagram* (IAU Symp. 80), Davis, Phillip A.G., and Hayes, D.S., Eds., Reidel, 1978, pp. 163–166.

[H8] Holmberg, E., "Magnitudes, Colors, Surface Brightness Intensity Distributions, Absolute Luminosities, and Diameters of Galaxies," in *Stars and Stellar Systems,* vol. 9: *Galaxies and the Universe,* Sandage, A., Sandage, M., and Kristian, J., Eds., University of Chicago Press, 1975, pp. 123–157.

[H9] Hubble, E.P., *The Realm of the Nebulae,* Dover, 1936.

[H10] Humphreys, R.M., "Studies of Luminous Stars in Nearby Galaxies. II—M Supergiants in the Large Magellanic Cloud," *Astrophys. J.,* suppl. ser., 1979, **39,** pp. 389–403.

[I1] Iben, I., Jr., "Stellar Evolution within and off the Main Sequence," *Ann. Rev. Astron. Astrophys.,* 1967, **5,** pp. 571–626.

[I2] Iben, I., Jr., "Stellar Ages—A Rough Review," in *L'age des étoiles* (IAU Coll. 17), Cayrel de Strobel, G., and Deplace, A.M., Eds., Observatoire de Paris-Meudon, 1972, pp. XI, 1–33.

[J1] Jenkins, L.F., *Yale Parallax Catalogue,* Yale University Press, 1952.

[K1] Kamp, P. van de, "Unseen Astrometric Companions of Stars," *Ann. Rev. Astron. Astrophys.,* 1975, **13,** pp. 295–333.

[K2] Kippenhahn, R., Thomas, H.-C., Weigert, A., "Sternentwicklung IV—Zentrales Wasserstoff- and Heliumbrennen bei einem Stern von 5 Sonnenmassen," *Z. Astrophys.,* 1965, **61,** pp. 241–267.

[K3] Kumar, C.K., "Reddening of HII Regions in M31, *Astrophys. J.,* 1979, **230,** pp. 386–389.

[K4] Kunth, D., "Primordial Helium and Emission-Line Galaxies," in *Cosmology and Particles,* Audouze, J., Crane, P., Gaisser, T., Hegyi, D., and Tran Thanh Van, J., Editions Frontières, Dreux, France, 1982, pp. 241–251.

[L1] Lee, T.A., "Photometry of High-Luminosity M-Type Stars," *Astrophys. J.,* 1970, **162,** pp. 217–238.

[L2] Lequeux, J., Peimbert, M., Rayo, J.F., Serrano, A., and Torres-Peimbert S., "Chemical Composition and Evolution of Irregular and Blue Compact Galaxies," *Astr. Astrophys.,* 1979, **80,** pp. 155–166.

[L3] Luyten, W.J., "Statistical HR Diagrams for One Hundred and Fifteen Thousand

Proper-Motion Stars," in *The HR Diagram* (IAU Symp. 80), Davis Phillip, A.G., and Hayes, D.S., Eds., Reidel, 1978, pp. 63–64.

[L4] Lynds, B.T., "Catalogue of Dark Nebulae," *Astrophys. J.*, suppl. ser., 1962, **7,** pp. 1–52.

[M1] Mihalas, D., and Binney, J., *Galactic Astronomy,* Freeman, 1981.

[M2] Mermilliod, J.C., "The Present Data Situation for Stars in Open Clusters" in *Star Clusters* (IAU Symp. 85), Hesser, J.E., Ed., Reidel, 1980, pp. 129–133.

[M3] Morgan, D.H., and Nandy, K., "Infrared Interstellar Extinction in the LMC," *Month. Not. R. Astron. Soc.,* 1982, **199,** pp. 979–986.

[O1] Olive, K.A., Schramm, D.N., Steigmann, G., Turner, M.S., and Yang, J., "Big-Bang Nucleosynthesis as a Probe of Cosmology and Particle Physics," *Astrophys. J.*, 1981, **246,** pp. 557–568.

[O2] Olsen, B.I., "On the Ratio of Total-to-Selective Absorption," *Publ. Astron. Soc. Pacific,* 1975, **87,** pp. 349–351.

[P1] Pagel, B.E.J., "Abundances of Elements of Cosmological Interest," *Philos. Trans. R. Soc. London,* ser. A, 1982, **307,** pp. 19–35.

[P2] Popper, D.M., "Stellar Masses," *Ann. Rev. Astron. Astrophys.,* 1980, **18,** pp. 115–164.

[R1] Richstone, D.O., and Graham, F.G., "A New Determination of the Luminosity Density of the Galaxy," *Astrophys. J.,* 1981, **248,** pp. 516–523.

[S1] Sandage, A., "Current Problems in the Extragalactic Distance Scale," *Astrophys. J.*, 1958, **127,** pp. 513–526.

[S2] Sandage, A., "Main-Sequence Photometry, Colour-Magnitude Diagrams, and the Globular Clusters, M3, M13, M15, and M92," *Astrophys. J.,* 1970, **162,** pp. 841–870.

[S3] Sandage, A., "The Redshift-Distance Relation. V—Galaxy Colors as Functions of Galactic Latitude and Redshift: Observed Colors Compared with Predicted Distributions for Various World Models," *Astrophys. J.,* 1973, **183,** pp. 711–730.

[S4] Sandage, A., "High-Latitude Reflection Nebulosities Illuminated by the Galactic Plane," *Astr. J.,* 1976, **81,** pp. 954–957.

[S5] Sandage, A., "The Oosterhoff Period Groups and the Age of Globular Clusters. III—The Age of the Globular Cluster System," *Astrophys. J.,* 1982, **252,** pp. 553–573.

[S6] Sandage, A., and Tammann, G.A., "Steps toward the Hubble Constant. VII—Distances to NGC 2403, M101, and the Virgo Cluster Using 21-Centimeter Line Widths Compared with Optical Methods. The Global Value of H_0," *Astrophys. J.,* 1976, **210,** pp. 7–24.

[S7] Savage, B.D., and Mathis, J.S., "Observed Properties of Interstellar Dust," *Ann. Rev. Astron. Astrophys.,* 1979, **17,** 73–111.

[S8] Schmidt, M., "The Mass of the Galactic Halo Derived from the Luminosity Function of High-Velocity Stars," *Astrophys. J.,* 1975, **202,** pp. 22–29.

[S9] Schneps, M.H., Lane, A.P., Downes, D., Moran, J.M., Genzel, R., and Reid, M.J., "Proper Motions and Distances of H_2O Maser Sources. III—W51NORTH," *Astrophys. J.,* 1981, **249,** pp. 124–133.

[S10] Schramm, D.N., "Nuclear Constraints on the Age of the Universe," in *Highlights of Astronomy,* vol. 6, West, R.M., Ed., Reidel, 1983, pp. 241–253.

[S11] Spitzer, L. Jr., *Physical Processes in the Interstellar Medium,* Wiley, 1978.

[S12] Symbalisty, E.M.D., and Schramm, D.N., "Nucleocosmochronology," *Rep. Prog. in Phys.,* 1981, **44,** pp. 293–328.

[T1] Tayler, R.J., *The Stars, Their Structure and Evolution,* Taylor and Francis, 1970.

[T2] Tayler, R.J., *Origin of the Elements,* Taylor and Francis, 1972.

[T3] Tinsley, B., and Larson, R.B., Eds., *Evolution of Galaxies and Stellar Populations,* Yale University Press, 1977.

[V1] VandenBergh, D.A., "Star Clusters and Stellar Evolution. I—Improved Synthetic Color-Magnitude Diagrams for the Oldest Clusters, *Astrophys. J.,* suppl. ser., 1983, **51,** pp. 29–66.

[V2] Vaucouleurs, G. de, "The Extragalactic Distance Scale. I—A Review of Distance Indicators: Zero Point Errors of Primary Indicators," *Astrophys. J.,* 1978, **223,** pp. 351–363.

[V3] Vaucouleurs, G. de, Vaucouleurs, A. de, and Corwin, H.G., Jr., *Second Reference Catalogue of Bright Galaxies,* University of Texas Press, 1976.

[W1] Wilson, O.C., and Bappu, M.K.V., "H and K Emission in Late-Type Stars: Dependance of Line Width on Luminosity and Related Topics," *Astrophys. J.,* 1957, **125,** pp. 661–683.

CHAPTER THREE

PRIMARY EXTRAGALACTIC DISTANCE INDICATORS

We now try to take the first steps up the distance ladder outside our Galaxy. In this chapter we describe those methods of measuring the distances to external galaxies which can be calibrated from observations in our Galaxy or from theoretical considerations. We call these the *primary* distance methods. Then when we have established the distances to nearby galaxies with primary methods, we use them to calibrate the less secure *secondary* methods (see Chap. 4), which have to be used to find the distances to remote galaxies.

The four major primary distance indicators are Cepheids, supernovae, novae, and RR Lyrae variables, described in Secs. 3.1 to 3.4. Several other less well-developed methods, using W Virginis stars, red giants, blue supergiants, Mira variables, and eclipsing variables, are also described in Sec. 3.4.

The most thoroughly studied and probably most fundamental of the primary indicators is the Cepheid variable star, both because it is luminous and readily (if laboriously) observable in nearby galaxies and because the period–luminosity–color relation is on a secure theoretical basis. Infrared observations hold out the promise of an enormous improvement in the precision of distances based on Cepheid variables.

The supernova method has the greatest potential of the primary methods, because supernovae are in principle observable to enormous distances (~ 1000 Mpc). But in practice there are still considerable complexities in the theoretical understanding of supernovae. It also seems unfair that although statistical studies in other galaxies imply that a supernova should go off in our Galaxy every 50 years or so on the average, none has been clearly seen since Kepler's supernova of 1604.

Novae show a considerable range in their observed properties. In recent years there has been great improvement in the theoretical understanding of the classical nova phenomenon. Novae have the potential to determine distances out to the Virgo cluster, but they have only been applied to the determination of the distances of our satellites, the Magellanic Clouds, and the Andromeda nebula.

The low luminosity of RR Lyrae variables means that they are hard to observe much beyond the Magellanic Clouds. However, they provide a useful check on the other distance indicators. Finally in Sec. 3.5 we give a summary of the galaxies whose distances have been measured with primary indicators, and discuss how far out we can reach with primary methods.

Before discussing these primary distance methods it is worthwhile to try to get a picture of our cosmic neighborhood. It is only in the present century that we have realized with certainty that the universe extends far beyond our own Milky Way galaxy. The nearest galaxies to our own, the Large and Small Magellanic Clouds, must, however, have been familiar nighttime objects to the peoples of the southern hemisphere for thousands of years, for even to the unaided eye they are a spectacular sight, large nebulous patches of light. They are essentially companions to our own Galaxy, in relatively close orbit round it. Modern estimates place them at about 150,000 lt-yr from us, only just beyond the most distant globular clusters in the galactic halo. Other, smaller companions to our Galaxy have been found with the aid of modern telescopes, the Sculptor, Draco, and Ursa Minor dwarf spheroidal galaxies, with distances in the range of 150,000 – 300,000 lt-yr. These have been studied much less thoroughly than the Magellanic Clouds.

At a much greater distance, about 2 million lt-yr away, lies the Andromeda galaxy, M31, dimly discernible with the naked eye and, as we remarked in Chap. 1, known to the Arab astronomers of the tenth century A.D. It is rather similar in size and structure to our own Galaxy and has several dwarf satellite galaxies as well as the more substantial companion, the spiral galaxy M33 in the Triangulum. Our Galaxy, M31, and their companions make up a group of 30 or so galaxies called the Local Group of galaxies.

To find the next groups of galaxies in our cosmic neighborhood, we have to travel out about 6 – 10 million lt-yr, to the Sculptor and M81 groups. Each is similar to the Local Group in that it comprises two or three large galaxies and a number of smaller ones. At this range of distance we are already reaching the limit for most of the primary distance methods. At about 15 – 20 million lt-yr we reach the M101 group, important because it contains the giant spiral galaxy M101, the Pinwheel. Such galaxies can be observed out to very great distances. Finally, to complete our look at our cosmic neighborhood, at about 40 – 60 million lt-yr we find the Virgo cluster of galaxies, a vast agglomeration of thousands of galaxies spread across the northern sky, centered on the constellation of Virgo, and first noticed by William Herschel at the end of the eighteenth century. This has played a dominant, almost obsessive role in the controversy about the cosmological distance ladder, since astronomers had hoped to calibrate the Hubble velocity – distance law, and hence to determine the size of the universe, using the Virgo cluster. As we shall see in Chaps. 5 and 6, this is probably a misguided hope.

3.1 Cepheid variable stars

When massive stars have exhausted the hydrogen in their cores and begin to evolve rapidly back and forth across the HR diagram, they tend to become pulsationally unstable as they cross a narrow zone known as the instability strip (see Sec. 2.7). The visual luminosity of the stars undergoes regular large-amplitude variations with a period in the range of 2–150 days. Since massive stars are short-lived, the Cepheid variables are found among young Population I stars in open clusters and OB associations.

The most important property of Cepheids for distance determination is the correlation between their mean absolute magnitude and their period, discovered by Henrietta Leavitt in 1907. Fig. 3.1 shows a more recent plot by Allan Sandage of Leavitt's period–luminosity law for Cepheids in our Galaxy and other Local Group

FIGURE 3.1

Period–luminosity relation for Cepheids in the Galaxy, Large and Small Magellanic Clouds, M31, and NGC 6822. (a) Absolute blue magnitude M_B *at mean light, corrected for interstellar extinction, versus log* P. *(b) Absolute visual magnitude* M_V *at mean light versus log* P. *(From [S3].)*

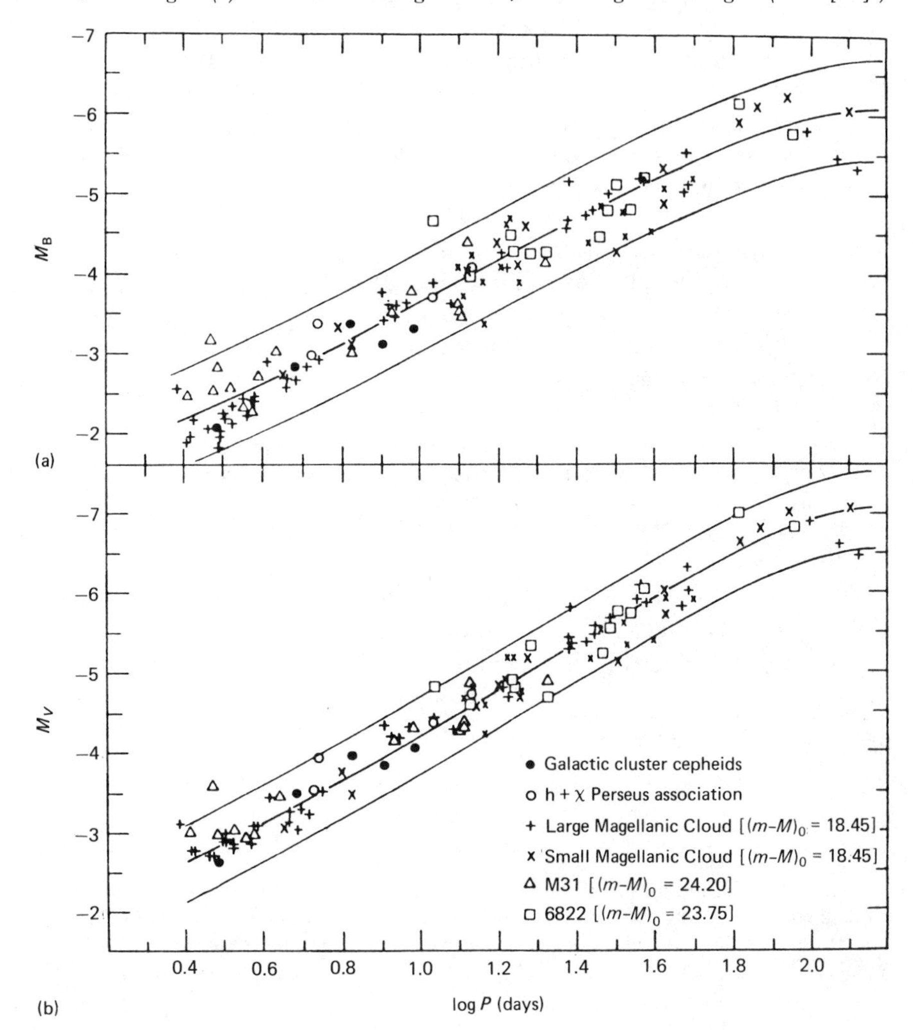

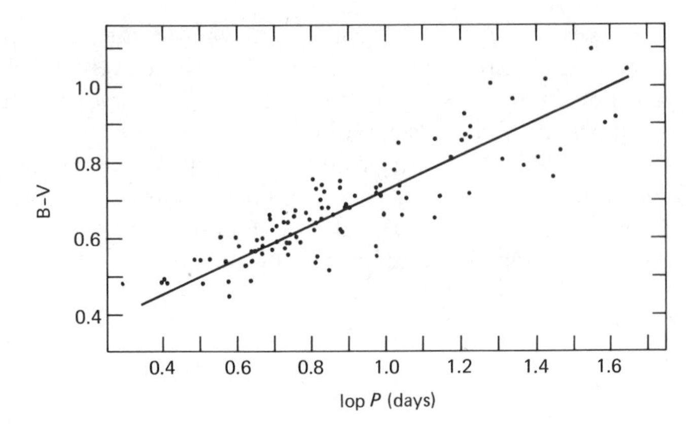

FIGURE 3.2
B-V color at mean light, corrected for interstellar reddening, versus log P *for galactic cepheids. (From [D2].)*

galaxies. The significance of this period–luminosity relation is, as mentioned in Chap. 1, that simply by measuring the period of a Cepheid, its luminosity can be estimated and its distance found from the inverse-square law. Because the instability strip is inclined to the vertical in the HR diagram (see Fig. 2.30), there is a correlation between mean absolute magnitude and color for stars in the instability strip, and hence also between period and color (Fig. 3.2).

In 1958 Sandage found that the displacement of the points in the period–luminosity diagram from the average line is correlated with the mean color of the star, measured by the blue-visual color index B-V [S1]. The most accurate way to measure distances with Cepheids is therefore through a period–luminosity–color (PLC) relation.* This relationship can be explained as follows. The pulsation period P for Cepheids will be related to the natural oscillation period of a star, which can be shown to be proportional to $\rho^{-1/2}$ where ρ is the mean density of the star. This natural oscillation period is essentially the time it would take for the star to collapse if pressure support were suddenly removed. For the sun this time is about 1 hour. For Cepheids we therefore expect

$$P\rho^{1/2} = Q \tag{3.1}$$

where Q is a constant called the pulsational constant, which may be expected to be the same for all Cepheids pulsating in the same manner. Now

$$\rho = \frac{3M}{4\pi R^3} \qquad \text{and} \qquad L = 4\pi R^2 \sigma T_{\rm eff}^4 \tag{3.2}$$

* Sandage and Tammann [S6] have also used the amplitude of Cepheid light variations in place of the B-V color, but later work has cast doubt on the uniqueness of the relationship between amplitude and color. (See, for example, the discussion by Tammann, Sandage, and Yahil [T3].)

where M, R, L, and T_{eff} are the mass, radius, luminosity, and effective temperature of the star and σ is the Stefan–Boltzmann constant. Over the mass range of Cepheids we may expect the following:

1 A relationship between mass and luminosity of the form

$$\log M = hM_{bol} + \text{constant} \tag{3.3}$$

where M_{bol} is the absolute bolometric magnitude of the star
2 The bolometric correction to be proportional to B-V, so

$$M_{bol} = M_V + d(\text{B-V}) + \text{constant} \tag{3.4}$$

where M_V is the absolute visual magnitude
3 A temperature–color relationship,

$$\log T_{eff} = a(\text{B-V}) + \text{constant} \tag{3.5}$$

where h, d, and a are constants which have to be determined observationally or from theory

Combining Eqs. (3.1)–(3.5), we obtain

$$\log P + (0.5h + 0.3)M_V + (0.5hd + 0.3d + 3a)(\text{B-V}) = \text{constant}. \tag{3.6}$$

An equation of this form was first derived by Sandage in 1958 and applied by him to the Cepheids in the Magellanic Clouds. In 1972 Icko Iben and R.S.Tuggle [I1] used detailed pulsation and evolution theory for Cepheids to determine the constants in this formula and found

$$M_V \simeq -2.61 - 3.76 \log P + 2.60(\text{B-V}) \tag{3.7}$$

$$= n - m \log P + q(\text{B-V}) \tag{3.8}$$

where M_V and B-V are magnitudes corresponding to the average luminosity in the relevant wavelength band over a cycle (mean light) and have already been corrected for the effects of extinction by dust. A recent and thorough investigation of Cepheids in the Large Magellanic Cloud by W.L. Martin, P.R. Warren, and M.W. Feast [M3] yielded the following empirical relation at mean light:

$$V = 16.41(\pm 0.14) - 3.80(\pm 0.13) \log P + 2.70(\pm 0.23)(\text{B-V}) \tag{3.9}$$

The coefficients of log P and B-V being in excellent agreement with Eq. (3.7). An improvement over most earlier work is the use of the infrared I (0.9 μm) magnitude, in addition to the usual B and V magnitudes, to help determine the reddening by dust in the line of sight to each Cepheid.

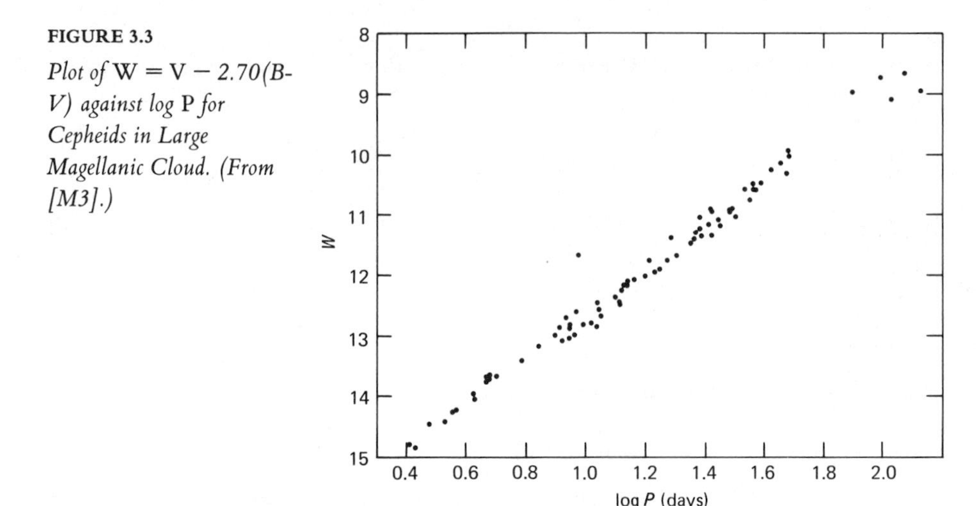

FIGURE 3.3
Plot of W = V − 2.70(B-V) against log P for Cepheids in Large Magellanic Cloud. (From [M3].)

Now by coincidence the coefficient of the B-V term in Eq. (3.7) is close to the 3.0 expected for the interstellar reddening relation $A_V = 3.0E(\text{B-V})$, where E(B-V) is the color excess (see Box 2.4). Thus if we subtract q(B-V) from the observed magnitude V, where q is the coefficient of B-V in Eq. (3.8), we will take out not only the color term in the Cepheid period–luminosity–color relation, but also most of the effect of interstellar extinction. The quantity $W = V - q(\text{B-V})$ will therefore depend only on the period and will hardly be affected by reddening due to dust at all. Fig. 3.3 shows the plot by Martin et al. of $W = V - 2.70(\text{B-V})$ against log P for Cepheids in the Large Magellanic Cloud. Different workers use slightly different values for q (see Table 3.2). Sidney van den Bergh [B7] has compared observations of Cepheids in the Magellanic Clouds, M31 (Andromeda), and the Local Group irregular galaxy, NGC 6822, with lines of the form $W = V - 2.67(\text{B-V}) = A - 3.67 \log P$, and his study shows that the period–luminosity–color relation does not differ greatly from galaxy to galaxy. However, theoretical studies show that the coefficients of B-V and log P in Eq. (3.8) depend on the detailed structure of the stars, for example, on the chemical composition and surface gravity of the star and on the details of the microturbulence in the surface layers. In 1975 Iben and Tuggle considered the effect of composition on Eq. (3.8) and derived the more elaborate relation

$$\begin{aligned} M_V(1 + 2.37\Delta Z + 0.29\Delta Y) = {} & -2.77 + 0.29\Delta Y \\ & - 31.1\Delta Z(1 - 0.43\Delta Z + 0.3\Delta Y) - 3.64 \log P \\ & + 2.63(1 - 0.37\Delta Z + 0.4\Delta Y)(\text{B-V}) \end{aligned} \tag{3.10}$$

where $\Delta Y = Y - 0.28$, $\Delta Z = Z - 0.02$, and Y and Z are the star's abundances by mass in the form of helium and heavy elements, respectively. Applying this to the Magellanic Clouds and M31, they obtained the results shown in Table 3.1. Note the lower heavy-element abundances in the Large Magellanic Cloud (LMC) and, especially, the Small Magellanic Cloud (SMC) compared with those of our Galaxy. These agree with direct estimates of heavy-element abundances in Small Magellanic

TABLE 3.1

Distance moduli and abundances for Magellanic Clouds and M31 derived from Cepheids by Iben and Tuggle (1975).

	LMC	*SMC*	*M31*
Z	0.01	0.005	0.02
Y	0.29	0.29	0.29 ± 0.06
μ_0	18.73	19.25	24.40

Y, Z—abundances of helium and heavy elements by mass.
SOURCE: From [I2].

Cloud Cepheids [H2] and with estimates in gaseous nebulas in the Magellanic Clouds made by Bernard Pagel and co-workers [P1]. The dependence of the coefficients in Eq. (3.10) on Y is rather weak.

In light of the above discussion, some of the ways that the period–luminosity relation has been applied in practice to determine the distances to galaxies are less than satisfactory from a theoretical point of view. For example, it is almost always assumed that the coefficients of $\log P$ and B-V in the period–luminosity–color relation, Eq. (3.8), are the same for every galaxy. These coefficients (m and q) are usually determined using the Cepheids of the Magellanic Clouds, since there are too few Cepheids with known distances in our Galaxy.

To calibrate the period–luminosity relation absolutely from observations, that is, to determine the quantity n in Eq. (3.8), we then have to use the Cepheids in our Galaxy with well-determined distances. These are listed in App. A.9. Martin et al. [M3] also use an additional 36 Cepheids for which L.A. Balona has determined the radii by the Baade–Wesselink method (see Sec. 2.5). Table 3.2 summarizes the values of m, n, and q determined by different authors, and Table 3.3 shows various estimates of the distances to the Magellanic Clouds using Cepheids. The values of m and n found by Sandage and Tammann in 1969 [S5] are slightly lower than those found in most later work, but the resulting distance moduli for the Magellanic Clouds do not seem very different from those found by other workers. On the whole there is a reasonable consensus on the values of the parameters m, n, and q. However, this consensus is challenged by Victor Clube and J.A. Dawe [C4], who claim that $q = -1$, and by J.P. Brodie and Barry Madore [B17], who claim $q = 5$. These claims, which are not generally accepted, are criticized in [F3] and [S17]. Table 3.2 also summarizes the coefficients s and t found by different authors for the period–color relation

$$\text{B-V} = s \log P + t. \tag{3.11}$$

The method of statistical parallaxes (see Sec. 2.3) has been applied by several authors to the nearer Cepheids in our Galaxy (distance $\lesssim 1$ kpc) to try to establish their absolute magnitudes [J2],[H4],[W1],[C4]. The results are in broad agreement with the calibrations of galactic Cepheids given in Table 3.2, but are of insufficient accuracy to discriminate between the proposed calibrations. T.G. Barnes and co-workers [B3],[B4] have tried to establish an independent distance scale for galactic Cepheids by using a modification of the Baade–Wesselink method (see Sec. 2.5).

TABLE 3.2

Constants in period–luminosity–color and period–color relations for Cepheids.

$$M_V = -n - m \log P + q(\text{B-V})$$

$$\text{B-V} = s \log P + t$$

Authors	Location/method	n	m	q	s	t
Fernie (1967) [F4]	Theory	2.54*	2.79	2.0		
Fernie (1967) [F4]	Galaxy				0.49	0.24
Sandage and Tammann (1968) [S4]	Galaxy				0.26	0.37
	LMC, SMC				0.44	0.06
Sandage and Tammann (1969) [S5]		2.46	3.425	2.52		
Sandage and Tammann (1971) [S6]		2.47	3.53	2.65	0.32	0.29
Iben and Tuggle (1972) [I1]	Theory	2.61*	3.76	2.60		
van den Bergh (1975) [B7]	LMC, SMC, Galaxy	2.46	3.67	2.67		
Iben and Tuggle (1975) [I2]	Galaxy	2.77*	3.64	2.63		
	LMC	2.50*	3.71	2.70		
	SMC	2.37*	3.76	2.75		
	M31	2.77*	3.64	2.63		
van den Bergh (1977) [B8]		1.86	4.11	2.52	0.48	0.27
Dean et al. (1978) [D2]					0.46	0.27
Butler (1978) [B19]	SMC		3.39 ± 0.11	1.82 ± 0.24	0.40	0.31
Martin et al. (1979) [M3]	LMC, Galaxy	2.33 ± 0.07	3.80 ± 0.13	2.70 ± 0.23		
	Baade–Wesselink	2.42* ± 0.05				

† $\mu_{\text{Hya}} = 3.03$, except values with asterisks, which are independent of μ_{Hya}.

TABLE 3.3

*Distance moduli μ_0 to Magellanic Clouds using Cepheids.**

Authors	*LMC*	*SMC*
Sandage and Tammann (1971) [S6]	18.59 ($A_B = 0.32$)	19.27 ($A_B = 0.08$)
Gascoigne (1974) [G3]	18.74 ($A_B = 0.32$)	18.95 ($A_B = 0.08$)
van den Bergh (1975) [B7]	18.85 ($A_V = 0.18$)	19.46 ($A_V = 0.06$)
Iben and Tuggles (1975) [I2]†	18.73 ($A_V = 0.24$)	19.25 ($A_V = 0.06$)
van den Bergh (1977) [B8]		
Period-luminosity-color	19.24 ($A_V = 0.15$)	19.54 ($A_V = 0.09$)
Period-luminosity	18.55 ($A_V = 0.15$)	18.81 ($A_V = 0.09$)
Feast (1977) [F1]		19.00 ($A_B = 0.12$)
Butler (1978) [B19]	18.78 ($A_B = 0.20$)	18.90 ($A_B = 0.48$)
de Vaucouleurs (1978) [V3]‡	18.28 ($A_B = 0.43$)	18.70 ($A_B = 0.31$)
Martin et al. (1979) [M3]	18.69 ($A_B = 0.20$)	
Madore (1983) [M2]§	18.65 ($A_H = 0.05$)	19.07 ($A_H = 0.03$)

* Based on $\mu_{Hya} = 3.03$, except where noted.
† Theoretical analysis, independent of μ_{Hya}.
‡ $\mu_{Hya} = 3.29$, correction of $\Delta\mu_0 = -0.15$ for line blanketing, based on period-luminosity relation only.
§ $\mu_{Hya} = 3.29$.

Using stars whose diameters have been measured (by interferometric methods), they have found an empirical correlation between the visual surface brightness of these stars and their visual-red (V-R) color index. From this correlation the angular diameter variation of Cepheids can be deduced. The radial velocity variations can then be used to infer the radii and hence the distances of the stars. The results are again consistent with the calibrations of Table 3.2, within the uncertainties of the method. The Baade-Wesselink method has also been applied to Cepheids by R.B. Stothers [S18], again with results consistent with those of Table 3.2.

Gerard de Vaucouleurs argues that little confidence can be placed in the period-luminosity-color relation until its shape and dependence on color have been more securely established [V2]. Essentially he is arguing that the parameter q in the period-luminosity-color relation [Eq. (3.8)] is too poorly determined for it to be worthwhile to use the color correction. He uses instead the absolute magnitude at a standard period, $\log P = 0.8$, with no inclusion of a color term. In view of the rather strong empirical and theoretical evidence for a color term, and the excellent agreement between the two, this is a surprising decision. For Cepheids with periods different from $\log P = 0.8$, he interpolates to this standard value, assuming $d \log M_V / d \log P = -2.8$. This method is therefore equivalent to using an M_V-$\log P$ relation in which the slope is assumed to be known a priori. De Vaucouleurs's scepticism about the period-luminosity-color method is also shared by M.J. Stift [S17].

The net result of de Vaucouleurs's assumptions, including the use of larger corrections for interstellar extinction, is that he obtains significantly lower distances to the Magellanic Clouds than the other workers listed in Table 3.3. For example, he finds a distance of 46 kpc to the Large Magellanic Cloud compared with 51–70 kpc found by other workers, and 55 kpc to the Small Magellanic Cloud compared with 58–78 kpc by other workers. The four estimates by van den Bergh shown in this table confirm that lower distances are obtained by the period–luminosity method, with no account taken of the color term, than by the period–luminosity–color or *W* methods. In this book I shall make use of distance estimates based on observations of Cepheids in the optical band only if they take account of the color term.

An exciting recent development is the use of infrared magnitudes in place of visual ones, an idea first investigated by J.D. Fernie in 1975. Infrared measurements offer the tremendous advantage of a great reduction in the effects of interstellar extinction and of the heavy-element abundance of the stars. A further advantage of infrared measurements of Cepheids, emphasized by Madore and co-workers [M2],[M8], is that the amplitude of the light variations is also much reduced. In fact these authors show that for a sample of 40 Cepheids in the Large Magellanic Cloud, the scatter in the period–luminosity relation in the H band (1.65 μm) for observations *at random phase* is only 60% of the scatter of the corresponding relation for time-averaged observations in the B band (0.44 μm). A single infrared measurement of a Cepheid can yield a distance estimate accurate to $\pm 10\%$. This demonstrates the immense potential of the infrared for Cepheid studies, though it should be noted that searches for new Cepheids and the determination of their periods will still have to be done in the visible band. Madore and co-workers have now applied the infrared method to the Large and Small Magellanic Clouds, M31, M33, NGC 6822, and IC 1613,* and will soon extend it to NGC 300 in the Sculptor group, and NGC 2403 in the M81 group.

Distances to Local Group galaxies using Cepheids are given in Tables 3.3 and 3.8 (Magellanic Clouds), 3.10 (M31), and 3.11 (other Local Group galaxies). In a major recent study Sandage and G. Carlson [S11] have established the distance to M33 using Cepheids, finding a distance modulus 0.7 mag greater than the old value due to Hubble. Sandage and Carlson also report that work is in progress on Cepheids in NGC 3109, IC 5152, and the Wolf–Lundmark, Sextans B, Leo A, and Pegasus galaxies [S10]. J.A. Graham has reported preliminary results on the distance to NGC 300 in the Sculptor group using Cepheids [G7]. He finds that the blue-magnitude distance modulus μ_B (uncorrected for the effects of interstellar and internal extinction) for NGC 300 is 7.36 mag greater than that of the Large Magellanic Cloud. Correcting for the effects of interstellar and internal extinction in the Large Magellanic Cloud and in NGC 300 (our adopted formulas from Sec. 2.8 give $A_B = 0$ and $A_i = 0.27$ for the latter galaxy), the true distance modulus for NGC 300 is 7.45 mag greater than that of the Large Magellanic Cloud, corresponding to a distance 31 times greater than that of the latter.

The most distant galaxy in which Cepheids have been studied is NGC 2403 in the M81 group. Fig. 3.4 shows the 1968 observations by Gustav Tammann and Sandage of some of the Cepheids they found. Because the stars are so faint, the full variability cycle cannot be followed, and for this reason Tammann and Sandage

* IC stands for index catalog, a supplement to the NGC catalog of nebulae.

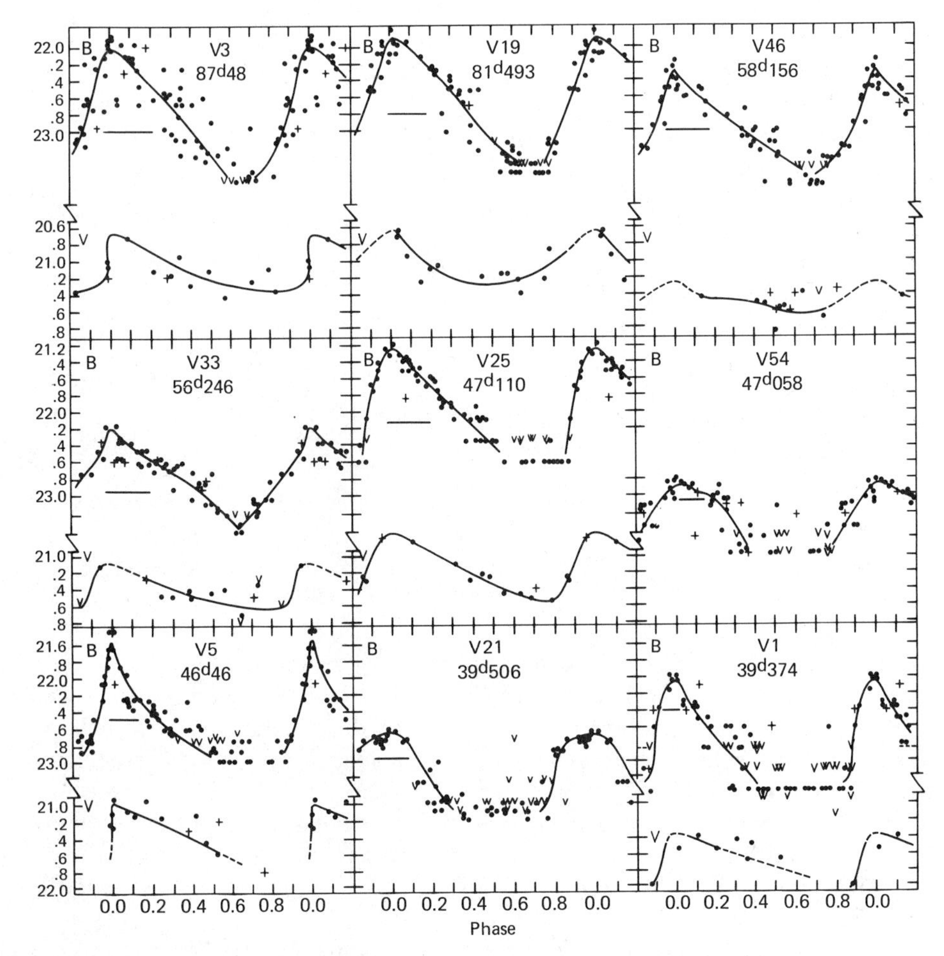

FIGURE 3.4

Light curves in B and V for nine Cepheids in NGC 2403. At minimum light the B curves lie below the limit of observation. The horizontal lines denote the limit of accurately calibrated magnitudes. (From [T2].)

worked in terms of the brightness of the stars at maximum light, instead of at mean light. They calibrated this using the stars of the Magellanic Clouds and our Galaxy, as before.

The distance deduced by Tammann and Sandage for NGC 2403 is 3.25 Mpc [T2]. Madore, however, believes that the distance is actually 2.2 Mpc and suggests that Tammann and Sandage have been led to an overestimate for two reasons [M1]. First, Madore argues that there is significant extinction within NGC 2403 itself. Second, he finds that if he removes those stars that are in crowded fields, and therefore most susceptible to observational error, the mean period–luminosity–color relation is significantly shifted toward brighter magnitudes and, hence, a

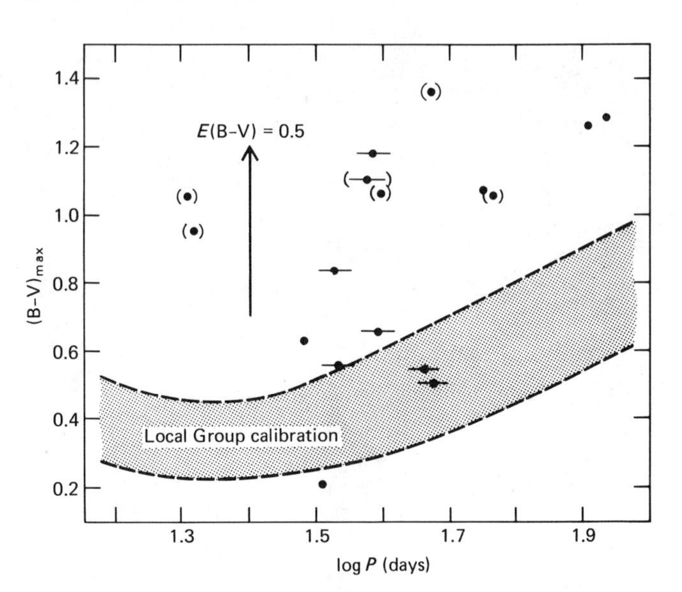

FIGURE 3.5
B-V color at maximum light $(B\text{-}V)_{max}$ versus log P for Cepheids in NGC 2403. Cepheids on dense backgrounds or crowded by nearby stars are marked with horizontal bars, those with uncertain colors are in parentheses. The shaded area is the Local Group calibration. The arrow shows the effect of 0.5 mag of reddening. (Adapted from [M1], [T2].)

lower distance. Fig. 3.5 shows Madore's plot of B-V against log P for NGC 2403 Cepheids. Stars in crowded fields are marked with horizontal bars, and it can be seen that most of the remaining points lie well above the calibration based on Local Group galaxies, apparently implying considerable reddening by dust. Madore estimates a reddening of $E(\text{B-V}) = 0.4$ mag for the Cepheids in this galaxy, considerably more than the estimate of 0.06 given by Tammann and Sandage based on the reddening of foreground stars in the same direction in the sky.

Madore uses the correlation of $W = V_{\text{max}} - 3.1(\text{B-V})_{\text{max}}$ with log P for the "best" NGC 2403 Cepheids to deduce a distance modulus $\mu_0 = 26.7$ mag ($d =$ 2.2 Mpc), using the same Hyades modulus as Tammann and Sandage, $\mu_{\text{Hya}} =$ 3.03 mag. Madore's conclusions, however, should be viewed in light of the following comments:

1 The foreground interstellar reddening predicted by the Fisher–Tully law (adopted in Sec. 2.8) is 0.055, in good agreement with the observations of Tammann and Sandage of 0.06 ± 0.015. The total internal reddening in NGC 2403 predicted by the de Vaucouleurs law (adopted in Sec. 2.8) is 0.085. (The assumption by Tammann and Sandage that internal reddening in NGC 2403 is negligible does not seem plausible and is inconsistent with the procedure adopted by them in work on other galaxies of this type [S8].) The total reddening due to interstellar dust in our Galaxy and to internal dust in NGC 2403 is therefore predicted to be 0.15 mag, far below Madore's estimate of 0.40.

2 Madore noted the uncertainty underlying observations of stars in crowded fields, and eliminated these stars in order to select the "best" Cepheid data. But there are also several NGC 2403 Cepheids whose colors Tammann and Sandage note to be uncertain. (These are shown in Fig. 3.5 in parentheses.) The omission of these stars

from the calculations would work in the opposite direction to the selection procedure adopted by Madore. It would appear to be as reasonable to suppose that the redder colors of the NGC 2403 Cepheids are due to observational errors (as assumed by Tammann and Sandage) as to assume that they are due to very high internal reddening in NGC 2403. This could be settled by infrared observations of these stars.

3 Madore's definition of W would correctly remove the effects of reddening by interstellar dust, but overcorrects for the intrinsic color effect in Cepheids. The value of q found by Martin et al. [M3] was 2.70, and most values in Table 3.2 are lower than this.

In conclusion, we believe that Tammann and Sandage's distance modulus for NGC 2403, based on B (blue) magnitudes at maximum light, should be corrected for internal extinction by $4 \times 0.085 = 0.34$ mag (see Table 3.9), that is, to 27.22 mag (using the old Hyades modulus, $\mu_{\rm Hya} = 3.03$), corresponding to a distance of 2.8 Mpc. Also, Madore's distance modulus should be increased by $(3.1 - 2.7) \times [(\text{B-V})^0_{\max} - 0.085]$, where $(\text{B-V})^0_{\max}$ is the mean B-V color at maximum light corrected for interstellar reddening, to allow for effect 3 above. For the "best" Cepheids used by Madore, $(\text{B-V})^0_{\max} = 0.91$. Thus Madore's estimate is increased to 27.03 mag, or a distance of 2.55 Mpc. Adopting the mean of these two values as the best compromise, correcting for the new Hyades modulus of 3.30 mag, which we adopted in Sec. 2.4, and applying van den Bergh's correction of -0.15 mag for line blanketing (see Sec. 2.4), we obtain a revised distance modulus for NGC 2403 from Cepheids of 27.24 ± 0.25* ($d = 2.8 \pm 0.3$ Mpc).

Some of the problems in using Cepheids to estimate the distances of external galaxies have been reviewed by Tammann, Sandage, and A. Yahil in their 1979 Les Houches lectures [T3] and by van den Bergh at the IAU Colloquium No. 37 in 1977 [B8]. These problems may be summarized as follows:

1 The difficulty of determining the correction for extinction in the galaxy observed (quite apart from the uncertainty in the extinction due to dust in our Galaxy, see Sec. 2.8). This problem can be overcome by using infrared observations to help determine the reddening to each star, as in the work of Martin et al. discussed before, or better still by working entirely in terms of infrared fluxes, as in the work of Madore and his collaborators.

2 The dependence of the period–luminosity–color relation on the heavy-element abundance. This affects the calibration of galactic Cepheids, since the abundances in the open clusters and associations are not accurately known, as well as that of Cepheids in external galaxies. Both the luminosities and the colors of Cepheids are modified by abundance variations [see Eq. (3.10)]. It has been realized for some time that the Magellanic Cloud Cepheids are bluer than their counterparts in our Galaxy, and we have seen that this can be interpreted as implying lower heavy-element abundances in the Magellanic Clouds than in our Galaxy.

* The uncertainties we have assigned to each distance method are summarized in Table 4.12.

3 The pulsation constant Q may not, as was assumed in the derivation of the period–luminosity–color relation [Eq. (3.6)], be independent of the period and the mean density of the star. This would lead to a disagreement between theoretical and empirical period–luminosity–color relations. The curvature in the period–luminosity relation for longer period Cepheids claimed by Sandage and Tammann (see Fig. 3.1), if it is real, could be due to variation of Q with period. The possible breakdown of the linear relation would be unfortunate since it is the longer period, more luminous Cepheids that are more easily observable in external galaxies. Van den Bergh [B7], Madore [M1], and Martin et al. [M3] do not find any deviation from linearity after correction for the effects of reddening.

4 Since the evolutionary tracks of stars form loops in the HR diagram which cross the instability strip more than once (see Fig. 2.25), the instability strip may not be uniformly populated with Cepheids, and systematic deviations from the true mean line may be introduced. Also a star may have a different mass-to-light ratio at each crossing, and this could undermine the concept of a unique period–luminosity relation. The seriousness of this problem is not really known at present.

5 The lower heavy-element abundances of the clusters in which Cepheids are found, compared with the Hyades, means that a line-blanketing correction should be applied, which van den Bergh has estimated to be $\Delta\mu_0 = -0.15$ mag, assuming that the clusters and associations in which Cepheids are found have approximately the same heavy-element abundance as the sun [B8]. De Vaucouleurs has emphasized that the magnitude of this correction may be rather uncertain [V3]. Harris [H1] has estimated the heavy-element abundances of galactic Cepheids, and for 16 of the stars in App. A.9 the average value of

$$[\mathrm{Fe/H}] = \log \{[n(\mathrm{Fe})/n(\mathrm{H})]_{\mathrm{star}}/[n(\mathrm{Fe})/n(\mathrm{H})]_{\mathrm{sun}}\}$$

which measures the iron abundance in the stars compared with that of the sun, is -0.02 ± 0.07, supporting van den Bergh's assumption that the galactic Cepheids used to calibrate the period–luminosity–color relation have approximately the same heavy-element abundance as the sun.

Despite these problems, many of which are being overcome in modern work, the Cepheid distance indicator remains powerful and important. The limit to which Cepheids can be observed with ground-based telescopes is about 5 Mpc, and there are several groups of galaxies (the Sculptor group, and M81 group, and IC 342 group, and the NGC 5128 group) within that range according to the distance scale adopted in this book. With the Hubble Space Telescope, due to be launched in the late 1980s, Cepheids may be observable out to distances of 20 Mpc or more, which will enormously strengthen the cosmological distance ladder.

3.2 Supernovae

Over the past two millenia, astronomers have from time to time been startled to see a new star appear in the sky and brighten rapidly to become, for a few months, one of the brightest stars in the sky. The ancient Chinese astronomers, who watched the sky systematically each night over a period of at least 1500 years, recorded several such events. Tycho Brahe and Kepler, two of the greatest astronomers of the Renaissance, each saw one. During the outburst, the star brightens by a factor of a billion or more and up to 10^{51} ergs of energy are generated, mainly in the form of the kinetic energy of the debris from the exploding star. App. A.10 lists some of the most reliable historical reports of supernovae in our Galaxy. The most famous of these is the supernova of 1054 A.D., the debris of which can still be seen as the Crab Nebula. The remnants of many other much older supernova events can be seen throughout the Milky Way, especially at radio wavelengths.

In the past century many supernovae have been observed in other galaxies, particularly following the pioneering work of Fritz Zwicky in the 1930s. The one of 1885 in the Andromeda galaxy, given the variable star label S Andromedae, was at first assumed to be a nova (see below). This mistake prevented astronomers from realizing the great distances of the galaxies. App. A.10 and Table 3.5 give the best observed supernovae seen in external galaxies. Two types of supernova outbursts, called type I and type II and having distinctive spectra and evolutionary characteristics, have been recognized. Type I supernovae occur in both spiral and elliptical galaxies and are generally believed to be associated with older, relatively low-mass

FIGURE 3.6

The 1972 type I supernova in NGC 5253. The supernova can be seen to the lower right of the galaxy image on the right. (Courtesy of Mount Palomar Observatory.)

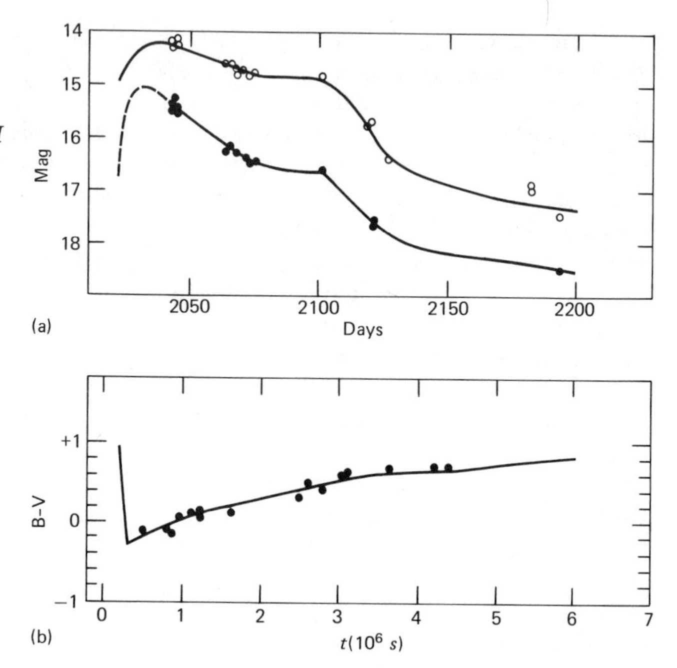

FIGURE 3.7

(a) Light curves in B (solid dots) and V (open circles) for the 1973 type II supernova in NGC 3627. (From [R2].) (b) Variation of B-V color, corrected for reddening by dust, with time for type II supernova 1969l in NGC 1058, compared with predictions of a theoretical model. (From [S14].)

stars. Type II supernovae are seen only in spiral galaxies, concentrated to the spiral arms, so are believed to take place in massive stars.

The sequence of events in type II supernova explosions is now fairly well understood. These supernovae are the last stages of the evolution of stars of mass greater than $8M_\odot$, following the formation of an iron core (see Sec. 2.7). The core collapses, attaining very high density and generating a strong shock wave, which propogates through the extended outer layers of the star, driving them off with velocities in the range of 3000 – 10,000 km s^{-1}. The compressed core is thought to form a neutron star. The expanding envelope consists of $1M_\odot$ or more of new heavy elements, together with a larger mass of unprocessed hydrogen from the star's original outer layers. The spectra of type II supernovae are dominated by strong hydrogen emission lines. The observed properties of type II supernovae can be modeled by calculating the effects of propogating a strong shock wave through a $5-10M_\odot$ red supergiant envelope of approximately solar composition. The star, initially at a temperature of about 3000 K at its surface, heats up to 10,000 K or more, and the color changes suddenly from red, before outburst, to blue. The characteristic observed variation of the light output and color during the outburst are shown in Fig. 3.7. The variation of color predicted by the shock model taken from the work of Stuart Schurmann, David Arnett, and Sidney Falk [S14], is also shown in Fig. 3.7(b). (See [R1] for recent reviews of supernovae models and observations.)

In 1974 Robert Kirschner and John Kwan first applied the Baade-Wesselink method (see Sec. 2.5) to the determination of the distances of supernovae [K3]. The method had been used earlier to compare observed and predicted expansion velocities in supernovae by Halton Arp, in 1961 [A5]. The basic idea of this method is that theoretical models (in the simplest case, the assumption that the source is a blackbody) are used to deduce the surface brightness variation in the source from observations of its color variation. The brightness variation can then be combined with this inferred surface brightness variation to deduce how the angular size of the source varies with time. Observations of the radial velocity in the source, deduced from the Doppler shifting of its spectral lines, can then be used to deduce the linear extent of the source and hence its distance. Schurmann et al. used their detailed models for the evolution of type II supernovae to deduce distances to several galaxies by this method. They found distances of 13.7 Mpc for NGC 1058, using supernova SN 1969l,* and 7.3 Mpc for M101, the Pinwheel galaxy in Ursa Major, using SN 1970g, with an uncertainty of 10–15% [S14]. Two other less well observed type II supernovae occurred in NGC 7331 (SN 1959d), for which the best models give $d = 13.4-14$ Mpc, and in M99, a spiral in Coma Berenices, (SN 1972q), for which models give d between 4.8 and 5.9 Mpc. David Branch and co-workers have also applied the Baade–Wesselink method to the type II supernova 1979c in M100, a bright spiral galaxy in the Virgo cluster, and obtained a distance of 23 Mpc [B16]. Fig. 3.8 shows the variation of temperature with time that Branch et al. derived for this event from a variety of color measurements at wavelengths ranging from the far untraviolet to the infrared. Because type II supernovae occur in young massive stars,

* Supernovae are given a number which consists of the year in which they were first observed, followed by a letter.

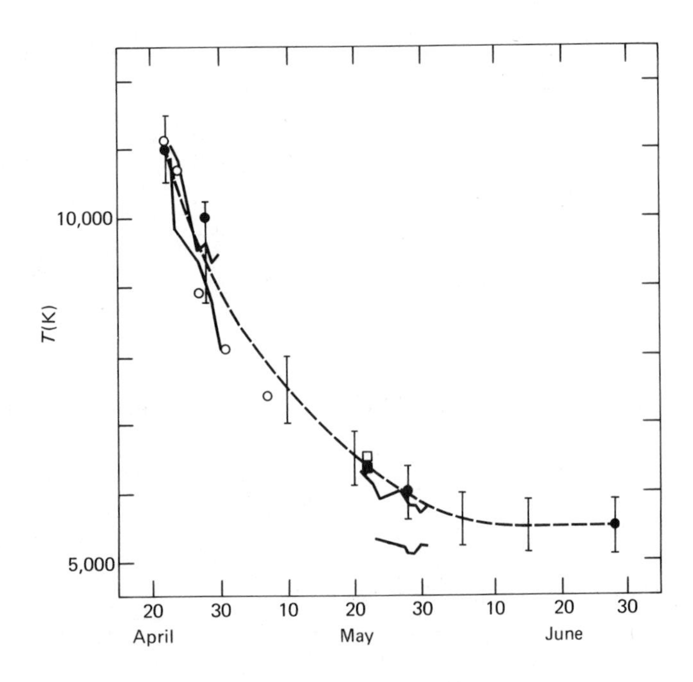

FIGURE 3.8
Variation of color temperature, calculated from spectroscopy and U, B, V, R, I photometry, with time for the type II supernova 1979c in M100. (From [B16].)

they may take place in or near clouds of gas and dust from which the stars are forming. An essential part of determining the distance to type II supernovae is therefore an estimate of the reddening E(B-V) and of visual extinction A_V to the supernova. These are given, for the cases discussed above, in Table 3.4. The contributions from interstellar and internal extinction predicted from the formulas adopted in Sec. 2.8 are also given. The Baade–Wesselink method should eventually be applicable to type II supernovae in very distant galaxies (1000 Mpc or more).

Type I supernovae, the other main type of supernova, show a remarkable homogeneity in their light and color variations with time, and a composite curve can be constructed from the observations of many such events (Fig. 3.9). After the rapid rise to maximum light and an initial rapid fall, the light curve settles to a steady exponential decline with time, with a half-life of 60 days. The homogeneity of type I supernova events is strongly supported by the work of Jay Elias et al., who have shown that the light curves for two type I supernovae in NGC 1316, the brightest galaxy in the Fornax cluster, and one in NGC 4536, probably in the Virgo cluster, are almost identical in the infrared J (1.25 μm), H (1.65 μm), and K (2.2 μm) bands [E1].

The most likely explanation of type I supernovae is that they result from the explosion of a white dwarf composed of carbon and oxygen, as a result of accretion of matter from a binary companion. At first the accreted hydrogen burns to helium in a shell near the surface of the star, and the helium burns to carbon and oxygen. But when the white dwarf mass grows to the Chandrasekhar limit of $1.4M_\odot$, the degenerate carbon at the center of the white dwarf ignites and the star is completely disrupted. Another model, which could arise from single-star evolution, is a star in the $6-8M_\odot$ range, which has lost its hydrogen during the core helium-burning phase, leaving a helium star of about $2M_\odot$. Such a star will develop a carbon–oxygen core, which eventually detonates and disrupts the star. In either case, type I supernovae are the result of explosive burning of about $1M_\odot$ of carbon and oxygen. The main product of this is the radioactive isotope nickel 56, which rapidly decays

TABLE 3.4

Distances to type II supernovae.

Supernova	*Galaxy*	*E(B-V)*	A_V	*Model*	*d(Mpc)*	*Reference*	*Predicted A_V (Sec. 2.8)* Interstellar	*Internal*
1969l	N1058	0.15	0.45	A	13.7	[S14]	0.33	0.14
1970g	M101	0.15	0.44	D	7.3	[S14]	0.0	0.14
1959d	N7331	0.5	1.5	E, F	13.7	[S14]	0.33	0.38
1972q	M99	1.2	3.6	A–G	5	[S14]	0.0	0.16
1979c	M100	0.1	0.3		15	[B16]	0.0	0.16
					±15%			

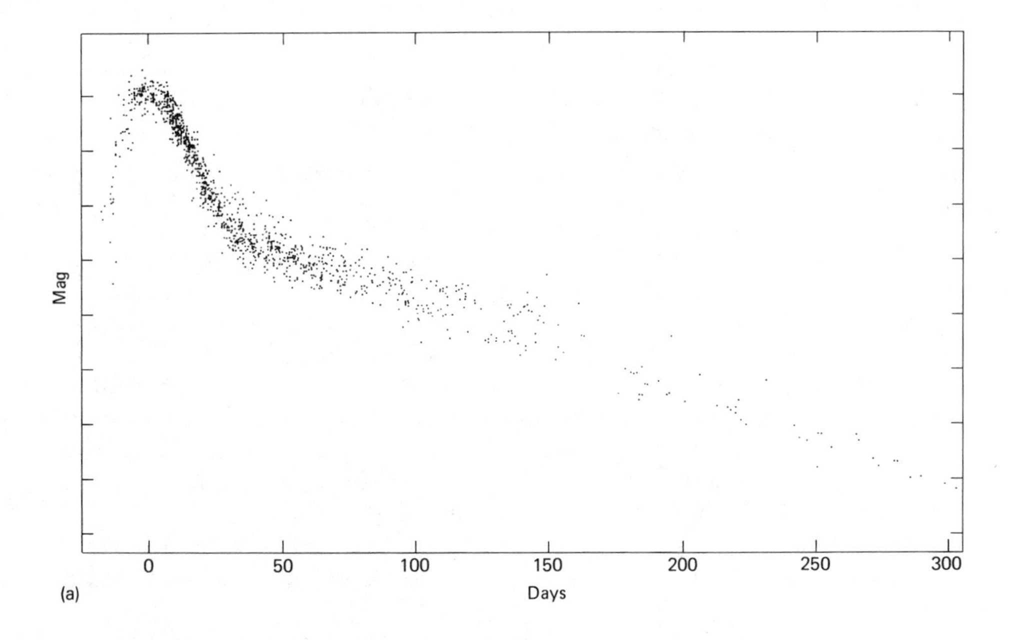

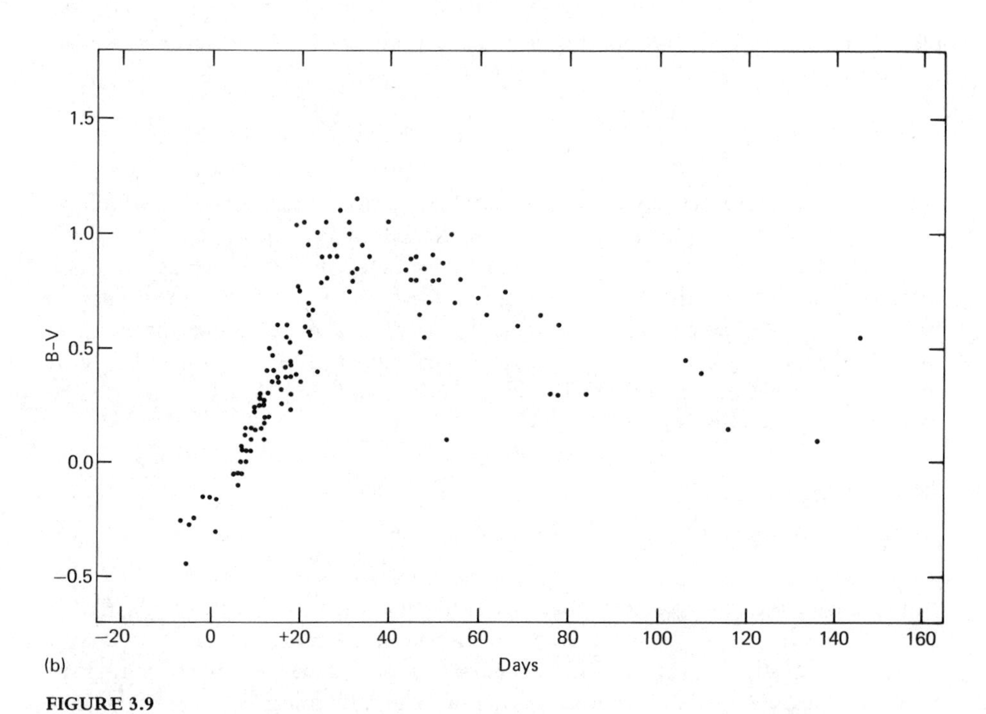

FIGURE 3.9

(a) Average blue light curve of type I supernovae. (b) Average B-V color, corrected for reddening by dust, as a function of time for type I supernovae. (From [B2].)

to cobalt 56 and finally to iron 56. This exponential radioactive decay explains the observed decline in the light curve. The expanding remnant consists of about $1M_\odot$ of iron, initially blanketed by smaller amounts of other elements like helium or carbon. No hydrogen is detected in the spectra of type I supernovae, in contrast to type II supernovae, whose spectra are dominated by strong emission lines of hydrogen.

Branch has shown that the distances to type I supernovae can also be estimated by the Baade–Wesselink method [B12],[B13]. If R_0 is the radius of the supernova photosphere at t_0, the time of maximum light, then from the variation of magnitude and color with time (the color can be translated to the effective temperature assuming that the spectrum is approximately a blackbody), the quantity $d[R(t)/R_0]/dt$, where $R(t)$ is the radius of the photosphere at time t, can be deduced. This can be combined with the radial velocity deduced from red-shifted spectral lines, which is found to be $10{,}900 \pm 700$ km s^{-1}, to give $R_0 = 1.4 \pm 0.14 \times 10^{15}$ cm. When this radius is combined with the effective temperature, the luminosity of the supernova can be deduced from Eq. (3.2). In his 1979 paper Branch found that

$$M_B(t_0) = -19.62 \pm 0.57. \tag{3.12}$$

Observations of the blue magnitude at maximum light $B(t_0)$ for any particular type I supernova, corrected for interstellar extinction and, in the case of spirals, internal obscuration, yields the distance. According to Branch and C. Bettis [B15], the photographic magnitude of supernovae at maximum light is related to the blue magnitude by

$$m_{pg} \simeq B - 0.32$$

and this relation allows us to determine distances to type I supernovae for which we have only the photographic magnitude at maximum light [from $M_{pg}(t_0) = -19.94$].

Arnett has developed a detailed model for type I supernovae for the case in which material deposited on the surface of a white dwarf causes the star to undergo a thermonuclear explosion, and he has used this model to determine distances to individual supernovae and to the "average" type I supernova, which he takes to be represented by SN 1975a in the spiral galaxy NGC 2207 [A3],[A4]. For the latter he deduces that at maximum light

$$\begin{aligned} M_B(t_0) &= -19.12 \pm 0.3 \\ M_{pg}(t_0) &= -19.5 \pm 0.3 \end{aligned} \tag{3.13}$$

where the uncertainty quoted is that of the mean value (zero-point error). Arnett estimates the dispersion about this mean relation to be 0.4 mag. The agreement with the kinematic estimate, Eq. (3.12), is encouraging, but given the lack of a consensus about the nature of type I supernovae, I shall use the calibration (3.12) in preference to Eq. (3.13), since the former is not dependent on which supernova model is correct.

TABLE 3.5

*Well-observed type I supernovae in external galaxies and their distance moduli.**

Supernova	*Galaxy*	$m_{pg}(max)$	A_B	A_i	μ_0
1885a	N224(M31)	6.0:	0.36	0.53	(25.05)
1895b	N5253	8.0	0.32	0.42	(27.20)
1919a	N4486	12.3:	0.0	0.0	(32.24)
1937c	IC4182	8.4	0.0	0.19	28.15
1937d	N1003	12.8	0.53	0.49	31.72
1939a	N4636	12.2	0.04	0.0	32.10
1939b	N4621	11.9	0.0	0.0	31.84
1954a	N4214	9.8	0.0	0.27	29.47
1954b	N5668	12.3	0.08	0.19	31.97
1956a	N3992	12.3	0.0	0.32	31.92
1957b	N4374	12.5	0.0	0.0	32.44
1959c	A1308 + 03	13.6	0.05	0 ?	33.49
1960f	N4496	11.6	0.03	0.26	31.25
1960r	N4382	11.83	0.0	0.25	31.52
1961d	A1248 + 28	$V = 16.5$	0.0	0.0	35.97
1961h	N4564	11.2	0.0	0.0	31.14
1961p	A0232 + 37	14.3	0.43	0.32	33.49
1962a	A1304 + 28	15.6	0.0	0 ?	35.54
1962j	N6835	13.6	0.46	0.62	32.46
1963c	A1255 + 28	15.7	0.0	0.0	35.64
1963j	N3913	13.7	0.0	0.19	33.45
1963p	N1084	$B = 14.0$	0.10	0.39	33.13
1964l	N3938	13.3	0.0	0.20	33.04
1967c	N3389	13.0	0.03	0.37	32.54
1968h	A1255 + 27	16.6	0.0	0 ?	36.54
1969c	N3811	$V = 13.7$	0.0	$A_V = 0.20$	32.97
1971l	N6384	$V = 13.0$	$A_V = 0.35$	$A_V = 0.29$	31.83
1972e	N5253	$B = 8.5$	0.31	0.42	27.39
1972j	N7634	$B = 14.2$	0.13	0.22	33.47
1973n	N7495	$B = 15.4$	0.15	0.20	34.67
1974g	N4414	$B = 12.5$	0.0	0.35	31.77
1975a	N2207	$B = 14.6$	0.56	0.31	33.35
1975b	A0316 + 41	15.5	0.76	0 ?	34.68

* List from [V4]; corrections for interstellar extinction A_B and internal obscuration A_i as adopted in Sec. 2.8; distance modulus μ_0 based on $M_{pg} = -19.94$, $M_V = -19.47$ and $M_B = -19.62$ (see text).

Tammann has also tried to calibrate the absolute magnitude at maximum light from ancient observations of historical supernovae in our Galaxy [T1], but this cannot be given very much weight as we have no real idea what the accuracy of these observations was.

The supernova method has tremendous potential for measuring distances, both because of its independence of any galactic distance scale and because the huge luminosities of supernovae make them detectable out to very great distances. However, there are still some difficulties with the method:

1 Type II supernovae take place in stars with a wide range of masses ($10-100M_\odot$) and show a considerable range of observed properties. For example, according to the review by Tammann [T1], the three type II supernovae that have occurred in the nearby spiral galaxy NGC 6946 (SN 1939c, 1948b, and 1980k) show a range of 1.7 mag in $m_B(t_0)$. Although the shock models of Schurmann et al. (see above) take some account of the individuality of type II supernovae, the model used to fit the observations of a particular supernova is probably not uniquely determined by these observations, and the accuracy of the method is likely to have been overstated.
2 The estimation of effective temperatures from B-V and other color indices is not a reliable process, particularly for supernovae whose spectra are strongly affected by broad emission and absorption lines. The precise definition of the photospheric radius depends on a deep understanding of supernova atmospheres, which is not really available yet.
3 In the absence of a consensus on the nature of type I supernovae it may be risky to assume that they represent a homogeneous population. Some anomalous events are known, and until these are understood, it is hard to be confident of the very small dispersion in the absolute magnitude at maximum light implied by Eqs. (3.12) and (3.13). In a recent paper Branch has argued that a possible correlation between the decline rate of type I supernovae and the absolute magnitude at maximum light (in the sense that less luminous supernovae decline faster) should be allowed for in estimating distances to them [B14], and this may be an indication that the type I supernova phenomenon is not quite as simple and homogeneous as is often assumed.
4 Observations of both supernova types are affected by extinction by dust in our Galaxy and, for late-type galaxies, in the galaxy where the supernova is occurring. The latter is a very serious problem for type II supernovae, which may take place in the vicinity of dense clouds of dust and molecules out of which new massive stars are forming. In this book I shall take the line that type II supernovae can only be used for distance determination if the extinction to the supernova has been directly measured. For type I supernovae the total extinction correction adopted in Sec. 2.8 will be applied for late-type galaxies. Distances to galaxies using type I supernovae are given in Table 3.5.

De Vaucouleurs, and Sandage and Tammann, prefer to use supernovae as secondary distance indicators, calibrating their absolute magnitudes at maximum light from supernovae in nearby galaxies for which distance estimates have been made by other methods. This will be discussed in Sec. 4.3.

3.3 Novae

Novae are explosive variable stars which brighten up by a factor of tens or hundreds of thousands during an outburst. The increase to maximum light is very rapid, sometimes lasting only a day. This is followed by a much slower decline over a number of years. Historically novae were the first objects to be used to try to estimate extragalactic distances, by Heber Curtis in 1917 and Knut Lundmark in 1919 (see Sec. 1.4). They are often called classical novae to distinguish them from the much less powerful dwarf nova phenomenon, in which a star brightens up by a factor of 10 – 100 every few days or weeks.

Before and after the outburst the spectra of the stars show that they are hot subdwarfs, and it was shown by Robert Kraft in 1964 that novae are probably all located in binary systems. At maximum light they reach an absolute visual magnitude M_V of -5 to -10, about 10,000 times fainter than a supernova, but still bright enough to be among the brightest stars in a galaxy, brighter than Cepheids, for example. Most novae show fluctuations during decline and even after they have returned to their normal state. Some novae have been observed to recur, and it is generally assumed that most would do so if observed for long enough. During an

FIGURE 3.10
Nova Persei of 1901. (Courtesy of Royal Astronomical Society.)

outburst a shell of matter is ejected from the star, a shell containing about 10^{-6}–$10^{-4} M_{\odot}$ and moving with a velocity of several hundred kilometers per second. A schematic light curve is shown in Fig. 3.11, labeled with some of the different stages that are observed. After the very rapid initial brightening there is a premaximum halt, believed to be due to the changing opacity of the expanding shell. There is then a slower rise to maximum, followed by a decline. Some novae show strong fluctuations during decline, others show a smoother decrease in brightness. Slow novae often have a deep dip in their light curve during decline, now known to be due to the condensation of dust grains in the expanding gas shell.

The origin of novae is believed to lie in gas captured gravitationally by a hot white dwarf from its binary companion, a red main-sequence star. The gas, ejected as a wind by the red star, falls onto the surface of the white dwarf and is compressed and heated to a high temperature. Eventually the temperature is high enough for nuclear reactions to commence, and a hydrogen-burning shell forms on the surface of the white dwarf. As more hydrogen is accreted, the hydrogen-burning shell becomes unstable, and the resulting huge increase in energy generation causes a pulsational instability of the whole star, the nova event. Geoff Bath and G. Shaviv [B5] showed in 1976 that the light curve of a nova can be modeled by assuming that a dense wind of gas is driven radially outward from the white dwarf by radiation pressure, at a constant velocity. The total bolometric luminosity is assumed to remain constant, but as the ejected material thins out, we see further in toward the star, and the observed temperature of the shell increases. The bulk of the energy output therefore shifts toward the ultraviolet region of the spectrum. The constancy of the total bolometric luminosity means that there has to be a decline in the amount of radiation observed in the visible band. The model also predicts that the more

FIGURE 3.11

Schematic visual light curve of a nova. The time scale has been magnified during the early stages of the outburst. The dip often seen during the decline from maximum light is due to condensation of dust grains in the expanding shell of gas thrown off during the nova outburst. (From [C3].)

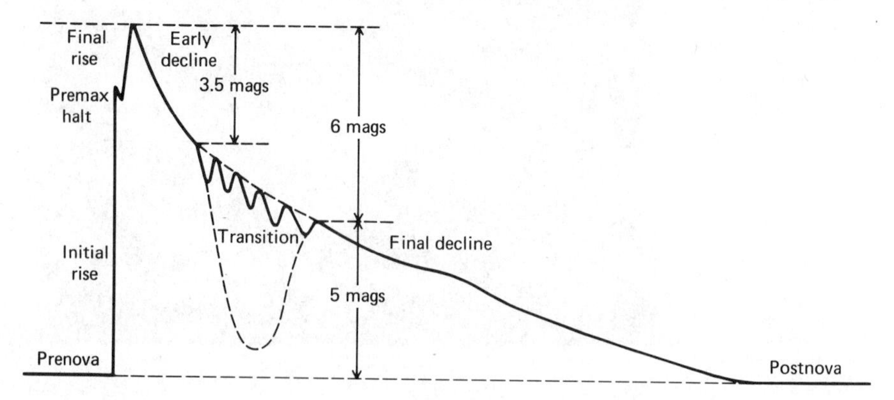

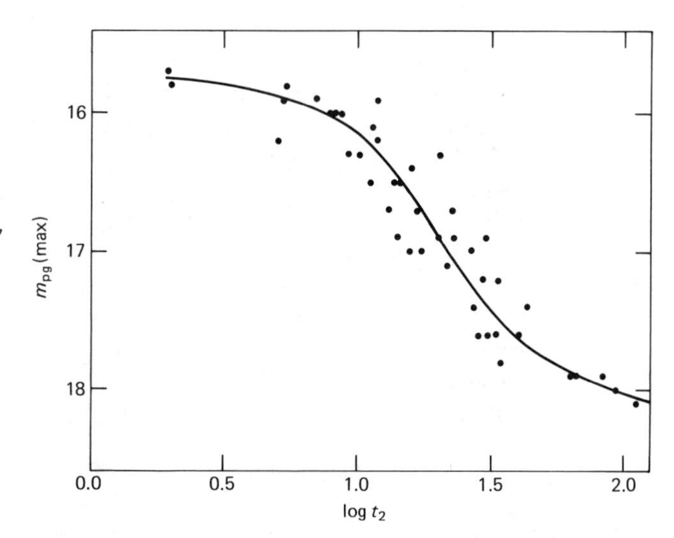

FIGURE 3.12

Photographic magnitude at maximum light m_{pg}*(max) versus rate of decline log* t_2 *for well-observed novae in M31.* t_2 *is the time in days for the brightness to drop by 2 mag. (From [B7].)*

luminous the nova is at maximum, the more rapid will be its subsequent decline at visual wavelengths. This is because the more luminous novae correspond to more massive white dwarfs, for which the thermonuclear runaway can get started in a lower mass envelope, which can then be more easily ejected.

The maximum absolute magnitude reached by novae shows a big range, but it is indeed quite well correlated with the rate of decline. Fig. 3.12 shows a plot by van den Bergh of the magnitude at maximum for novae in M31 compared with t_2, the time for the brightness to drop by 2 mag. The solid curve is the mean line adopted by van den Bergh. Fig. 3.13 shows the absolute magnitude at maximum for novae in

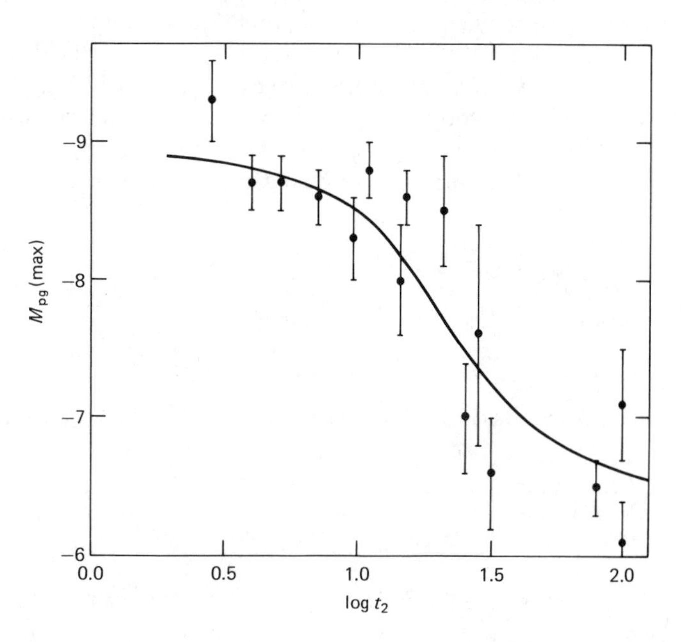

FIGURE 3.13

Calibration of the relation between absolute photographic magnitude at maximum light M_{pg}*(max) and the rate of decline log* t_2 *for galactic novae.* t_2 *is the time in days for the brightness to drop by 2 mag. The smooth curve was taken from the M31 observations shown in Fig. 3.12. (From [B7].)*

our Galaxy against t_2, the smooth curve being taken from the M31 data. The distances to the novae in our Galaxy are derived by several different methods: (1) combining measurements of the rate of angular expansion of the ejected shell $\dot{\theta}$ with the radial expansion velocity V to give $d = V/\dot{\theta}$ (the expansion parallax); (2) measuring the amount of interstellar extinction in the line of sight and then estimating the distance, assuming an average amount of extinction per kiloparsec; (3) kinematic distances (see Sec. 2.9); and (4) globular cluster distances (see Sec. 2.6). Van den Bergh [B7] used comparisons like those between Figs. 3.12 and 3.13 to derive distances to the Small and Large Magellanic Clouds and M31. F.D.A. Hartwick and J.B. Hutchings [H3] have used the Bath and Shaviv model for nova outbursts described above to fit the M_B(max) versus t_2 diagrams for our Galaxy and M31, and derived a distance to the latter (see Table 3.6).

In his application of novae to the distance scale, de Vaucouleurs assumes a linear relationship between the absolute magnitude at maximum light M_0(pg) and log t_3, where t_3 is the time until the luminosity has dropped by 3 mag:

$$M_0(\mathrm{pg}) = b \log t_3 - a \tag{3.14}$$

where a and b are constants, which de Vaucouleurs determines as $a = 11.3$ and $b = 2.4$. He uses Eq. (3.14) to estimate distances to novae in other galaxies [V2]. He also suggests that the luminosity of all novae 15 days after maximum light is the same,

$$M_{15}(\mathrm{pg}) = -5.5 \pm 0.15.$$

M.M. Shara [S15],[S16] has given a theoretical justification for these assumptions, based on the Bath and Shaviv nova model. However, the existence of a tight correlation between absolute magnitude at maximum light and the rate of decline is questioned by H.W. Duerbeck [D4], who has derived distances for 35 galactic novae. These show far more scatter than is seen in Fig. 3.12, suggesting that we should not rely too heavily on the nova method for estimating distances.

Table 3.6 gives the distance estimates to some nearby galaxies using novae. The method could in principle reach out to several nearby groups of galaxies. Novae are

TABLE 3.6

Galaxies with distance moduli μ_0 from novae.

Authors	*LMC*	*SMC*	*M31*
van den Bergh (1975) [B7]	19.5 ($A_V = 0.18$)	19.36 ($A_V = 0.06$)	24.01 ($A_V = 0.33$)
van den Bergh (1977) [B8]	19.08	19.27	24.03 ($A_V = 0.30$)
Hartwick and Hutchings (1978) [H3]			24.24 ($A_B = 0.36$)
de Vaucouleurs (1978) [V3]	18.46 ($A_B = 0.43$)	18.59 ($A_B = 0.31$)	23.95 ($A_B = 0.41$)

* van den Bergh (1977) [B8] also uses one nova in M33 to determine $\mu_0 \simeq 23.31$ ($A_V = 0.18$).

now being studied in M81 and M101, and it is hoped that they may be used eventually to reach out to the Virgo cluster [G7]. An interesting development in recent years has been detailed infrared observations of novae, which show that dust condenses out in the expanding envelope. (See, for example, the review by Gallagher and Starrfield [G1].) Future applications of the nova method should therefore include infrared measurements to monitor the extent to which the visible light is being absorbed by this circumstellar dust.

3.4 RR Lyrae variables and other methods

We have already encountered RR Lyrae variable stars in Sec. 2.6, where we saw that they vary their light output regularly on a time scale of 0.4 – 1 day. They are Population II stars, located mainly in the halo of the Galaxy, for example in globular clusters. They are found in the HR diagram where the horizontal branch in globular clusters intersects the instability strip (see Fig. 2.17) and have a blue color (spectral type A). They are stars of $1 - 2M_\odot$, which have exhausted their hydrogen, undergone a helium flash, and are now evolving to the right, back toward the red giant branch (see Fig. 2.28). While there is quite a range in estimates for their mean absolute magnitude (see Table 2.3) and a suggestion of dependence on heavy-element abundance, most estimates of the mean absolute magnitude lie within a few tenths of a magnitude of:

$$\overline{M_{\mathrm{V}}} = 0.6, \qquad \overline{M_{\mathrm{B}}} = 0.8 \tag{3.15}$$

so they are effective distance indicators. Because they are so much fainter than novae or Cepheids, RR Lyrae variables have been studied only in galaxies within 100 kpc, like the Magellanic Clouds. They are a useful test of the Magellanic Cloud distances though, because their calibration from statistical parallaxes, from the Baade – Wesselink method, and from globular cluster main-sequence fitting, is independent of the Hyades distance modulus. Table 3.7 summarizes different estimates of the distances to the Magellanic Clouds using RR Lyrae variables. For the dwarf ellipticals of the local group, RR Lyrae stars are the best distance indicators we have at present. Distances to Draco, Leo II, Ursa Minor, and Sculptor based on RR Lyrae stars are given in Table 3.11.

Some other indicators which have been used to estimate extragalactic distances are:

W VIRGINIS STARS

These are also called type II Cepheids and were originally confused with classical (type I) Cepheids. W Virginis stars are considerably less luminous than classical Cepheids and define a period – luminosity relation with a good deal of scatter. With

TABLE 3.7

Distance moduli μ_0 to Magellanic Clouds from RR Lyrae and other methods.

Authors	*Method*	*LMC*	*SMC*
van den Bergh (1977) [B8]	RR Lyrae	18.28 ($A_V = 0.15$)	18.80 ($A_V = 0.09$)
	Supernova remnants*	18.21	
	Globular cluster giants	18.35	18.81
	H_β and H_γ lines *	18.05	18.35
	Spectral gradients*	18.43	
	Color index C versus H_β*		18.98
	MK classification*	18.55	
de Vaucouleurs (1977) [V3]	RR Lyrae	18.17 ($A_B = 0.43$)	18.49 ($A_B = 0.31$)
	AB supergiants†	18.32	18.65
	Eclipsing binaries	18.36	18.74
Tammann et al. (1979) [T3], [G6]	RR Lyrae	18.61 ($A_B = 0.20$)	19.12 ($A_B = 0.08$)
Crampton (1979) [C5]	Spectral classification and H_γ widths of O,B stars*	18.63 ± 0.2	
Crampton and Greasley (1982) [C6]	Spectral classification of O,B stars*		19.1 ± 0.25
Glass and Evans (1982) [G5]	Mira variables	18.5	

* $\mu_{Hya} = 3.03$ with no correction for line blanketing, which is equivalent to $\mu_{Hya} = 3.18$ with this correction.
† $\mu_{Hya} = 3.29$.

further study they could provide a useful check on the distances to M31 and the Magellanic Clouds.

THE BRIGHTEST RED GIANTS IN GLOBULAR CLUSTERS

The advantage of these over RR Lyrae stars is that they are 3 mag brighter. The disadvantage is that their absolute magnitude depends sensitively on heavy-element abundance.

MIRA VARIABLES

Glass and Evans [G5] have obtained a period–luminosity relation for Mira variable stars in the Large Magellanic Cloud and then compared this with Miras in our galaxy to estimate the distance of the Large Magellanic Cloud.

O, B, AND A SUPERGIANTS

Crampton has used spectroscopic parallaxes to O and B supergiants to estimate the distance of the Large Magellanic Cloud [C5], and with Greasley he has applied the same method to the Small Magellanic Cloud [C6]. These estimates are in good agreement with distances from other methods.

The strengths of the Balmer H_β and H_γ lines of hydrogen in the brightest O and B stars in the Magellanic Clouds have been used to estimate the distances of the Clouds [B8],[C5]. In the past this method has tended to give lower distances than other methods, perhaps because of the presence of supergiants with anomalously strong hydrogen lines [G6]. The distances derived by Crampton [C5] for the Large Magellanic Cloud using H_γ line strength is, however, consistent with other estimates.

Divan and de Vaucouleurs have used a three-parameter spectral classification system, based on the detailed form of the star's blue continuum, to compare supergiants of types A and B in the Magellanic Clouds with those in our Galaxy and hence estimate the distance of the Clouds [D3],[V3].

ECLIPSING BINARIES

This is essentially the method described in Sec. 2.2. The temperatures of the stars are estimated from spectra and colors, and the eclipse durations give the stellar radii in terms of the size of the orbit. Using Kepler's law and the mass–luminosity relation, the relative contributions of the stars to the total brightness and the distance of the system can be estimated. Dworak and de Vaucouleurs have applied the method to 10 eclipsing variables in the Large Magellanic Cloud, 19 stars in the Small Magellanic Cloud and 6 stars in M31 [D5],[V3].

The latter two methods are given a similar weight by de Vaucouleurs to the Cepheid, nova, and RR Lyrae methods, though their potential seems yet to be fully realized. Some distance estimates made using the five methods summarized above are given in Table 3.7, but I have not used any of these five methods in trying to determine the distances to the Magellanic Clouds and M31 because they do not yet seem to be reliable enough.

3.5 *Galaxies with distances from primary indicators*

As we have seen, the nearest galaxies to our own Galaxy are the Large and Small Magellanic Clouds whose existence was first reported to Europe by Magellan's circumnavigating expedition of 1519–1521. They have been the subject of intensive study for more than a century, and it was as a result of her study of variable stars in the Magellanic Clouds that in 1908 Henrietta Leavitt announced the crucial period–luminosity law for Cepheids. They are both small irregular galaxies, with a much higher proportion of gas and Population I material than our Galaxy. Both show some evidence of a barred spiral structure; this is clearest for the Large Magellanic Cloud (Fig. 3.14).

Almost all modern distance estimates to the Magellanic Clouds place them at 40–80 kpc from us, only just beyond the outskirts of our own Galaxy. This proximity to our Galaxy results in a strong tidal interaction between our Galaxy and its two small companions. As the Magellanic Clouds orbit around the Galaxy, gas is drawn away from them into a long thin stream, which stretches from the Clouds down toward the Milky Way (Fig. 3.15).

Table 3.8 summarizes the best available estimates of the distances of the Magellanic Clouds, taken from Tables 3.3, 3.6, and 3.7. I have corrected all Cepheid distances for an assumed Hyades distance modulus of 3.30 (see Sec. 2.4) and applied a line-blanketing correction of −0.15 (see Sec. 3.1). There is still some dispute about what Hyades distance should be used. Sandage and Tammann suggest that the good agreement between their estimates of the distances of the Magellanic Clouds from Cepheids based on the old (pre-1960) Hyades distance modulus ($\mu_0 = 3.03$) with that using RR Lyrae stars supports the old Hyades modulus [T3],[S9]. In the work of Martin, Warren, and Feast discussed in Sec. 3.1, they find the distance to the Large Magellanic Cloud using Cepheids calibrated via open-cluster main-sequence fitting and via the Baade–Wesselink method and again prefer the old Hyades modulus. However, we saw in Sec. 2.4 that the distance modulus to the Hyades cluster is almost certainly in the range of 3.2–3.4 and a value of 3.03 is improbable in light of the modern data.

The mean distances to the Magellanic Clouds using Cepheids, adopted in Table 3.8, do not differ significantly from the mean distances using novae and RR Lyrae stars, and as emphasized by van den Bergh [B9], the latter two distance indicators belong to Population II and are not affected by the Hyades modulus. The external errors of the weighted means using all three methods do not differ significantly

FIGURE 3.14

Some galaxies with distances from primary distance indicators. (a) Large Magellanic Cloud. (Courtesy of J.B. Whiteoak.) (b) Small Magellanic Cloud (Courtesy of Royal Observatory, Edinburgh.) (c) Andromeda nebula M31 (Courtesy of Mount Palomar Observatory.) (d) M33. (Courtesy of Mount Palomar Observatory.) (e) NGC 2403 (Courtesy of Mount Palomar Observatory.)

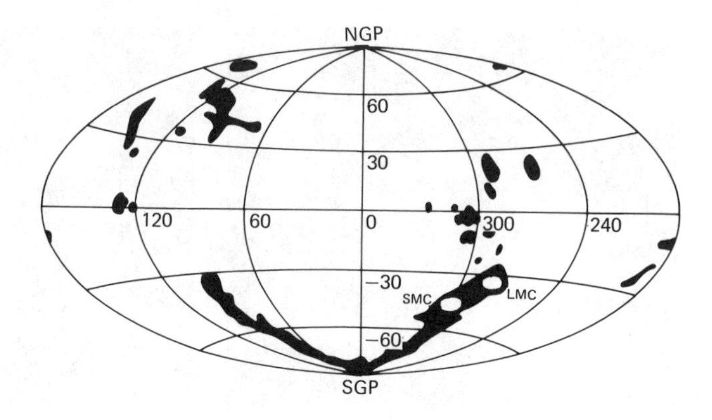

FIGURE 3.15

High-velocity hydrogen gas is plotted on the sky in galactic coordinates, with the galactic equator running horizontally along the middle, and the north (NGP) and south (SGP) galactic poles situated at the top and bottom. The lines are labeled with galactic latitude b *and longitude* l. *The Magellanic Stream is a ribbon of gas which spans the gulf between our Galaxy and its two satellites, the Large and Small Magellanic Clouds (marked as crosses). (From [M4].)*

from the internal errors (see App. A.17), so there is no statistical evidence for systematic errors.

One of the main problems in establishing an accurate extragalactic distance scale is the estimate of the extinction by dust, both within our Galaxy (A_B, interstellar) and in the galaxies being studied (A_i, internal). In published distance estimates the calculated uncertainties do not always include the uncertainty in estimating the extinction. Table 3.9 gives various observations of foreground reddening due to dust in our Galaxy toward Local Group galaxies, and of reddening to Population I objects like Cepheids and young supergiants in the galaxies. Averaged together, these observations can be used to estimate the interstellar extinction A_B, the total (interstellar + internal) blue extinction $A_B + A_i$, and hence A_i itself, together with the uncertainties in these quantities. Also given in Table 3.9 are the predictions of the models for interstellar and internal extinction adopted in Sec. 2.8. The adopted models agree well with observations in Local Group galaxies, except that the observed internal extinction in the Magellanic Clouds is smaller than the predicted values, perhaps due to the low heavy-element abundances of these galaxies.

In trying to combine the distance-modulus estimates by different methods, we need to take account of (1) the uncertainty in the mean calibrating relation being used σ_{zp} (the zero-point error), and (2) the dispersion about this in a single observation σ, which is reduced by a factor of $N^{1/2}$ if there are N independent observations (see App. A.17 on the theory of errors). This dispersion has three components: one due to the fact that successive measurements of the same property will scatter around the mean calibrating relation (due to both measurement error and differences in the calibrating property from one object to another), one due to the

TABLE 3.8

*Summary on distance moduli of the Magellanic Clouds.**

Authors	LMC μ_0	LMC W/Σ^2	SMC μ_0	SMC W/Σ^2
Period–luminosity–color for Cepheids				
Sandage and Tammann (1971) [S6]	18.86		19.54	
Gascoigne (1974) [G3]	19.01		19.22	
Iben and Tuggles (1975) [I2]	18.73		19.25	
van den Bergh (1977) [B8] (mean of $W - \log P$ and period–luminosity–color relation)	19.29		19.79	
Feast (1977) [F1]			19.27	
Butler (1978) [B19]	19.05		19.47†	
Martin et al. (1979) [M3]	18.77			
Madore (1983) [M2]	18.65		19.07	
Mean for Cepheids	18.91		19.37	
Corrected for line blanketing	18.76	138.9	19.22	138.9
RR Lyrae				
Graham (1975) [G5], using $M_B = 0.75$ [T1]	18.61		19.12	
van den Bergh (1977) [B8]	18.28		18.80	
de Vaucouleurs (1978) [V3]	18.36‡		18.69§	
Mean for RR Lyrae	18.42	43.8	18.87	43.8
Novae				
van den Bergh (1977) [B8]	19.08		19.27	
de Vaucouleurs (1978) [V3]	18.65‡		18.79§	
Mean for novae	18.86	29.7	19.03	29.7
Overall weighted mean	**18.70**	212.4	**19.12**	212.4
Internal error	0.14		0.14	
External error	0.10		0.10	

* Corrected to $\mu_{Hya} = 3.30$ where appropriate.
† $A_B + A_i = 0.18$ (Table 3.9)
‡ $A_B = 0.24$ (Table 3.9)
§ $A_B = 0.11$ (Table 3.9)

uncertainty in the interstellar extinction, and one due to the uncertainty in the internal extinction (for late-type galaxies). The zero-point error σ_{zp} will include the uncertainties in the distance scale in our Galaxy for indicators calibrated via galactic sources. In addition I believe that a weighting W should be applied as follows:

$W = 1$ for methods based on empirical correlations between observable properties

$W = 2$ for methods which are supported not only by empirical evidence, but also

TABLE 3.9

Interstellar and internal extinction in Local Group galaxies and NGC 2403.

	LMC		*SMC*		*M33*	
Foreground reddening, E_0(B-V)	0.07 ± 0.04	[F2]	0.02 ± 0.02	[T5]	0.03 ± 0.02	[M7]
	0.06 ± 0.02	[S12]	0.02	[G2]	0.11 ± 0.02	[H9]
	0.06 ± 0.01	[B10]	0.04 ± 0.03	[A6]		
	0.07 ± 0.01	[B18]	0.019 ± 0.004	[M9]		
	0.08	[G2]	0.04	[C6]		
	0.03	[D1]				
	0.07	[C5]				
	0.034	[M9]				
Adopted	0.06 ± 0.02		0.03 ± 0.02		0.07 ± 0.04	
Adopted interstellar extinction, $A_B = 4E_0$(B-V)	0.24 ± 0.08		0.11 ± 0.08		0.28 ± 0.16	
Predicted A_B						
de Vaucouleurs	0.43		0.35		0.31	
Sandage	0.10		0.05		0.11	
Fisher and Tully	0.23		0.15		0.25	
Reddening to Cepheids and young supergiants, E_1(B-V)	0.08	[G2]	0.02	[G2]	0.08 ± 0.03	[S13]
	0.12 ± 0.01	[B18]	0.07	[C6]	0.27 ± 0.02	[H9]
	0.04	[M3]				
	0.11	[H6]				
	0.10	[C5]				
Adopted	0.09 ± 0.02		0.045 ± 0.025		0.175 ± 0.095	
Adopted total interstellar plus internal extinction, $A_B + A_i = 4E_1$(B-V)	0.36 ± 0.08		0.18 ± 0.10		0.70 ± 0.38	
Adopted total internal extinction, $A_i = 4[E_1\text{(B-V)} - E_0\text{(B-V)}]$	0.12 ± 0.11		0.07 ± 0.13		0.42 ± 0.20	
Predicted A_i						
de Vaucouleurs	0.23		0.40		0.33	
Sandage and Tammann	0.18		0.27		0.45	
Adopted correction to face-on magnitude	0		0		0.16	

* Observations refer to variable star field IV, toward the edge of the galaxy

NGC 6822		*IC 1613*		*M31**		*NGC 2403*	
0.27	[K1]	0.03	[S2]	0.11 ± 0.02	[B7]	0.60 ± 0.015	[T2]
0.19 ± 0.03	[B7]	0.03	[H8]	0.07 ± 0.03	[H7]		
0.34 ± 0.05	[H8]						
0.27 ± 0.04		0.03 ± 0.02		0.09 ± 0.02		0.06 ± 0.015	
1.08 ± 0.16		0.12 ± 0.08		0.36 ± 0.08		0.24 ± 0.06	
0.79		0.21		0.41		0.38	
0.26		0.00		0.22		0.13	
0.71		0.09		0.41		0.22	
0.27 ± 0.03	[K1]	0.08 ± 0.02	[H8]	0.16 ± 0.03	[B1]	0.06 ± 0.015	[T2]
0.42 ± 0.05	[H8]			0.14	[H7]	0.4	[M1]
0.345 ± 0.075		0.08 ± 0.02		0.15 ± 0.02		(0.145)	
1.38 ± 0.30		0.32 ± 0.08		0.60 ± 0.08		(0.58)	
0.30 ± 0.20		0.20 ± 0.12		0.24 ± 0.12		(0.34)	
0.20		0.20		0.53		0.34	
0.16		0.16		1.20		0.46	
0.03		0.03		0.36		0.17	

by quantitative theoretical arguments (such as methods using Cepheids, novae, RR Lyrae stars, and supernovae).

The total root-mean-square uncertainty is then

$$\Sigma = \left(\sigma_{zp}^2 + \frac{\sigma^2}{N}\right)^{1/2} \tag{3.16}$$

and the adopted mean distance modulus for m measurements $\mu_{0,i}$, each with uncertainty Σ_i and weight W_i, is

$$\bar{\mu}_0 = \left(\sum_{i=1}^{m} \frac{W_i \mu_{0,i}}{\Sigma_i^2}\right) \Big/ \sum_{i=1}^{m} \frac{W_i}{\Sigma_i^2}. \tag{3.17}$$

The internal error in this estimate is

$$\sigma_{\text{int}} = \left(\sum_{i=1}^{m} \frac{W_i^2}{\Sigma_i^2}\right)^{1/2} \Big/ \left(\sum_{i=1}^{m} \frac{W_i}{\Sigma_i^2}\right) \tag{3.18}$$

while the external error is

$$\sigma_{\text{ext}} = \left[\frac{1}{m-1} \sum_{i=1}^{m} \frac{W_i(\bar{\mu}_{0,i} - \bar{\mu}_0)^2}{\Sigma_i^2} \Big/ \sum_{i=1}^{m} \frac{W_i}{\Sigma_i^2}\right]^{1/2}. \tag{3.19}$$

The weighted mean distance moduli to the Magellanic Clouds are given in Table 3.8. The weights and errors adopted in this book for different methods are summarized in Table 4.12.

My adopted distance moduli for the Magellanic Clouds are 18.70 mag for the Large Magellanic Cloud ($d = 55$ kpc) and 19.12 mag for the Small Magellanic Cloud ($d = 67$ kpc). These are 0.4–0.5 mag greater than those arrived at by de Vaucouleurs. The reasons for this are as follows:

1 His estimates of dust extinction are larger than those obtained from reddening observations (for example, to Cepheids he adopts a total extinction $A_B + 0.5A_i =$ 0.54 and 0.55 to the Large and Small Magellanic Clouds, respectively, compared with 0.36 and 0.18 from Table 3.9).

2 His treatment of the Cepheid period–luminosity relation differs from the period–luminosity–color relation used in most modern work.

3 I did not adopt his proposal that novae, RR Lyrae stars, and other methods should be given the same weight as the Cepheid method.

Points (2) and (3) also explain the 0.3 mag lower distance moduli adopted by van den Bergh for the Magellanic Clouds in his 1977 review [B8]. Neglect of the color term for Cepheids also explains why Stothers [S18] obtains slightly lower distance

TABLE 3.10

Distance moduli to M31.

Authors	μ_B	$A_B + A_i$	μ_0	W/Σ^2
*Period–luminosity–color for Cepheids**				
van den Bergh (1975) [B7], W-method			24.60	
van den Bergh (1977) [B8], period–luminosity method		$(A_V = 0.21)$	(24.37)	
de Vaucouleurs (1978) [V3], period–luminosity method	24.50	0.41	(24.09)	
Tammann et al. (1979) [T3]	25.07	0.64	24.43	
Mean for Cepheids			24.51	
Corrected for line blanketing			24.36	67.8
Novae				
van den Bergh (1977) [B8]		$(A_V = 0.30)$	24.03	
de Vaucouleurs (1978) [V3]	24.36	0.41	23.95	
Hartwick and Hutchings (1978) [H3]	24.60	(0.36)	24.24	
Mean for novae			24.07	41.7
Other Methods				
W Virginis van den Bergh (1977) [B8]		$(A_V = 0.30)$	(23.62)	0
Population II giants van den Bergh (1975) [B7]		$(A_V = 0.33)$	(24.07)	0
de Vaucouleurs (1978) [V3]	(24.7)	0.41	(24.3)	0
Eclipsing binaries de Vaucouleurs (1978) [V3]	(24.57)	0.41	(24.16)	0
Overall weighted mean§			**24.25**	109.5
Internal error			0.19	
External error			0.14	

* Corrected to $\mu_{Hya} = 3.30$.
§ Excluding bracketed values.

moduli to the Magellanic Clouds than those found in Table 3.8. (His average distance moduli based on Cepheids and RR Lyrae stars are 18.52 and 18.82 mag for the Large and Small Magellanic Clouds, respectively. These estimates are based on the old Hyades modulus, without correction for line blanketing.) However, support for a lower value for the distance modulus of the LMC, perhaps as low as 18.4, has come from recent work by R. A. Schommer, E. W. Olszewski and M. Aaronson, fitting the Hyades main sequence to LMC star clusters (Astrophysics Journal 1985).

After the Magellanic Clouds, the next best studied galaxy is the Andromeda nebula M31, which under good observing conditions is discernible to the naked eye as a fuzzy patch of light. M31 and our Galaxy are the dominant members of a group of 30 or so nearby galaxies, known as the Local Group of galaxies. The distance estimates to M31 from primary indicators are summarized in Table 3.10, and those

TABLE 3.11

Distances to other Local Group galaxies by primary methods.

Galaxy	μ_0	*Remarks*
NGC 6822	23.77	Based on μ_B (N6822) = 25.03 from Cepheids [T3],[K1] (μ_{Hya} = 3.03), $A_B + A_i$ = 1.38 (Table 3.9), corrected to μ_{Hya} = 3.30 and for line blanketing ($\Delta\mu_0 = -0.15$)
	(23.32)	From infrared observations of nine cepheids [M5], using A_H = 0.20, corrected for line blanketing
M33	24.65	Based on μ_B (M33) = 25.35 from Cepheids [S15], $A_B + A_i$ = 0.70 (Table 3.9)
	(24.19)	From infrared observations of fifteen Cepheids [M2], using A_H = 0.13, corrected for line blanketing
IC 1613	24.29	Based on μ_B (IC 1613) = μ_B (SMC) + 5.31 for Cepheids [S2], $A_B + A_i$ = 0.32 for IC 1613, 0.18 for SMC (Table 3.9)
	(24.14)	From infrared observations of ten Cepheids, [M6], using A_H = 0.04, corrected for line blanketing
Sextans A	25.37	Based on μ_B (Sex A) = μ_B (LMC) + 6.66 for Cepheids [S10], A_B = 0.07 [S10] and A_i = 0.24 (Box 2.6) for Sextans A, $A_B + A_i$ = 0.32 for LMC (Table 3.9)
Fornax	20.70	From comparison of horizontal branch [B8], A_V = 0.21
Sculptor	19.4	From RR Lyrae stars [B8], A_V = 0.18
Leo II	20.3	From RR Lyrae stars [B8], A_V = 0.21
Draco	19.0	From RR Lyrae stars [B8], A_V = 0.30
Ursa Minor	19.0	From RR Lyrae stars [B8], A_V = 0.24

to other members of the Local Group in Table 3.11. The properties of Local Group galaxies are given in Table 3.12, distances in parentheses being estimates by secondary methods. It can be seen that the Sculptor, Draco and Ursa Minor dwarf spheroidal galaxies are also close satellites of our Galaxy. Most of the Local Group galaxies are dwarf irregulars or dwarf spheroidal galaxies, and the only substantial galaxies are the giant spirals M31 and our own Galaxy, the Sc spiral M33, and the Large Magellanic Cloud.

Table 3.13 compares the distance moduli adopted here for Local Group galaxies with those given by Sandage and Tammann, van den Bergh, de Vaucouleurs, and Humphreys. The distance estimates quoted by de Vaucouleurs for M33 and the Local Group irregular galaxies NGC 6822 and IC 1613 are averages of values from both primary and secondary methods, which rather undermines the logical consistency of his distance scale, though the effect on the final distances of including the secondary methods is small. In Table 3.13 only primary indicators have been included. On average the de Vaucouleurs distance moduli in the Local Group are 0.29 mag less than our adopted values, while those of Sandage and Tammann are 0.03 mag greater. The latter difference is increased to 0.15 mag if account is taken of the lower Hyades modulus used by Sandage and Tammann, and with allowance for the effects of line blanketing.

Apart from supernovae, which we will discuss further below, the only primary indicator that has been studied in galaxies outside the Local Group are the Cepheids,

TABLE 3.12

Properties of Local Group galaxies.

Galaxy	*Other names*	*Type*	*Distance (kpc)*	M_V	*Diameter D(0) (kpc)*	*Radial velocity (km s^{-1})*
M31	NGC 224	Sb	710	−21.2	30	−299
Our Galaxy		(Sb-Sbc)		(−20.5)	(30)	
M33	NGC 598	Sc	850	−19.1	14	−183
LMC		Irr I,SBm	55	−18.8	10	260
SMC		Irr I,SBm	67	−16.8	4.9	150
NGC 205		E5	(710)	−16.5	3.0	−239
M32	NGC 221	E2	(710)	−16.5	1.8	−217
NGC 6822	DDO 209	Irr I	570	−16.0	1.7	−56
NGC 185		dE3	(710)	−15.4	2.3	−252
IC 10		Irr I	(1300)	−15.3	1.9	−345
NGC 147		dE4	(710)	−15.0	2.4	−245
IC 1613		Irr I	710	−14.7	2.6	−238
Wolf–Lundmark	DDO 221	Irr I	(1300)	−14.6	3.1	−120
IC 5152		Irr I	(1600)	−14.3*	1.9	83
Sextans A	DDO 75	Irr I	1180	−13.8*	1.6	326
Fornax		dE3	(140)	−12.6	0.8	53
And II		dE0	(710)	−10.8	0.7	
And III		dE0	(710)	−10.8	0.9	
UKS 1927-177		Irr	(1100)	−10.5*	1.0	
UKS 2323-326		Irr	(1300)	−10.4*	0.6	
And I		dE0	(710)	−10.4	0.5	
Sculptor		dE3	(75)	−10.4	1	
Leo I	DDO 74	dE3	(110)	−10.1	0.3	
Sag DIG		Irr	(460)	−9.3*	1	−58
LGS 3		Irr I	(740)	−9.0*	0.6	−280
Leo II	DDO 93	dE0	(105)	−8.3	0.4	
Draco	DDO 208	dE0	(63)	−8.2:	0.7	
Ursa Minor	DDO 199	dE4	(63)	−7.9:	0.5	
And IV			(710)		0.2	
Sextans B			(1350)			

* Blue magnitude.
† Other possible members: DDO 187, GR 8, NGC 3109, Pegasus, Leo A, DDO 210.
SOURCE: Compiled from [A2],[V1],[B6],[L1],[V2],[Y1],[C2],[T4].

TABLE 3.13

Comparison of Local Group distance moduli.

Authors	*LMC*	*SMC*	*M31*	*N6822*
Adopted μ_0 ($\mu_{Hya} = 3.30$)*	**18.70**	**19.12**	**24.25**	**23.77**
d, kpc	55	67	710	570
Sandage and Tammann (1974) [S7] ($\mu_{Hya} = 3.03$)	18.59	19.27	24.16	23.95
van den Bergh (1977) [B8] ($\mu_{Hya} = 3.03$)	18.43	18.82	24.02	23.64
de Vaucouleurs (1978) [V3] ($\mu_{Hya} = 3.29$)	18.31	18.62	24.07	23.73
Humphreys (1983) [H10] ($\mu_{Hya} = 3.03$)	18.6	19.0		23.2

* Corrected for line blanketing, $\Delta\mu_0 = -0.15$.

which have been studied in NGC 2403 in the M81 group, and in NGC 300 in the Sculptor group. The work of Tammann and Sandage on NGC 2403 was described in Sec. 3.1 together with a critique of this by Madore. De Vaucouleurs has cast doubt on whether this determines the distance of the M81 group by suggesting that NGC 2403 (and also NGC 2366) are foreground galaxies, appreciably nearer to us than the other members of the group [V1]. While the question of group or cluster membership can be problematic, these doubts about the M81 group do not seem to be supported by most other workers (see, for example, [A1],[S8],[K2],[H3]). However it does seem reasonable to insist that at least two members of a group or cluster have their distances measured before a believable distance to the group can be assigned, and therefore to leave the question of the M81 group distance open. In this I am therefore supporting the view of de Vaucouleurs and van den Bergh that the distance to the M81 group is not yet established by primary methods. The same argument applies to the Sculptor group.

Supernovae give us, in principle, primary distances to galaxies much further out than these nearby groups (see Tables 3.4 and 3.5). We argued in Sec. 3.2 that this method still has considerable uncertainty associated with it, and it would, of course, be very valuable if the supernova method could be tied to more classical methods by finding a supernova in the local or M81 groups. Not much weight can be placed on the observations of the type I supernova of 1885 in M31, since the extinction is unknown and the classification uncertain. The supernovae that have been observed in this century in groups within 10 Mpc are given in Table 3.14. From the calibration given in Eq. (3.12) for type I supernovae we can infer distances to NGC 5253 in the NGC 5128 group of $\mu_0 = 27.38 \pm 0.8$ ($d = 3.0$ Mpc) and to the Canis Venatici I cloud of $\mu_0 = 28.81 \pm 0.7$ ($d = 5.8$ Mpc). However, in view of the uncertainties in the supernova method it seems preferable to combine these esti-

Authors	*M33*	*IC 1613*	*Mean* ($\mu_{0,\text{adopted}} - \mu_{0,\text{other}}$)
Adopted μ_0 ($\mu_{\text{Hya}} = 3.30$)*	**24.65**	**24.29**	0
d, kpc	850	720	
Sandage and Tammann (1974) [S7] ($\mu_{\text{Hya}} = 3.03$)	24.56	24.43	−0.03
Van den Bergh (1977) [B8] ($\mu_{\text{Hya}} = 3.03$)	24.06	24.41	0.23
De Vaucouleurs (1978) [V3] ($\mu_{\text{Hya}} = 3.29$)	24.30	24.02	0.29
Humphreys (1983) [H10] ($\mu_{\text{Hya}} = 3.03$)		24.3	0.20

mates with others from secondary methods (see Sec. 4.9). The type II supernova method of Schurmann et al. gave a distance to M101 of 7.3 Mpc, but further evidence is needed to establish the distance of this group. The Schurmann et al. method has not been applied to the other type II supernovae listed in Table 3.14 due to insufficiently detailed observations.

Doubt has been cast by several authors [J1],[B11],[C1] on whether M101 is associated with the other galaxies which make up the M101 group, following an earlier suggestion by Eric Holmberg [H5]. To see that this is probably not a very substantial objection, let us examine the distribution on the sky of nearby galaxies, defined as galaxies with recession velocity, corrected for the sun's motion around the galaxy, $V_0 \leq 500$ km s^{-1} (Fig. 3.16). To eliminate galaxies of the Local Group and most of those of the M81 and Sculptor groups, galaxies with corrected recession velocity $V_0 \leq 300$ km s^{-1} are omitted. Nearby groups of galaxies have been circled. Although there are a number of galaxies that do not seem to be part of any group, the

TABLE 3.14

Supernovae in galaxies within 10 Mpc.

Type I		*Type II*	
1937c	IC 4182, in Canes Venatici I cloud	1923a	NGC 5236, in N5128 group
1954a	NGC 4214, in Canes Venatici I cloud	1957a	NGC 2841, in N2841 group
1972e	NGC 5253, in NGC 5128 group	1968b	NGC 5236, in N5128 group
		1970g	NGC 5457, in M101 group

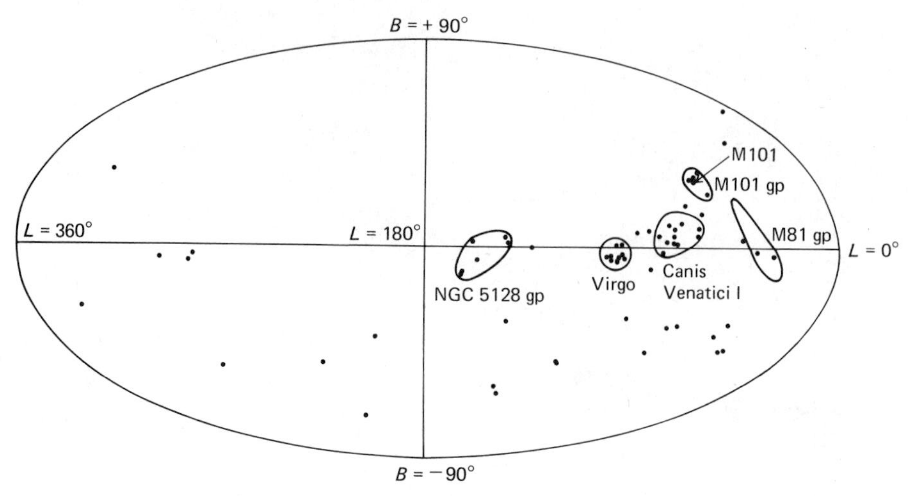

FIGURE 3.16

Plot on sky in supergalactic coordinates L *and* B *of galaxies with recession velocity, corrected for galactic rotation, in the range of* $300 \leq v_0 \leq 500$ *km* s^{-1}. *Some nearby groups and clusters and the location of the galaxy M101 are indicated. It does not appear very plausible that M101 is a chance superposition on the other galaxies of the M101 group. (Data from [K4].) Supergalactic coordinates, defined by de Vaucouleurs et al. [V5], are simply a rotation of galactic coordinates so that the plane* $B = 0°$ *coincides with the plane of the Local Supercluster.*

galaxies of the M101 group form a tight grouping, and the idea that M101 is a chance superposition on the other galaxies of the group is hard to accept.

The most noticeable feature of the distribution of bright galaxies on the sky is the Virgo cluster [Fig. 3.17(a)]. It is now believed that most galaxies within about 20 Mpc of the core of the Virgo cluster, including the Local Group of galaxies, form part of an immense disk-shaped supercluster, the Local Supercluster [Fig. 3.17(b)]. Several other examples of superclusters are known, and their extent may range up to 100 Mpc. Most galaxies probably belong to groups or clusters ranging in size from small groups like the Local Group or the M81 group to rich clusters like Virgo.

FIGURE 3.17

(a) Core of Virgo cluster. (Courtesy of Royal Observatory, Edinburgh.) (b) Projection of the distribution of bright galaxies onto the plane perpendicular to the plane of the Local Supercluster, defined by de Vaucouleurs et al. [V5]. The concentrated band of points across the middle of the figure is the Local Supercluster. The two straight lines enclose the zone of avoidance near the Milky Way, in which few galaxies can be seen because of the effect of extinction by dust. (From [T6].)

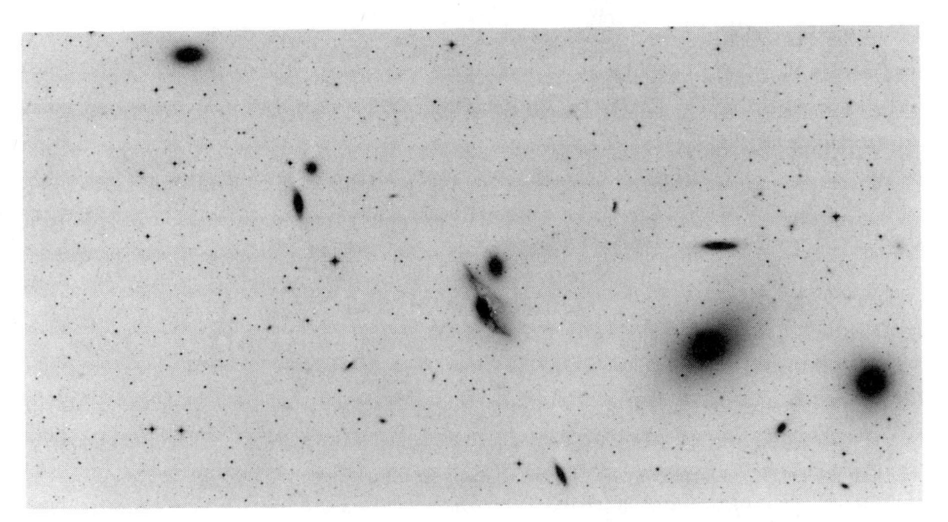
a

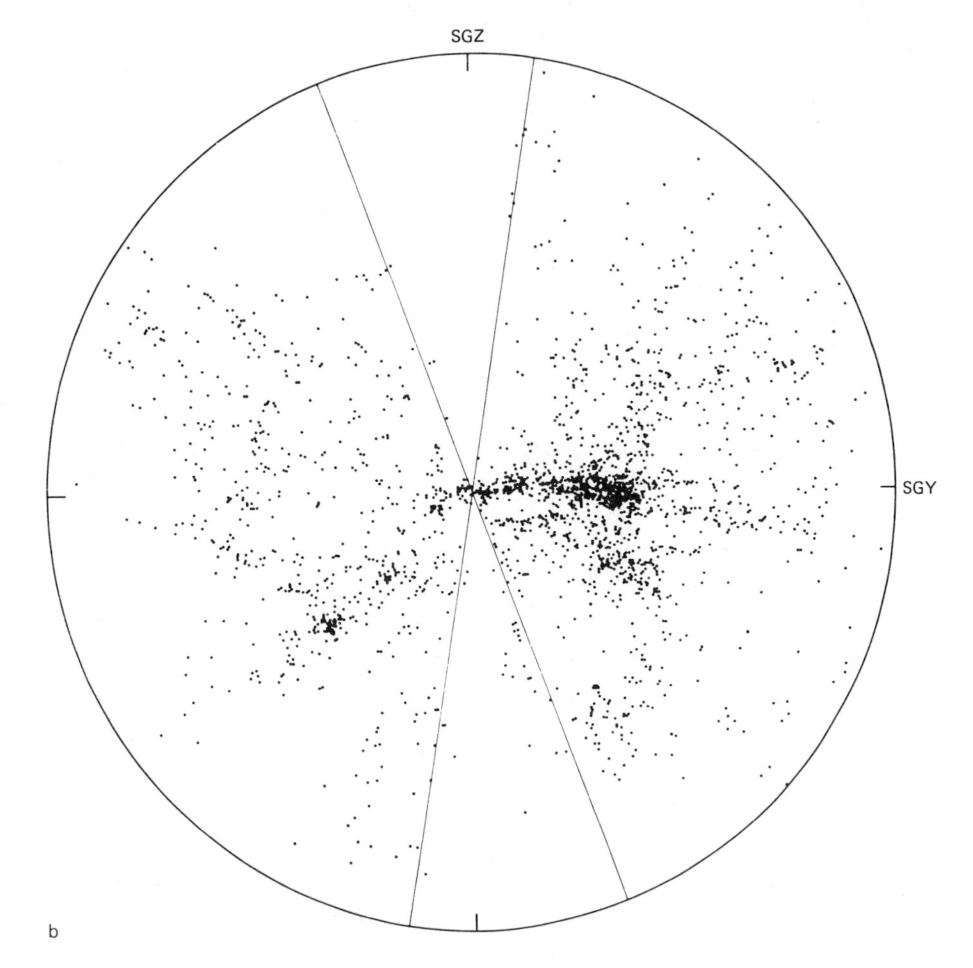

b

BIBLIOGRAPHY

[A1] Aaronson, M., Mould, J., and Huchra, J., "A Distance Scale from the Infrared Magnitude/HI Velocity Width Relation. I—The Calibration," *Astrophys. J.,* 1980, **237,** pp. 655-665.

[A2] Allen, C.W., *Astrophysical Quantities,* 3d ed., Athlone Press, London, 1973.

[A3] Arnett, W.D., "The Cosmic Distance Scale: Methods for Determining the Distance to Supernovae," *Astrophys. J.,* 1982, **254,** pp. 1-7.

[A4] Arnett, W.D., "Type I Supernovae. I—Analytical Solutions for the Early Part of the Light Curve," *Astrophys. J.,* 1982, **253,** 785-795.

[A5] Arp, H.C., "U-B and B-V Colors of Blackbodies," *Astrophys. J.,* 1961, **133,** pp. 874-882.

[A6] Azzopardi, M., and Vigneau, J., "The Small Magellanic Cloud. I—A Study of the Structure Revealed by the Supergiants," *Astr. Astrophys.,* 1977, **56,** pp. 151-161.

[B1] Baade, W., and Swope, H.H., "Variable Star Field 96′ South Proceeding the Nucleus of the Andromeda Galaxy," *Astr. J.,* 1963, **68,** pp. 435-469.

[B2] Barbon, R., Ciatti, F., and Rosino, L., "On the Light Curve and Properties of Type I Supernovae," *Astron. Astrophys.,* 1973, **25,** pp. 241-248.

[B3] Barnes, T.G., Evans, D.S., and Parsons, S.B., "Stellar Angular Diameters and Visual Surface Brightness. II—Early and Intermediate Spectral Types," *Month. Not. R. Astron. Soc.,* 1976, **174,** pp. 503-512.

[B4] Barnes, T.G., Dominy, J.F., Evans, D.S., Kelton, P.W., Parsons, S.B., and Stover, R.J., "The Distances of Cepheid Variables," *Month. Not. R. Astron. Soc.,* 1977, **178,** pp. 661-674.

[B5] Bath, G.T., and Shaviv, G., "Classical Novae—A Steady State, Constant Luminosity, Continuous Ejection Model," *Month. Not. R. Astron. Soc.,* 1976, **175,** pp. 305-322.

[B6] van den Bergh, S., "Search for Faint Companions to M31," *Astrophys. J.,* 1972, **171,** L31-L33; "The Dwarf Spheroidal Companions to the Andromeda Nebula," *Astrophys. J.,* 1974, **191,** pp. 271-272.

[B7] van den Bergh, S., "The Extragalactic Distance Scale," in *Stars and Stellar Systems,* vol. 5: Galaxies and the Universe," Sandage, A., Sandage, M., and Kristian, J., Eds., University of Chicago Press, 1975, pp. 509-593.

[B8] van den Bergh, S., "The Distance Scale within the Local Group," in *Decalages vers le rouge et expansion de l'univers—l'évolution des galaxies et ses implications cosmologiques* (IAU Colloq. 37), Balkowski, C., and Westerlund, B.E., Eds., Editions du CNRS, Paris, 1977, pp. 13-22.

[B9] van den Bergh, S., "The Distance to the Hyades Cluster and the Extragalactic Distance Scale," *Astrophys. J.,* 1977, **215,** pp. L103-L105.

[B10] van den Bergh, S., and Hagen, G.L., "UBV Photometry of Star Clusters in the Magellanic Clouds," *Astr. J.,* 1968, **73,** pp. 569-578.

[B11] Bottinelli, L., and Gouguenheim, L., "The Luminosity Classes of Galaxies and the Hubble Constant," *Astr. Astrophys.,* 1976, **51,** pp. 275-282.

[B12] Branch, D., "Type I Supernovae and the Value of the Hubble Constant," *Month. Not. R. Astron. Soc.,* 1977, **179,** pp. 401-408.

[B13] Branch, D., "On the Use of Type I Supernovae to Determine the Hubble Constant," *Month. Not. R. Astron. Soc.,* 1979, **186,** pp. 609-616.

[B14] Branch, D., "The Hubble Diagram for Type I Supernovae," *Astrophys. J.,* 1982, **258,** pp. 35-40.

[B15] Branch, D., and Bettis, C., "The Hubble Diagram for Supernovae," *Astr. J.,* 1978, **83,** pp. 224-227.

[B16] Branch, D., Falk, S.W., McCall, M.L., Rybski, P., Uomoto, A.K., and Wills, B.J., "The Type II Supernova 1979c in M100 and the Distance to the Virgo Cluster," *Astrophys. J.,* 1981, **244,** pp. 780-804.

[B17] Brodie, J.P., and Madore, B.F., "The Period-Luminosity Relation for Cepheids. I—Numerical Simulations," *Month. Not. R. Astron. Soc.,* 1980, **191,** pp. 841-854.

[B18] Brunet, J.P., "UBV Photometry for Supergiants of the Large Magellanic Cloud," *Astr. Astrophys.,* 1975, **43,** pp. 345-358.

[B19] Butler, C.J., "Photometry of Cepheid Variables in the Large Magellanic Cloud," *Astr. Astrophys.,* suppl. ser., 1978, **32,** pp. 83-126.

[C1] Cappacioli, M., and Fasano, G., "On the Distance of the Giant Spiral Galaxy M101," *Astr. Astrophys.,* 1980, **83,** pp. 354-362.

[C2] Cesarsky, D.A., Laustsen, S., Lequeux, J., Schuster, H.E., and West, R.M., "Two New Faint Stellar Systems Discovered on ESO Schmidt Plates," *Astr. Astrophys.,* 1977, **61,** pp. L31-L33.

[C3] Chiu, H.Y., and Muriel, A., *Stellar Evolution,* MIT Press, 1972.

[C4] Clube, S.V.M., and Dawe, J.A., "Statistical Parallaxes and the Fundamental Distance Scale. II—Applications of the Minimum Likelihood Technique to RR Lyrae and Cepheid variables," *Month. Not. R. Astron. Soc.,* 1980, **190,** pp. 591-610.

[C5] Crampton, D., "The Distance to the Large Magellanic Cloud from OB stars," *Astrophys. J.,* 1979, **230,** pp. 717-723.

[C6] Crampton, D., and Greasley, J., "OB Stars in the Small Magellanic Cloud," *Publ. Astron. Soc. Pacific,* 1982, **94,** pp. 31-35.

[D1] Dachs, J., "Photometry of Bright Stars in the Large Magellanic Cloud," *Astr. Astrophys.,* 1972, **18,** pp. 271-286.

[D2] Dean, J.F., Warren, P.R.N., and Cousins, A.W.J., "Reddening of Cepheids Using BVI Photometry," *Month. Not. R. Astron. Soc.,* 1978, **183,** pp. 569-583.

[D3] Divan, L., "Calibration en magnitudes absolues de la classification BCD. Application à la détermination du module de distance du Grand Nuage de Magellan," in *Problems of Calibration of Absolute Magnitudes and Temperatures of Stars* (IAU Sympo. 54), Hauck, B., and Westerlund, B.E., Eds., Reidel, 1973, pp. 78-85.

[D4] Duerbeck, H.W., "Light Curve Types, Absolute Magnitudes, and Physical Properties of Galactic Novae," *Pub. Astron. Soc. Pacific,* 1981, **93,** pp. 165-175.

[D5] Dworak, T.Z., "Determination of the Distances of the Nearest Galaxies by the Method of Parallax of Eclipsing Binaries," *Acta Cosmol.,* 1974, **2,** pp. 13-19.

[E1] Elias, J.H., Frogel, J.A., Hackwell, J.A., and Persson, S.E., "Infrared Light Curves of Type I Supernovae," *Astrophys. J.,* 1981, **251,** pp. L13-L16.

[F1] Feast, M.W., reported by G. Tammann, ref. [T1].

[F2] Feast, M.W., Thackeray, A.D., and Wesselink, A.J., "The Brightest Stars in the Magellanic Clouds," *Month. Not. R. Astron. Soc.,* 1960, **121,** pp. 337-385.

[F3] Feast, M.W., and Balona, L.A., "On the Determinations of the Color Terms in the P-L-C Relation for Cepheids," *Month. Not. R. Astron. Soc.,* 1980, **192,** pp. 439-443.

[F4] Fernie, J.D., "A Recalibration of the Period-Luminosity Color Relation for Classical Cepheids," *Astr. J.,* 1967, **72,** pp. 1327-1334.

[F5] Fernie, J.D., "A Preliminary Study of the Classical Cepheid P-L Relation in the R, I Photometric System," in *Variable Stars and Stellar Evolution* (IAU Symp. 67), Sherwood, V.E., and Plaut, L., Eds., Reidel, 1975, pp. 185-190.

[G1] Gallagher, J.S., and Starrfield, S., "Theory and Observations of Classical Novae," *Ann. Rev. Astron. Astrophys.,* 1978, **16,** pp. 171-214.

[G2] Gascoigne, S.C.B., "Further Observations of Magellanic Cloud Cepheids," *Month. Not. R. Astron. Soc.,* 1969, **146,** pp. 1-36.

[G3] Gascoigne, S.C.B., "Metal Abundances and the Luminosities of Cepheids," *Month. Not. R. Astron. Soc.,* 1974, **166,** pp. 25-27.

[G4] Geyer, U., "The Zero-point of the Period-Luminosity Relation for Cepheids," *Astr. Astrophys.,* 1970, **5,** pp. 116-126.

[G5] Glass, I.S., and Evans, T.L., "A Period-Luminosity Relation for Mira Variables in the Large Magellanic Cloud," *Nature,* 1981, **291,** pp. 303-304.

[G6] Graham, J.A., "The RR Lyrae Stars in the Small Magellanic Cloud," *Publ. Astron. Soc. Pacific,* 1975, **87,** pp. 641-682; "The RR Lyrae Stars of the Large Magellanic Cloud," *Pub. Astron. Soc. Pacific,* 1977, **89,** pp. 425-465.

[G7] Graham, J.A., "Distance Indicators in the Magellanic Clouds," in *Highlights of Astronomy,* vol. 6, West, R.M., Ed., Reidel, 1983, pp. 209-216.

[H1] Harris, H.C., "Photometric Abundances of Classical Cepheids and the Gradient in the Galactic Disk," *Astr. J.,* 1981, **86,** pp. 707-718.

[H2] Harris, H.C., "Photometric Abundances and the Colors of Cepheids in the Small Magellanic Cloud," *Astr. J.,* 1981, **86,** pp. 1192-1199.

[H3] Hartwick, F.D.A., and Hutchings, J.B., "Classical Novae: A Time-Dependent Optically Thick Wind Model for the Postmaximum Phase," *Astrophys. J.,* 1978, **226,** pp. 203-209.

[H4] Heck, A., and Jung, J., "On the Method of Statistical Parallaxes," *Astr. Astrophys.,* 1975, **40,** pp. 323-326.

[H5] Holmberg, E., "A Study of External Galaxies," *Arkiv Astronomi,* 1964, **3,** pp. 387-438.

[H6] Humphreys, R.M., "Studies of Luminous Stars in Nearby Galaxies. II—M Supergiants in the Large Magellanic Cloud," *Astrophys. J.,* suppl. ser., 1979, **39,** pp. 389-403.

[H7] Humphreys, R.M., "Studies of Luminous Stars in Nearby Galaxies. IV—Baade's Field IV in M31," *Astrophys. J.,* 1979, **234,** pp. 854-860.

[H8] Humphreys, R.M., "Studies of luminous Stars in Nearby Galaxies. V—The Local Group Irregulars NGC 6822 and IC 1613," *Astrophys. J.,* 1980, **238,** pp. 65-78.

[H9] Humphreys, R.M., "Studies of Luminous Stars in Nearby Galaxies. VI—The Brightest Supergiants and the Distance of M33, *Astrophys. J.,* 1980, **241,** pp. 587-597.

[H10] Humphreys, R.M., "The Brightest Stars as Extragalactic Distance Indicators," *Astrophys. J.,* 1983, **269,** pp. 335–351.

[I1] Iben, I. Jr., and Tuggle, R.S., "Comments on a PLC Relationship for Cepheids and on the Comparison between Pulsation and Evolution Masses for Cepheids," *Astrophys. J.,* 1972, **178,** pp. 441–453.

[I2] Iben, I., Jr., and Tuggle, R.S., "On the Intrinsic Properties of Cepheids in the Galaxy, in Andromeda, and the Magellanic Clouds," *Astrophys. J.,* 1975, **197,** pp. 39–54.

[J1] Jaakkola, T., and Le Denmat, G., "Remarks on the Low Value Obtained for the Hubble Constant," *Month. Not. R. Astron. Soc.,* 1976, **176,** pp. 307–313.

[J2] Jung, J., "On the Distance Scale of nearby Classical Cepheids," *Astr. Astrophys.,* 1970, **6,** pp. 130–137.

[K1] Kayser, S.E., "Photometry of the Nearby Irregular Galaxy, NGC 6822," *Astr. J.,* 1967, **72,** pp. 134–148.

[K2] Kennicutt, R.C., Jr., "HII Regions as Extragalactic Distance Indicators II—Applications of Isophotal Diameters," *Astrophys. J.,* 1979, **228,** pp. 696–703.

[K3] Kirschner, R.P., and Kwan, J., "Distances to Extragalactic Supernovae," *Astrophys. J.,* 1974, **193,** pp. 27–36.

[K4] Kraan-Korteweg, R.C., and Tammann, G.A., "A Catalogue of galaxies within 10 Mpc," *Astron. Nachr.,* 1979, **300,** pp. 181–194.

[L1] Longmore, A.J., Hawarden, T.G., Webster, B.L., Goss, W.M., and Mebold, U., "Two New Dwarf Galaxies in the Local Group: UKS 1927–177 and UKS 2323–326," *Month. Not. R. Astron. Soc.,* 1978, **183,** pp. 97–100.

[M1] Madore, B.F., "The Distance to NGC 2403," *Month. Not. R. Astron. Soc.,* 1976, **177,** pp. 157–165.

[M2] Madore, B.F., "Cepheids in External Galaxies," in *Highlights in Astronomy,* vol. 6, West, R.M., Ed., Reidel, 1983, pp. 217–224.

[M3] Martin, W.L., Warren, P.R., and Feast, M.W., "Multicolour Photoelectric Photometry of Magellanic Cloud Cepheids. II—An analysis of BVI Observations in the LMC," *Month. Not. R. Astron. Soc.,* 1979, **188,** pp. 139–157.

[M4] Matthewson, D.S., Cleary, M.N., and Murray, J.D., "The Magellanic Stream," *Astrophys. J.,* 1974, **190,** pp. 291–296.

[M5] McAlary, C.W., Madore, B.F., McGonegal, R., McLaren, R.A. and Welch, D.L., "The distance to NGC 6822 from Infrared Photometry of Cepheids," *Astrophys. J.,* 1983, **273,** pp. 539–543.

[M6] McAlary, C.W. Madore, B.F., and Davis L.E., "The Distance to IC 1613 from Infrared Photometry of Cepheids," *Astrophys. J.,* 1984, **276,** pp. 487–490.

[M7] McClure, R.D., and Racine, R., "The Reddening of M3, M13, M31, and M33 from Photometry of Late-Type Field Stars," *Astr. J.,* 1969, **74,** pp. 1000–1007.

[M8] McGonegal, R., McLaren, R.A., McAlary, C.W., and Madore, B.F., "The Cepheid Distance Scale: A New Application for Infrared Photometry," *Astrophys. J.,* 1982, **257,** pp. L33–L36.

[M9] McNamara, D.H., and Feltz, K.A., Jr., "The Galactic Foreground Reddening in the Direction of the Magellanic Clouds," *Publ. Astron. Soc. Pacific,* 1980, **92,** pp. 587–591.

[P1] Pagel, B.E.J., Edmunds, M.G., Fosbury, R.A.E., and Webster, B.L., "A Survey of Chemical Composition of HII Regions in the Magellanic Clouds," *Month. Not. R. Astron. Soc.,* 1978, **184,** pp. 569–592.

[R1] Rees, M.J., and Stoneham, R.J., Eds., "Supernovae: A Survey of Current Research," Reidel, 1982.

[R2] Rosino, L., "Observations of Supernovae at the Astrophysical Observatory of Asiago," in *Supernovae,* Schramm, D.N., Ed., Reidel, 1977, pp. 1–11.

[S1] Sandage, A., "Current Problems in the Extragalactic Distance Scale," *Astrophys. J.,* 1958, **127,** pp. 513–526.

[S2] Sandage, A.R., "The Distance Scale," in *Problems of Extragalactic Research,* McVittie, G.C., Ed., Macmillan, 1962, pp. 359–378.

[S3] Sandage, A., "Classical Cepheids: Cornerstone to Extragalactic Distances?" *Quart. J. R. Astron. Soc.,* 1972, **13,** pp. 202–221.

[S4] Sandage, A., and Tammann, G.A., "A Composite Period-Luminosity Relation for Cepheids at Mean and Maximum Light," *Astrophys. J.,* 1968, **151,** pp. 531–545.

[S5] Sandage, A., and Tammann, G.A., "The Double Cepheid CE Cassiopeiae in NGC 7790: Tests of the Theory of the Instability Strip and the Calibration of the Period–Luminosity–Color Relation," *Astrophys. J.,* 1969, **157,** pp. 683–708.

[S6] Sandage, A., and Tammann, G.A., "Absolute Magnitudes of Cepheids. III—Amplitude as a Function of Position in the Instability Strip: A Period–Luminosity–Amplitude Relation," *Astrophys. J.,* 1971, **167,** pp. 293–310.

[S7] Sandage, A., and Tammann, G.A., "Steps toward the Hubble Constant. I—Calibration of the Linear Sizes of Extragalactic HII Regions," *Astrophys. J.,* 1974, **190,** pp. 525–538.

[S8] Sandage, A., and Tammann, G.A., "Steps toward the Hubble Constant. VII—Distances to NGC 2403, M101 and the Virgo Cluster Using 21 Centimeter Line Widths Compared with Optical Methods: The Global Value of H_0," *Astrophys. J.,* 1976, **210,** pp. 7–24.

[S9] Sandage, A., and Tammann, G.A., "H_0, q_0, and the Local Velocity Field," in *Astrophysical Cosmology,* Bruck, H.A., Coyne, G.V., and Longair, M.S., Eds., Pontifica Academia Scientiarum, 1982, pp. 23–81.

[S10] Sandage, A., and Carlson, G., "Distance and Absolute Magnitudes of the Brightest Stars in the Dwarf Galaxy Sextans A," *Astrophys. J.,* 1982, **258,** pp. 439–456.

[S11] Sandage, A., and Carlson, G., "The Distance to M33 from a New Study of its Cepheids," *Astrophys. J.,* 1983, **267,** pp. L25–L28.

[S12] Sanduleak, N., and Davis Phillip, A.G., "A Stellar Group in Line of Sight of the Large Magellanic Cloud," *Astr. J.,* 1968, **73,** pp. 566–568.

[S13] Schmidt-Kahler, Th., "Reddening and the Distance of M33 as Derived from Open Clusters," *Astr. J.,* 1967, **72,** pp. 526–531.

[S14] Schurmann, S.R., Arnett, W.D., and Falk, S.W., "Type II Supernovae: Nonstandard Candles as Extragalactic Distance Indicators," *Astrophys. J.,* 1979, **230,** pp. 11–25.

[S15] Shara, M.M., "On the Constancy of the Absolute Magnitude $M_B(15)$ of a Classical Nova 15 Days after Maximum Light, *Astrophys. J.,* 1981, **243,** pp. 268–270.

[S16] Shara, M.M., "A Theoretical Explanation of the Absolute Magnitude–Decline

Time (M_B–t_3) Relationship for Classical Novae," *Astrophys. J.*, 1981, **243,** pp. 926–934.

[S17] Stift, M.J., "The Cepheid Period–Luminosity–Colour Relation: A Most Unsuitable Distance Indicator," *Astr. Astrophys.*, 1982, **112,** pp. 149–156.

[S18] Stothers, R.B., "A New Calibration of the Extragalactic Distance Scale Using Cepheids and RR Lyrae Stars," *Astrophys. J.*, 1983, **274,** pp. 20–30.

[T1] Tammann, G.A., "A Progress Report on Supernova Statistics," in *Supernovae,* Schramm, D.N., Ed., Reidel, 1977, pp. 95–116.

[T2] Tammann, G.A., and Sandage, A., "The Stellar Content and Distance of the Galaxy NGC 2403 in the M81 Group," *Astrophys. J.*, 1968, **151,** pp. 825–860.

[T3] Tammann, G.A., Sandage, A., and Yahil, A., "The Determination of Cosmological Parameters," in *Physical Cosmology,* Balian, R., Audouze, J., and Schramm, D.N., Eds., North-Holland, 1979, pp. 53–125.

[T4] Thuan, T.X., and Martin, G.E., "A New Dwarf Galaxy in the Local Group," *Astrophys. J.*, 1979, **232,** pp. L11–L16.

[T5] Tifft, W.G., "Magellanic Cloud Investigations. I—The region of NGC 121," *Month. Not. R. Astron. Soc.*, 1963, **125,** pp. 199–125.

[T6] Tully, R.B., "The Local Supercluster," *Astrophys. J.*, 1982, **257,** pp. 389–422.

[V1] Vaucouleurs, G. de, "The Extragalactic Distance Scale. III—Secondary Distance Indicators," *Astrophys. J.*, 1978, **224,** pp. 14–21.

[V2] Vaucouleurs, G. de, "The Extragalactic Distance Scale. I—A Review of Distance Indicators: Zero Points and Errors of Primary Indicators, *Astrophys. J.*, 1978, **223,** pp. 351–363.

[V3] Vaucouleurs, G. de, "The Extragalactic Distance Scale. II—Distances of the Nearest Galaxies from Primary Indicators," *Astrophys. J.*, 1978, **223,** pp. 730–739.

[V4] Vaucouleurs, G. de, "The Extragalactic Distance Scale. VI—Distances of 458 Spiral Galaxies from Tertiary Indicators," *Astrophys. J.*, 1979, **227,** pp. 729–755.

[V5] Vaucouleurs, G. de, Vaucouleurs, A. de, and Corwin, H.G., Jr., *Second Reference Catalogue of Bright Galaxies,* University of Texas Press, 1976.

[V6] Vaucouleurs, G. de, Peters, W.L., and Corwin, H.G., Jr., "Possible New Members of the Local Group of Galaxies from Solar Motion Solutions," *Astrophys. J.*, 1977, **211,** pp. 319–323.

[W1] Wielen, R., "Space Velocities of Nearby Classical Cepheids," *Astr. Astrophys.*, suppl. ser., 1974, **15,** pp. 1–15.

[Y1] Yahil, A., Tammann, G.A., and Sandage, A., "The Local Group: The Solar Motion Relative to Its Centroid," *Astrophys. J.*, 1977, **217,** pp. 903–915.

CHAPTER FOUR

SECONDARY DISTANCE INDICATORS

We now turn to distance indicators which depend for their calibration on our knowing of the distance to some representative nearby galaxies through primary distance indicators. In a classic series of papers in the early 1970s Allan Sandage and Gustav Tammann set out to provide a thorough calibration of the brightest HII regions and stars in galaxies (Secs. 4.1 and 4.3) and then used these secondary distance indicators to estimate distances of galaxies beyond the Local Group and the M81 group. Gerard de Vaucouleurs challenged their calibration of these methods and arrived at distances generally lower than those found by Sandage and Tammann. Robert Kennicutt Jr. has attempted to put the method using the brightest HII regions in galaxies on a more objective basis, but has also found, unfortunately, that the method is not capable of very high precision. Beautifully detailed work by Roberta Humphreys has enormously increased our knowledge of the brightest stars in nearby galaxies, but problems in trying to use them as distance indicators have been highlighted by David Hanes. Nevertheless Sandage and Tammann, in their latest work, use the brightest red stars to determine the distances of several galaxies, including two in the Canes Venatici I cluster in which type I supernovae have occurred. (These supernovae can then be used to calibrate the absolute magnitude of type I supernovae at maximum light.)

Another secondary distance indicator, which has been used by Sandage and Tammann, de Vaucouleurs, and, especially, Hanes, is a galaxy's globular cluster population (Sec. 4.2), assuming that the distribution of globular cluster luminosities is the same in all galaxies. However, some galaxies have many more globular clusters of the same luminosity than others, and the reason for these anomalies is not understood.

Brent Tully and Richard Fisher have discovered a method of estimating distances of galaxies which has attracted an immense amount of effort in recent years. This is based on a correlation found in spiral galaxies between the width of the 21-cm emission line of neutral atomic hydrogen and the absolute magnitude of the galaxy (Sec.4.4). This indicator has been applied by Tully and Fisher and others to well over 1000 galaxies. The method is relatively easy to apply and has the potential to reach to distances of 100 Mpc or more. There are some difficulties associated with the determination of the inclination of the disk of the galaxy to the line of sight and with estimating the internal extinction. The latter problem is greatly reduced by the use of infrared magnitudes. There is also controversy about the extent to which the Tully – Fisher correlation depends on galaxy type. However, apart from supernovae, the 21-cm line-width method probably holds the greatest hope of pushing the distance scale out to cosmological distances.

Most of the above methods apply only to spiral galaxies. In early-type galaxies (ellipticals and lenticulars) there is a correlation between color and luminosity which can be used as an approximate distance indicator (Sec. 4.5).

Secs. 4.6 to 4.8 deal with what de Vaucouleurs calls *tertiary* distance indicators, where the calibration of the indicator depends (at present) on galaxies whose distances are determined by secondary indicators. The philosophy of this book is that the need to use such galaxies as calibrators is a temporary deficiency only. The potential of primary indicators is far from fulfilled, and in the course of the next few years all the methods of this chapter should be calibrated by galaxies with distances determined from primary indicators. The nomenclature of de Vaucouleurs is, however, a fair description of the present situation.

The most widely used of these tertiary indicators is the spiral galaxy luminosity class, particularly for late-type spirals (Sc and Sd). This indicator is based on a correlation between the luminosity of spiral galaxies and their luminosity class, which in turn is based on the appearance of the galaxy (Sec. 4.6). There has been much controversy about the calibration of this relation, and in recent years it has become apparent, unfortunately, that these two properties are not nearly as well correlated as had been hoped and that this is therefore a rather approximate way of measuring distances.

The sizes of galaxies are also correlated with galaxy luminosity and luminosity class; so sizes, too, can be used as distance indicators (Sec. 4.7). But although a more objective quantity than the luminosity class, the size of a galaxy is not an easy quantity to define precisely, so this too is still a rather crude distance indicator.

Finally we come to the indicator which Sandage and others have used to push the distance scale out to truly cosmological distances ($>$1000 Mpc): the luminosity of the brightest galaxies in rich clusters of galaxies (Sec. 4.8). This can be calibrated through observations of relatively nearby clusters, like the Virgo cluster, which are accessible by several methods. Two factors that limit the usefulness of this method for very distant clusters are the evolution of galaxies with time (extremely distant galaxies are necessarily seen at considerably younger ages than nearby galaxies) and the tendency of luminous, massive galaxies to swallow up their neighbors through tidal interaction.

4.1 HII regions

The possibility that giant HII regions in galaxies could be used as distance indicators was first suggested by J.L. Sersic in 1960 [S19]. When a massive new O or B star (mass $M \gtrsim 10\ M_{\odot}$, say) is born in a dense molecular cloud, the copious flux of ultraviolet photons which it emits has a dramatic effect on the star's environment. Photons with wavelength less than 912 Å are capable of *ionizing* hydrogen by giving the electron in the atom sufficient energy to escape completely. The electron may later recombine with another ionized atomic nucleus or ion. (The ion of a hydrogen atom is just a proton.) When it does so, it usually finds itself in an excited state and then cascades down to lower energy levels, emitting photons with characteristic energies which can be observed as recombination lines in the spectrum. Transitions between very high energy levels lead to emission lines in the radio region of the spectrum—lines which have been extensively studied. Transitions down to the electron energy levels $n = 4$, 3, 2, and 1 (ground state) give spectral lines in the Brackett (1.46–4.05 μm), Paschen (0.84–1.875 μm), Balmer (0.3646–0.6563 μm), and Lyman (0.0912–0.1215 μm) series, respectively. However, the gas is normally so opaque to radiation in the Lyman lines that it is generally assumed that these photons are almost instantly reabsorbed by the gas or by dust grains present within the gas.

It can be shown that there is a sharp boundary between the completely ionized gas near the star (the HII region) and the neutral gas further out (the HI region). For a spherically symmetric geometry this boundary defines the Strömgren sphere, named after Bengt Strömgren who first discovered this phenomenon. The radius of this sphere r_s can be calculated by balancing the number of ionizing photons per second N_u with the total number of recombinations per second:

$$\frac{4\pi}{3} r_s^3 n_e n_H \alpha^{(2)} = N_u \qquad (4.1)$$

where n_H is the number density of hydrogen atoms per cubic centimeter, n_e is the number density of electrons ($= n_H$ for pure, completely ionized hydrogen), $\alpha^{(2)}$ is the recombination coefficient to the $n = 2$ level (transitions to the $n = 1$ level are excluded because they produce radiation which rarely escapes from the HII region), and N_u is the total number of ionizing photons (those with wavelength <912 Å) emitted by the star per second. App. A.13 gives the stellar temperature and radius, the number of ionizing photons, and the Strömgren radius for stars of different spectral type. For a typical interstellar gas density of 0.1–1 hydrogen atom per cubic centimeter the Strömgren radius for a star of spectral type O5 reaches the amazing dimensions of 100–500 pc, 10^7 times larger than the radius of the star. The Strömgren radius decreases sharply as we go to later spectral types, and the HII region associated with a B star is much less prominent.

Besides being copious emitters of line radiation, particularly in the lines of the Balmer series and in the Lyman α line, HII regions are strong sources of continuum radiation through the free–free (or thermal bremsstrahlung) mechanism. Free–free radiation is generated by the motion of the free electrons in the electrostatic field of the ions and can be observed at wavelengths from the radio to the visible bands. The brighter HII regions in galaxies are large, luminous objects, easily observable at great distances.

Two complications of the above picture should be mentioned. First the presence of heavy elements, and especially helium, complicates the discussion of the ionization structure (but also leads to a great variety of observable spectral lines). Second the presence of dust grains spread through the gas drastically cuts down the flow of ionizing photons during the early high-density phase in the evolution of an HII region.

This evolution can be outlined as follows. When an O star first forms, most of the ionizing photons are absorbed by dust and reemitted in the infrared. At this stage the HII region is very small, less than 0.1 pc in diameter. It is obscured by dust at visual wavelengths, and has OH and H_2O maser sources associated with it. As the HII region expands, because of its high internal pressure, it begins to become visible through the Balmer line (and associated continuum) radiation. The drop in density means that the ionizing photons can now penetrate to a greater mass of gas. As the HII region grows, it gradually eats into the molecular cloud from which the O star formed. If the star is not exactly in the center of the cloud, the HII region will eventually burst out of one side of the cloud, giving a "blister" geometry. In time the molecular cloud will be dispersed, and a fully developed Strömgren sphere will appear. The mass of ionized gas in the largest HII regions in galaxies can be as big as a hundred million solar masses [H1].

These giant HII regions can be modeled with two components, a central core embedded in a more extended halo, though this separation into two components is somewhat subjective.

Sandage and Tammann have described how photographic plates taken through a narrow interference filter centered on the Balmer H_α line at 6563 Å can be used to estimate core and halo diameters of the three brightest HII regions in a number of galaxies in the Local, M81, and M101 groups [S6]. They restricted themselves to late-type spiral and irregular galaxies (types Sc, Sd, and Im), since it was shown by Sersic that Sa and Sb galaxies have smaller HII regions than Sc galaxies. They found that the average of the core and halo diameters for the three brightest HII regions, $\langle D_H, D_C \rangle_3$, is correlated with the absolute photographic magnitude of the galaxy M^0_{pg}, corrected for interstellar extinction and the effects of inclination, by

$$\log \langle D_H, D_C \rangle_3 = -0.140\, M^0_{pg} - 0.202 \tag{4.2}$$

which is the straight line in Fig. 4.1. Unfortunately the slope of this correlation is close to the value of -0.2, which would make the relation between the HII region size and galaxy luminosity independent of distance. Sandage and Tammann therefore looked for a correlation between the HII region diameter and the galaxy luminosity class L_c and derived, for the same galaxies as above, a linear relation of the form

$$\langle D_H, D_C \rangle_3 = -96.5(\pm 14.4)L_c + 557(\pm 60) \quad [\text{pc}] \tag{4.3}$$

which can then be used to estimate the distances of galaxies for which the luminosity class is known and the HII region angular diameters have been measured.

This whole procedure has been criticized on a number of grounds:

1 The rather subjective method of measuring core and halo diameters on a photographic plate is liable to systematic errors, a point made by de Vaucouleurs [V2] and

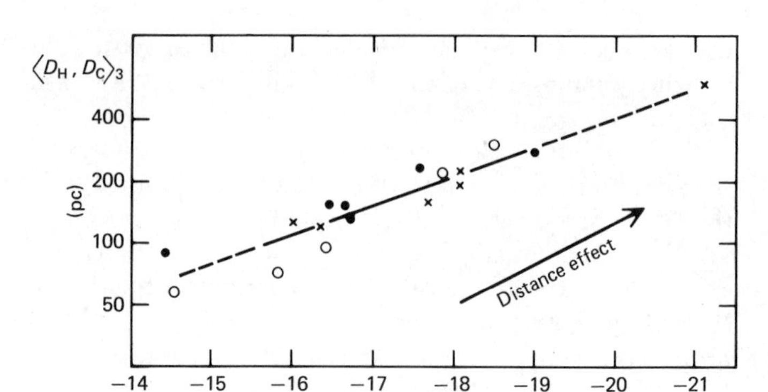

FIGURE 4.1

Correlation of $\langle D_H, D_C \rangle_3$*, the average of the core and halo diameters of the three largest HII regions in a galaxy, with the absolute blue magnitude* M^0_{pg}*, corrected for extinction by dust, for 17 calibrating galaxies. ○—members of Local Group; ●—members of M81–NGC 2403 group; ×—members of M101 group. The effect of arbitrarily increasing the distances of the galaxies is shown by an arrow. (From [S6].)*

confirmed by the work of Kennicutt [K1]. The latter has measured objectively defined isophotal diameters (diameters down to a particular intensity level, or isophotal contour) for the HII regions used by Sandage and Tammann. Tammann et al., in their 1979 Les Houches lectures [T4], appear to accept this criticism.

2 The adoption of the NGC 2403 Cepheid distance for the M81 group must be subject to considerable error, due to both uncertainties in the extinction correction (see Sec. 3.1) and a possible spread of the group in the line of sight.

3 The calibration was not done in a completely self-consistent way because galaxies from the M101 group were included in the solution, even though the distance to the M101 group was established essentially via the HII region method.

4 Van den Bergh has emphasized that the luminosity class L_c is a subjective classification and may be subject to distance-dependent errors [B5]. To this can be added the criticism that L_c is not a continuous variable, as is implied by Eq. (4.3), but rather a series of boxes with labels. In order to get a monotonic behavior from box to box it was already necessary to exclude certain boxes, namely, Sa and Sb galaxies. If the solution is not linear, this may only mean that some boxes are wider than others.

5 The use of D in Eq. (4.3) rather than log D means that errors which are a certain fraction of the distance have different weights for different values of D.

In 1982 Hanes redid the Sandage and Tammann analysis, correcting some of these points [H4]. However, he still used their data, so his work did not correct the errors in the core and halo diameter measurements. Also, he gave no weight at all to the distance estimate for NGC 2403 using the Cepheid method or for M101 using the supernova method, though these estimates must be superior to group distances derived through empirical relations such as Eq. (4.3). Finally Hanes still used L_c

as a continuous variable. He arrived at a distance modulus of 26.73 ± 0.13 ($d = 2.2$ Mpc) for the M81 group and of 28.95 ± 0.14 ($d = 6.2$ Mpc) for the M101 group, based on the Sandage and Tammann Local Group distances and on $\mu_{\mathrm{Hya}} = 3.03$. These distances for the M81 and M101 groups are considerably lower than those derived by Sandage and Tammann.

In this book I proceed as follows:

1 I use Kennicutt's isophotal HII region diameters.
2 I adopt a distance modulus of 27.24 ± 0.28* for NGC 2403, as discussed in Sec. 3.1, but leave the distances of the other galaxies in the M81 group as quantities to be found by comparison with Local Group galaxies.
3 I adopt the distance modulus to M101, $\mu_0 = 29.30 \pm 0.46$*, from the supernova method, but leave the distances of the other galaxies in the group as quantities to be determined by comparison with Local Group galaxies.
4 I treat the galaxy luminosity class L_c as a subdivision into boxes, not as a continuous variable.
5 I work in terms of log D.
6 The Local Group distances used are those adopted in Tables 3.8, 3.10, and 3.11.

The calibration of the method is then as given in line 1 of Table 4.1. There is one calibrating galaxy of each type, except for type IV-V for which there are two, and for types I-II, II, and IV for which there are none. We therefore have no estimate for the uncertainty associated with this calibration, a fact obscured by treating L_c as a continuous variable. We can estimate that the uncertainty must be about that associated with moving ± 1 bin in L_c, that is, ± 0.2 in log D or ± 1.0 in μ_0. In the absence of better information I adopt $\sigma(\mu_0) = 1.0$ as the dispersion about the calibration relation. Since the calibration is essentially determined by a single object in each luminosity class interval, the zero-point error will also be ± 1.0 in μ_0. For luminosity class IV I adopt the mean of log D for the two adjacent luminosity classes (III-IV and IV-V), which appears justified in light of the fairly steady increase in log D from $L_c = \mathrm{V}$ to I.

The method can now be applied to the galaxies of the M81 group (except NGC 2403). The mean distance modulus for these galaxies by this method is 26.59 ± 1.1 ($d = 2.1$ Mpc). Applying the method to the galaxies of the M101 group (excluding M101 itself), we find $\mu_0 = 28.55 \pm 1.1$ ($d = 5.1$ Mpc). The only part played by NGC 2403 and M101 themselves in these estimates is to lend plausibility to the estimate of $\sigma(\mu_0)$ and to the interpolation for luminosity class IV. These distance estimates are clearly of very low weight. Combined with the other distance estimates discussed in this chapter, however, much more reliable distances can be estimated (see Table 4.13). Using these, the galaxies of the M81 and M101 groups can be combined with those of the Local Group to give an improved calibration. This is given in line 2 of Table 4.1 and can be used to estimate the distances of other galaxies for which Kennicutt has measured the isophotal diameters of the three brightest HII regions [$\mu_0(D)$ column in Table 4.2].

* The uncertainties assigned to each distance method are summarized in Table 4.12.

TABLE 4.1

Calibration of diameters and fluxes of HII regions in galaxies.

	Luminosity class and L_c								
	I	*I–II*	*II*	*II–III*	*III*	*III–IV*	*IV*	*IV–V*	*V*
	1	*1.5*	*2*	*2.5*	*3*	*3.5*	*4*	*4.5*	*5*
Mean isophotal diameter D_3 of three largest HII regions in a galaxy, (pc)									
1 Calibration based on Local Group, N2403 ($\mu_0 = 27.24$), and M101 ($\mu_0 = 29.30$)									
Calibrating galaxies	M101	–	–	M33	N2403	LMC	–	SMC N6822	IC1613
$\langle \log D_3 \rangle$	2.86	–	–	2.39	2.52	2.41	(2.23)	2.055	1.77
2 Calibration based on Local Group, M81 group ($\mu_0 = 27.31$), and M101 group ($\mu_0 = 29.16$)									
Calibrating galaxies	M101	–	–	M33	N2403	LMC	N4236 N5204 N5474 N5585	SMC N6822 N2366 IC2574 Ho II N5477 Ho IV	IC1613 Ho I Ho I
$\langle \log D_3 \rangle$	2.83	–	–	2.39	2.54	2.41	2.31	2.205	1.76
Mean H_α luminosity F_3 of three brightest HII regions in a galaxy, erg s^{-1}									
3 As line 1									
Calibrating galaxies	M101	–	–	M33	N2403	LMC	–	SMC N6822	IC1613
$\langle \log F_3 \rangle$	40.29	–	–	39.08	39.35	39.26	(38.79)	38.325	37.45
4 As line 2									
Number of calibrating galaxies	1	–	–	1	1	1	4	7	2
$\langle \log F_3 \rangle$	40.23	–	–	39.08	39.37	39.26	38.78	38.62	37.55
5 As line 4, plus Virgo cluster ($\mu_0 = 31.32$)									
Number of calibrating galaxies	6	–	4	6	5	3	5	7	2
$\langle \log F_3 \rangle$	39.92	–	39.58	39.19	39.17	39.36	38.91	38.62	37.55

Kennicutt has proposed a further distance indicator, based on the luminosity of HII regions in the Balmer H_α line at 6563 Å. This is correlated with the absolute magnitude of the galaxy, the correlation taking the form

$$\log F_3 = -(0.32 \pm 0.04)M^0_{pg} + \text{constant} \tag{4.4}$$

where F_3 is the average H_α luminosity of the three brightest HII regions and M^0_{pg} is the absolute magnitude of the galaxy corrected for interstellar extinction and the effects of inclination [K2],[K3].

An analysis exactly parallel to that given for the diameters of the three brightest HII regions in galaxies can be applied to observations of the H_α fluxes of the three brightest HII regions, to give independent estimates of the distances of the M81 and

TABLE 4.2

Galaxies with isophotal diameters and H_α fluxes for three brightest HII regions.

NGC	$\log \theta_3$*	$\log f_3$†	*Luminosity class*	A_B‡	A_i‡	m_{pg}§	$\mu_0(D)$‖	$\mu_0(F_\alpha)$‖	$\mu_0(M_{pg})$‖
428	0.826	−12.90	III–IV	0.09	0.09	11.74	29.47	30.45	29.45
628	0.924	−12.43	I	0.14	0.02	9.74	31.08	30.68	30.75
672	0.908	−12.64	III	0.22	0.31	11.31	29.71	29.33	29.63
925	0.732	−12.76	II–III	0.34	0.17	10.53	29.84	29.68	29.28
1073	0.908	−13.09	II	0.12	0.02	11.43	–	31.48	31.25
1232	0.778	−12.84	I	0.10	0.04	10.46	31.81	31.70	31.45
2500	0.756	−13.12	IV	0.19	0.02	12.13	29.32	29.88	29.78
3184	0.908	−12.77	II	0	0.01	10.28	–	30.68	30.23
3631	0.763	−12.97	I	0	0.04	10.91	31.89	32.03	32.00
3810	0.820	−12.85	I	0	0.11	11.30	31.60	31.73	32.32
3938	0.732	−13.02	I	0	0.03	10.79	32.04	32.15	31.89
4214	1.033	−12.07	III–IV	0	0.10	10.12	28.44	28.38	27.91
4321	0.845	−12.79	I	0	0.04	10.07	31.48	31.58	31.17
4395	1.090	−12.38	IV–V	0	0.06	10.66	27.13	27.30	26.73
4449	1.228	−12.02	III	0	0.14	9.90	28.11	27.76	28.61
5068	0.986	−12.52	III–IV	0.19	0.03	(10.92)	28.76	29.50	28.59
5194	1.068	−12.39	I	0	0.12	8.88	30.36	30.58	29.89
6015	0.763	−12.94	II	0.07	0.29	11.69	–	31.10	31.29
6946	1.114	−12.23	I	0.87	0.04	9.67	30.13	30.18	29.89
7741	0.763	−12.80	II	0.22	0.12	11.63	–	30.75	31.25

* θ_3— mean angular diameter of three largest HII regions, in seconds of arc.
† f_3— mean H_α flux, in erg s^{-1} cm^{-2}.
‡ Number of magnitudes of extinction in blue band due to interstellar (A_B) and internal (A_i) dust.
§ (Uncorrected) apparent magnitude of galaxy.
‖ Distance moduli estimated from HII region diameters and fluxes and from the absolute magnitude of the galaxy as a function of luminosity class.

SOURCE: Compiled from [K1],[K2].

M101 groups. Kennicutt has given an estimate of the dispersion of log F_3 about the calibration relation, using galaxies of the Virgo cluster, and this corresponds to $\sigma(\mu_0) \simeq 1.0$. The calibration of log F_3 as a function of L_c is given in line 3 of Table 4.1, and the result is a distance modulus for the M81 group (excluding NGC 2403) of 26.46 ± 1.1 ($d = 1.95$ Mpc) and for the M101 group (excluding M101) of 28.92 ± 1.1 ($d = 6.1$ Mpc). These are consistent with the results given above, based on the diameters of HII regions, but are again of very low weight. Using the final best estimate of the M81 and M101 group distances from Table 4.13, the calibration of this method based on the galaxies of these groups and the Local Group is given in line 4 of Table 4.1.

This calibration can be used to estimate the distance of the Virgo cluster. Kennicutt [K3] has given the H_α fluxes of the three brightest HII regions in 21 galaxies of the Virgo cluster. Four of these are of luminosity class II, for which there is no calibration from nearby galaxies, and so give no distance information. The 17 remaining galaxies yield a mean distance modulus to the Virgo cluster of 31.51 ± 1.05 ($d = 20$ Mpc). For comparison, Kennicutt found, using a slightly different procedure,

$$\mu_0 = 31.4 \pm 0.25 \; (d = 19 \text{ Mpc}).$$

His lower error estimate arises because of his assumption that L_c can be treated as a continuous variable.

Again, my final best estimate of the Virgo cluster distance (Table 4.13) can be used to obtain an improved calibration for the H_α luminosity method based on galaxies of the Local Group, the M81 and M101 groups, and the Virgo cluster. This improved calibration of log F_3 as a function of L_c is given in line 5 of Table 4.1, and I have applied it to the other galaxies for which Kennicutt has measured the H_α fluxes of the three brightest HII regions [$\mu_0(F_\alpha)$ in Table 4.2].

Kennicutt and Paul Hodge [K6] have shown that the dependence of the brightest HII region diameters and H_α luminosities on the absolute magnitude of the galaxy can be predicted on the basis of a statistical model. They observed some 600 HII regions in the giant Sc galaxy NGC 628 (M74) and obtained the H_α luminosity and diameter distribution functions shown in Fig. 4.2. If the total HII region content of a late-type spiral or irregular galaxy is assumed to scale with the galaxy luminosity, the H_α luminosity and diameter of the ith brightest HII region, where $i = 1, 2, 3, \ldots$, can be shown to be

$$\log_{10} F_i = -(0.27 \pm 0.04) M^0_{pg} + \text{constant} \qquad (4.5)$$

$$\log_{10} D_i = -(0.13 \pm 0.02) M^0_{pg} + \text{constant} \qquad (4.6)$$

where the constants will depend on the value of i. These are in good agreement with the empirical relations, Eqs. (4.4) and (4.2).

De Vaucouleurs has applied two further HII region properties to the determination of distance. He suggests that HII "rings" can be easily recognized and show only a small range in their linear diameters [V3]. However, as these have been

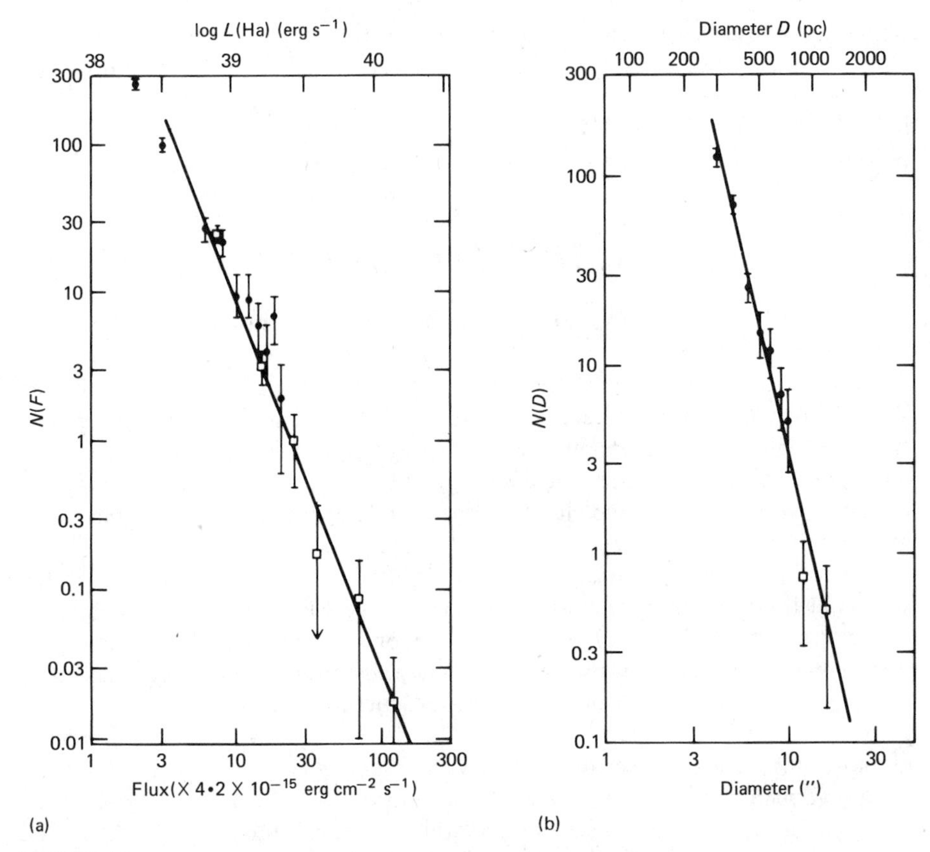

FIGURE 4.2

(a) Luminosity function for HII regions in NGC 628. Lower horizontal scale is the H_α flux from an HII region; upper scale shows the corresponding H_α luminosity F. *Vertical scale is* N(F), *where* N(F) dF *is the number of HII regions with H_α luminosities in the range of* F, F + dF. *(b) Diameter function for HII regions in NGC 628. Lower horizontal scale is the apparent angular diameter of an HII region; upper scale is the intrinsic angular diameter* D. *Vertical scale is* N(D), *where* N(D) dD *is the number of HII regions with diameters in the range of* D, D + dD. *(From [K6].)*

studied in only 11 galaxies (Large Magellanic Cloud, M31, M33, NGC 6822, IC 1613, NGC 2403, DDO 133, DDO 75, IC 342, and IC 10) and all but IC 10 are used in the calibration, the potential of this method would seem to have yet to be demonstrated. It has, however, been used by D.G. Lawrie and K.B. Kwitter [L1] to deduce a distance modulus to the M101 group of 28.18 ± 0.28 mag, which seems low compared with the other estimates given in Table 4.13.

A further distance indicator used by de Vaucouleurs is based on a correlation found by J. Melnick [M2] between the amount of spread in velocities in the HII region gas (called the *velocity dispersion*), measured by the widths of the spectral lines, with the absolute magnitude of the galaxy. The calibration uses galaxies of the Local Group and the M81 group, and the method has so far been applied to only two galaxies outside these groups: M101, for which the method depends on an extrapolation of the calibration, and NGC 300 in the Sculptor group. Again, more galaxies need to be studied to demonstrate the full potential of this method.

4.2 Globular clusters

The properties of the globular clusters in our Galaxy were described in Sec. 2.6. The virtue of their use as distance indicators is that their distances are independent of the Population I distance scale based on the Hyades. Reasonably detailed studies have been made of the globular cluster systems in only two other galaxies apart from our own, the Large Magellanic Cloud and M31. More fragmentary observations are available of globular clusters in six other members of the Local Group, for 15 Virgo cluster galaxies, and for a number of other galaxies in groups and clusters (see App. A.14). At distances well beyond the Virgo cluster, globular clusters have been detected around NGC 3311 in the Hydra I cluster, where the brightest globular cluster appears at the very limit of telescope detection, at blue magnitude $B = 23.5$. At distances beyond 5 Mpc, globular clusters have images which are typically smaller than 2 seconds of arc and can only be distinguished from foreground stars by statistical methods or by a combination of photometry and proper-motion studies. At distances between 0.5 and 5 Mpc, globular clusters may be confused with very distant background galaxies or with images of open clusters or other kinds of nebulosity.

The brightest globular clusters in large galaxies reach luminosities of $M_V \simeq -12$ and can be seen out to 100 Mpc. But reliable distances from globular clusters can only be obtained by observing a substantial number of the globular clusters in a galaxy, and this limits the range to about 10–20 Mpc.

The globular-cluster system in a galaxy can occupy a very large volume. The most remote cluster known in our Galaxy, NGC 2419, is about 100 kpc from the center of the Galaxy, twice as far as the Magellanic Clouds. The cluster system around giant ellipticals like M49 (NGC 4472) or M87 (NGC 4486), both in Virgo, extends to radii of at least 40–100 kpc, possibly up to 500 kpc. There is evidence that the density distribution of globular clusters around a galaxy is less concentrated toward the center of the galaxy than the light distribution, and this has been interpreted as implying that the globular clusters condensed out at an earlier stage than the other Population II stars, when the protogalaxy was more extended.

Brighter, more massive galaxies tend to have more globular clusters (see App. A.14). On the average the total number of globular clusters N_t is directly proportional to the galaxy luminosity L,

$$N_t \propto L^{1.0 \pm 0.3}. \tag{4.7}$$

However, there are dramatic exceptions to this rule. The bright Virgo elliptical M87, associated with the strong radio source Virgo A, has an enormous population

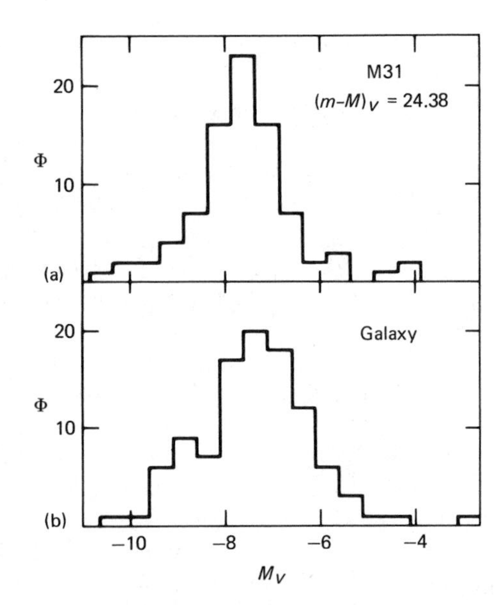

FIGURE 4.3

Luminosity distribution of globular clusters. (a) In M31. (b) In our Galaxy. Horizontal scale is the absolute visual magnitude M_V of a globular cluster; vertical scale is the number of globular clusters found in each bin of absolute magnitude. (From [R1].)

of 15,000 globular clusters, whereas the bright spiral galaxies NGC 891 and 4565 have far fewer than would be predicted by Eq. (4.7). J.A. Dawe and R.J. Dickens detected globular clusters in three large ellipticals of the Fornax I cluster, but failed to find them around several other galaxies of comparable luminosity [D1]. These anomalies mean that the globular cluster population of a galaxy may not be a very reliable distance indicator.

Fig. 4.3 shows the luminosity distributions of globular clusters in our Galaxy and M31. These distributions can be fitted with a Gaussian curve, and Table 4.3 gives the mean and the standard deviation for Gaussian fits to the globular-cluster

TABLE 4.3

Mean absolute visual magnitude of globular clusters in Local Group galaxies.

Galaxy	*M_V(galaxy)**	*$\overline{M}_V$(globular cluster)†*	*n*	*Distance modulus assumed in [R1]*
M31	−21.1	-7.59 ± 0.25	86	24.38
Galaxy	(−20)	-7.34 ± 0.18	103	-
M33	−18.9	-7.87 ± 0.40	6	24.38
LMC	−18.5	-7.43 ± 0.34	17	18.64
SMC	−16.9	-7.03 ± 0.37	13	18.95
N147, 185, 205	−15.5	-6.87 ± 0.34	15	24.38
Fornax	−13.6	-7.26 ± 0.53	5	20.90
Average		-7.26 ± 0.12		

* Absolute visual magnitudes of galaxy.
† Average absolute visual magnitude of globular clusters in galaxy, based on *n* globular clusters.
SOURCE: Compiled from [R1].

luminosity functions in Local Group galaxies. Despite the range of a factor of 1000 in galaxy luminosities, the globular-cluster luminosity distributions are consistent with being of the same form in all the galaxies. This suggests that the globular-cluster population in galaxies was formed with the same distribution of globular cluster masses (and hence luminosities) in all galaxies and that this distribution has not been modified by subsequent evolutionary effects.

Hanes used the Gaussian luminosity distribution to try to determine the distance modulus of the Virgo cluster by comparing the mean magnitude distribution for globular clusters in five Virgo galaxies with the luminosity distribution found in our Galaxy and M31 and obtained [H2]

$$\mu_0 = 30.27 \pm ^{0.49}_{0.43}.$$

With data reaching one magnitude deeper he derived $\mu_0 = 30.7 \pm 0.3$ [H3]. W.E. Harris and R. Racine, using essentially the same data, found $\mu_0 = 30.9 \pm 0.3$ [H5]. I therefore adopt 30.8 as the Virgo distance modulus by this method.

Not too much weight should be placed on the Gaussian form, though, since the bright end of the combined luminosity distribution for our Galaxy and M31 can also be fitted by a power-law

$$\Phi(M) \propto 10^{\alpha M} \tag{4.8}$$

with $\alpha = 1.2 \pm 0.2$, where $\Phi(M)\, dM$ is the number of globular clusters with absolute magnitude in the range of M to $M + dM$, while for M87 and five other giant elliptical galaxies in the Virgo cluster, the bright end of the luminosity distribution can be fitted by Eq. (4.8) with $\alpha = 0.8 \pm 0.2$ [R1]. A luminosity distribution of the form of Eq. (4.8) would not allow distances to be determined using globular clusters, since a shift in distance could be exactly compensated by a change on N_t, the total number of globular clusters.

Some efforts have been made to use the brightest globular cluster in a galaxy as a distance indicator. Assuming a universal Gaussian luminosity function for globular clusters, with mean absolute magnitude $\overline{M}_V = -7.3$ and standard deviation $\sigma = 1.1$, and assuming further that the total number of globular clusters in a galaxy N_t is related to the absolute blue magnitude of the galaxy $M_B^0(G)$ by

$$\log N_t = -0.3[M_B^0(G) + 11.0] \tag{4.9}$$

de Vaucouleurs calculated the absolute magnitude of the brightest globular clusters in the six brightest elliptical galaxies in Virgo [V1]. The apparent magnitude of the brightest globular clusters in these galaxies can then be used to give the distance modulus of the Virgo cluster, which de Vaucouleurs found to be $\mu_0 = 30.2 \pm 0.3$. This quoted uncertainty does not include the uncertainty in Eq. (4.9), which must be considerable, bearing in mind anomalous galaxies like M87 and NGC 891.

Sandage and Tammann [S12], on the other hand, using a similar method, derived $\mu_0 = 31.45 \pm 0.5$ for Virgo, the discrepancy from de Vaucouleurs merely reflecting the unreliability of using the brightest globular cluster in a galaxy as a distance indicator.

In his 1982 review J.A. Graham pointed out [G1] that in the Magellanic Clouds and other late-type galaxies, globular clusters are found which are bluer and younger than those known in our Galaxy and those observed around giant ellipticals. Thus the concept of a unique cluster luminosity function has to be used with care. I have given zero weight here to this method using the single brightest globular cluster in a galaxy.

A class of object that can be used as a distance indicator in a way similar to globular clusters, namely, by comparing the whole population in different galaxies, is that of planetary nebulae. These are shells of gas thrown off during the final stages in the life of a star like the sun, prior to its becoming a white dwarf (see Sec. 2.7). The star is very hot during this phase, ~ 100,000 K, and ionizes the gas, which then appears as a bright spherical halo around the star. About 700 planetary nebulae are known in our Galaxy, and the total population has been estimated by G.H. Jacoby to be $10,000 \pm 4000$ [J1]. The distances to planetary nebulae in our Galaxy are estimated by measuring the surface brightness of the nebula in H_α light and then assuming that all planetary nebulae have the same mass in ionized gas, a method introduced by Mike Seaton in 1968 [S18]. H.C. Ford and D.C. Jenner estimated that the distance of M81 is 3.5–4.0 times greater than that of M31, based on a comparison of the planetary nebulae in the two galaxies [F2]. This is in agreement with the distance scale adopted in this book [d(M81)/d(M31) = 4.1], but since the estimate is based on only eight planetary nebulae in M81, I have not used it here.

4.3 Brightest red and blue stars in galaxies

The galaxies in which the brightest stars have been studied best are the Magellanic Clouds. For our own Galaxy the effects of dust extinction mean that only the most luminous supergiants within 2 or 3 kpc of the sun have been studied in detail. Fig. 4.4(a) and (b) shows a comparison by Roberta Humphreys and K. Davidson of the HR diagrams (M_{bol} against spectral type) for the brightest stars in the neighborhood of the sun and in the Large Magellanic Cloud. Although the proportions of stars in different parts of the diagram are different for the two galaxies, the upper envelope seems to be similar in both, holding out the hope that the luminosities of the very brightest stars in a galaxy can be used as distance indicators.

However, when we turn to look at more distant galaxies, the problem of contamination by foreground stars starts to be acute. This is well illustrated by Fig. 4.4(c), in which Sandage and Tammann compare the color–magnitude diagram for an area of 0.079 square degrees centered on M101 with the color–magnitude diagrams found in two adjacent fields. At blue magnitudes fainter than $B \sim 19.5$, a strong concentration of blue stars (B-V ~ 0–0.6) can be seen in M101, presumably the upper main sequence. However, many of the brighter stars are foreground stars in the halo of our Galaxy and can only be distinguished from M101 stars by careful spectroscopy and photometry. Sandage and Tammann argued that few stars bluer

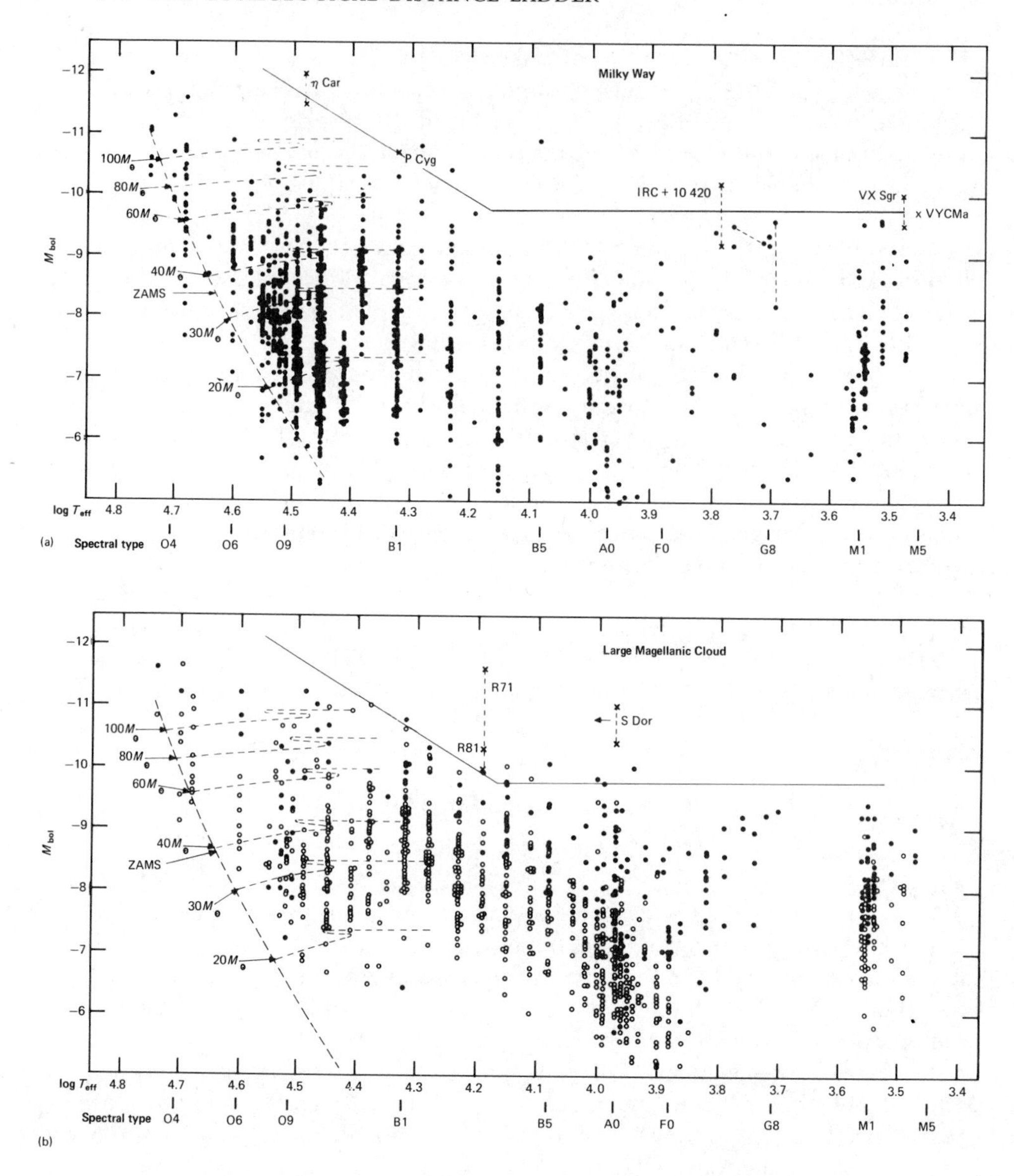

than B-V = 0.4 or redder than B-V = 2.0 are likely to be foreground halo stars and have identified the brightest very blue and very red stars in a number of late-type galaxies in the Local Group and in the M81 and M101 groups. Their 1982 summary of the absolute magnitudes of the brightest red and the brightest blue stars as well as the means of the brightest three red and the brightest three blue stars are given in Table 4.4. The latter quantities are shown in Fig. 4.5 as a function of the absolute magnitude of the galaxy. We see that while the brightest blue stars tend to be more luminous in the more luminous galaxies, the visual luminosity of the brightest red stars seems to be about the same in all galaxies.

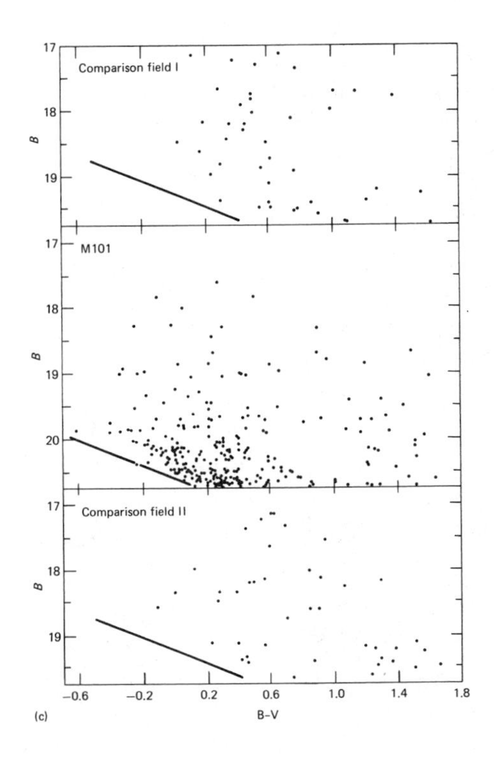

FIGURE 4.4

(a) HR diagram. Absolute bolometric magnitude M_{bol} versus logarithm of effective temperature T_{eff} for the most luminous supergiants in our Galaxy. Horizontal scale is also labeled with spectral type. The position of the zero-age main sequence (ZAMS) and the evolutionary tracks with mass loss are shown as broken curves. The solid line defines the approximate upper boundary of the supergiant luminosities. The names of a few of the very brightest stars of different types are given. (From [H11].) (b) As (a), but for supergiants in the Large Magellanic Cloud. (From [H11].) (c) Color–magnitude diagram. B versus B-V for the brightest stars in an area of 0.079 square degrees centered on M101 (middle), compared with the distribution in two control fields that have the same area near, but uncontaminated by, M101 itself. (From [S8].)

The possibility of using the absolute visual magnitude of the brightest very red stars in galaxies as a distance indicator has been actively investigated by Humphreys in her detailed studies of Local Group galaxies. She used infrared magnitudes to separate the foreground halo stars from the galaxy members and to obtain reliable estimates of the extinction to the stars. Sandage and Tammann assumed that the internal obscuration to these red and blue supergiants can be neglected, but this is not supported by the work of Humphreys. Her estimates of the mean internal extinction to supergiants in Local Group galaxies are given in Table 3.9 along with other estimates.

The explanation suggested by Humphreys and Davidson for the constancy of the absolute visual magnitude of the brightest very red supergiants in galaxies is that mass loss in supergiants, as they cross the HR diagram toward the red supergiant region, limits the stars to a fixed maximum mass [H1]. This would result, however, in a fixed maximum total luminosity, or absolute bolometric magnitude, for the stars, rather than a fixed absolute visual magnitude. This distinction is important in cool stars since their bolometric correction depends sensitively on their temperature (or spectral type). Humphreys pointed out that the lower number of luminous blue supergiant progenitors in lower luminosity galaxies (Fig. 4.5) could be expected to cause a corresponding deficiency of luminous red supergiants in these galaxies [H10]. However, she also pointed out that this effect seemed to be compensated for by the fact that stars in the low-luminosity galaxies tend to have lower heavy-ele-

TABLE 4.4

Mean apparent and absolute magnitudes of three brightest blue and red stars in nearby galaxies.

Galaxy	Luminosity class	Assumed μ_B[S14]	Brightest blue		Brightest red	
			$m_B(3)$	$M_B(3)$	$m_V(3)$*	$M_V(3)$
Solar neighborhood				−8.84		−7.97
LMC	III–IV	18.91	9.46	−9.45	11.07	−7.76
SMC	IV to IV–V	19.35	10.63	−8.72	11.77	−7.56
M33	II–III	24.68	15.72	−8.96	16.70	−7.95
NGC 6822	IV–V	25.03	16.89	−8.14	16.94	−7.82
IC 1613	V	24.55	16.68	−7.87	16.95	−7.57
Sex A		25.67	17.88	−7.79	18.09	−7.53
NGC 2403	III	27.80	18.27	−9.53	20.07	−7.67
NGC 2366	IV–V	27.75	18.97	−8.78	–	–
NGC 4236	IV	27.58	19.22	−8.36	–	–
IC 2574	IV–V	27.60	19.77	−7.83	(20.0)	(−7.6)
Ho II	IV–V	27.67	19.64	−8.03	(20.4)	(−7.2)
Ho I	V	27.63	19.73	−7.90	–	–
Ho IX		27.63	19.56	−8.07	(19.5)	(−8.1)
M101		29.2	18.99	−10.21	$\gtrsim$21.2	$\gtrsim$−8.0
NGC 5474		29.2	20.6	−8.6	–	–
NGC 5585		29.2	20.9	−8.3	–	–
Adopted mean [S14]						−7.72

* Mean of three brightest apparent visual magnitudes at maximum light, uncorrected for internal extinction.

SOURCE: Compiled from [S14].

ment abundances, and this has the effect that the red supergiants attain higher luminosities.

Jay Elias et al. have investigated whether the infrared (2.2 μm) luminosities of M supergiants could be a useful distance indicator [E1]. However, they found that the infrared absolute magnitude ranges from $M(2.2\ \mu\text{m}) = -12$ for the brightest M supergiants in the Milky Way and the Large Magellanic Cloud to $M = -11.4$ in the Small Magellanic Cloud, NGC 6822, and IC 1613 (Table 4.5). They therefore suggested that M_V is a better distance indicator. From the theoretical point of view, a maximum value of the absolute bolometric magnitude M_{bol} would be easier to justify, and the data given by Elias et al. show that the spread in M_{bol} for the brightest M supergiants is less than that for $M(2.2\ \mu\text{m})$. However, it would not be easy to rank observationally the brightest M stars in a galaxy in terms of M_{bol}.

Sandage and Tammann have used the mean magnitude of the three brightest very red stars in a galaxy to determine the distances to the nearby galaxies Wolf-Lundmark, Sextans B, Leo A, and Pegasus, and to NGC 4395, IC 4182, and NGC

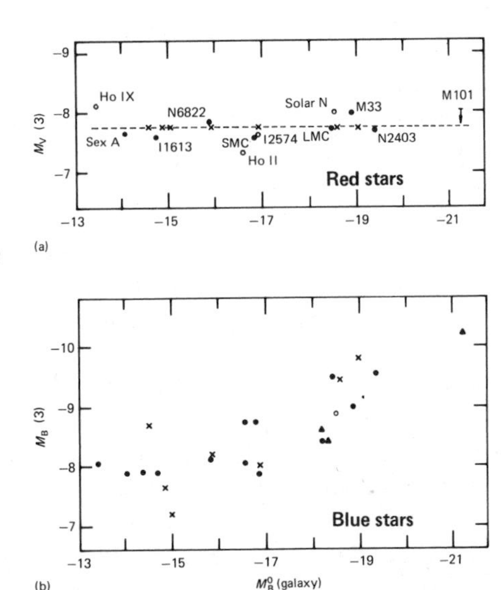

FIGURE 4.5

(a) Calibration of mean absolute visual magnitude of the three brightest red supergiants $M_V(3)$ in galaxies whose distances are known from Cepheids (or by other methods; in the case of the solar neighborhood, denoted by Solar N). ○—less certain values; ×—galaxies from Table 4.4, forced to be on the mean line defined by the calibration. (b) Mean absolute blue magnitude for the first three brightest blue supergiants in galaxies $M_B(3)$. The three triangles are galaxies in the M101 group. Horizontal axis is the absolute blue magnitude of the parent galaxy M_B^0(galaxy), corrected for extinction by dust. (From [S14].)

4214 in the Canes Venatici I cluster [S14]. The significance of the latter two galaxies is that type I supernovae have occurred in them (see Table 3.8), so Sandage and Tammann were then able to calibrate the absolute magnitude of type I supernovae at maximum light t_0, which they estimated to be

$$M_B(t_0) = -19.74 \pm 0.19 \quad \text{(zero-point error)}$$

in good agreement with Eq. (3.15). With the Hubble space telescope they hope to be able to measure the brightest red stars in Virgo cluster galaxies, thereby strengthening this calibration. Supernovae at great distances can then be used to determine the calibration of the Hubble velocity–distance law (see Sec. 6.1). De Vaucouleurs has also used supernovae in galaxies as secondary indicators, the luminosity at maximum light being calibrated from supernovae in galaxies with distances determined from the luminosity, size, and luminosity class of the galaxies (see Secs. 4.6

TABLE 4.5

Visual, infrared, and bolometric absolute magnitudes of brightest red variable stars at maximum light, in Local Group galaxies.

Galaxy	*Assumed* μ_0*[E1]*	*Brightest visual star* M_V	*Brightest infrared star* M_K	*Most luminous star* M_{bol}
Solar neighborhood		−8.2	−12.2	−9.4
LMC	18.6 ± 0.2	−8.1	−11.9	−9.3
SMC	19.0 ± 0.2	−8.3	−11.5	−9.1
NGC 6822	23.2 ± 0.2	−7.9	−11.2	−8.8
IC 1613	24.3 ± 0.1	−8.1	−11.3	−9.0

SOURCE: Compiled from [E1].

and 4.7). For type II supernovae the strong internal extinction seen toward some of these (see Sec. 3.2) means that they cannot be very reliable secondary indicators, except where this extinction can be measured directly. For type I supernovae de Vaucouleurs found that the mean absolute magnitude at maximum light is

$$M_{pg}(t_0) = -18.5 \pm 0.2$$

with no correction for internal obscuration [V4]. This is about 1 mag fainter than the prediction by the kinematic method, Eq. (3.15).

De Vaucouleurs has used the brightest red stars at maximum light in a galaxy to estimate distances to NGC 2403, IC 2574, and Holmberg II in the M81 group (Table 4.6). Corrected for the mean difference between the Local Group distance moduli of de Vaucouleurs and the values adopted in this book (see Table 3.13), they yield a mean distance modulus to the M81 group of $\mu_0 = 27.69$ mag ($d = 3.4$ Mpc), consistent within the uncertainties of the method with the mean modulus we adopt in Sec. 4.9 for this group, $\mu_0 = 27.31$ mag ($d = 2.9$ Mpc).

Humphreys and Steve Strom have found 90 red stellar objects in M101, candidates for being bright red supergiants in the galaxy [H13]. They estimated that the brightest stars in M101 appear at a visual magnitude $V = 20.9 \pm 0.2$. Using Humphreys's estimate that the mean absolute magnitude of the three intrinsically brightest red stars, after correction for internal extinction, is $M_V(3) = -8.0$ [H10], they deduced that the uncorrected visual distance modulus to M101 is $\mu_V = 28.9 \pm 0.3$ mag. They estimated that the total amount of extinction toward the red supergiants in M101 is $A_V = 0.3$, and so arrived at a distance modulus $\mu_0 \simeq 28.6$ mag for M101 ($d = 5.2$ Mpc), which is consistent with the mean modu-

TABLE 4.6

Distance moduli to galaxies using brightest red stars.

Galaxy	A_V	μ_0	*Reference*
Wolf–Lundmark	0.00	25.51	[S14]*
Sex B	0.05	25.61	[S14]*
Leo A	0.00	27.37	[S14]*
Pegasus	0.05	28.16	[S14]*
IC 4182	0.00	28.18	[S14]*
NGC 4395	0.00	28.79	[S14]*
NGC 4214	0.00	28.99	[S14]*
NGC 2403	0.28	27.79	[V4]†
IC 2574	0.23	27.46	[V4]†
Ho II	0.29	27.81	[V4]†
M101	0.3	28.8	[H13]‡

* $M_V(3) = -7.72$; no correction for internal extinction; corrected for mean Local Group difference $\Delta\mu_0 = -0.03$.
† $M_V(1) = -7.6$; no correction for internal extinction; corrected for mean Local Group difference $\Delta\mu_0 = 0.29$.
‡ $M_V(3) = -8.0$; corrected for internal extinction; corrected for mean Local Group difference $\Delta\mu_0 = 0.20$.

lus we adopt in Sec. 4.9, $\mu_0 = 29.16$ mag ($d = 6.8$ Mpc), though on the low side. As M101 is about 10 times more luminous than the most luminous of the Local Group galaxies with which the calibration of $M_V(3)$ was established, this extrapolation to M101 is questionable. The philosophy of Sandage and Tammann, that galaxies of the Local, M81, and M101 groups are needed to establish the constancy of $M_V(3)$, seems preferable. I have not used the Humphreys and Strom distance estimate to M101 from the brightest red stars, nor that by de Vaucouleurs for M81 group galaxies, in determining the distances of these groups.

In their earlier investigation of the brightest stars in galaxies, Sandage and Tammann found [S7] that the mean blue absolute magnitude of the three brightest blue stars in a galaxy $M_B(3)$ is correlated with the absolute photographic magnitude of the galaxy M^0_{pg}(galaxy) as

$$M_B(3) = 0.326\ M^0_{pg}(\text{galaxy}) - 3.09 \tag{4.10}$$

and that the absolute blue magnitude of the brightest blue star in a galaxy $M_B(1)$ satisfies

$$M_B(1) = 0.315\ M^0_{pg}(\text{galaxy}) - 3.48. \tag{4.11}$$

Such correlations can be understood as a statistical effect if all the galaxies have a common luminosity function for these bright blue stars, but the total number of such stars is proportional to the galaxy luminosity [S7]. A luminous galaxy is then assumed to have more bright blue stars, and the very brightest of these stars will therefore tend to have a higher luminosity than those in a less luminous galaxy. Humphreys [H10], however, argued that this correlation may be due to differences in the initial mass function for star formation in different galaxies.

The slopes of Eqs. (4.10) and (4.11) are not as unfavorable for distance determination as was the correlation for HII region diameters, Eq. (4.1), but they are still not really satisfactory. Thus Sandage and Tammann used a correlation with luminosity class L_c of the form

$$M_B(1) = (0.588 \pm 0.216)L_c - 11.16 \pm 0.90 \tag{4.12}$$

with $\sigma(M_{star}) \simeq 0.56$ mag, and

$$M_B(3) = (0.604 \pm 0.202)L_c - 11.01 \pm 0.84 \tag{4.13}$$

with $\sigma(M_{star}) \simeq 0.51$ mag. They used Eq. (4.12) to estimate the distances to NGC 5474 and 5585 in the M101 group [S8]. They also used an extrapolation of Eq. (4.12), which was determined for galaxies with luminosity class L_c in the range of II – III to V, to luminosity class I to estimate a distance to M101. As discussed in Sec. 4.1, the use of L_c as a continuous variable is unjustified, and to extrapolate an empirical relation like Eq. (4.12) outside the range over which it was determined is meaningless. Also Hanes has argued that the correlations given by Eqs. (4.12) and (4.13) are too poor to determine a distance to the M81 group [H4]. However,

Hanes's requirement that $M_B(1)$ and $M_B(3)$ be *correlated* with L_c implies that the latter is again being thought of, incorrectly, as a continuous variable.

Although distances using the brightest blue stars were a major part of the earlier Sandage and Tammann distance scale (and are also used to determine distances to several galaxies by de Vaucouleurs), Sandage and Tammann now appear to have abandoned this distance indicator [S14], and I have not used the method in Table 4.13.

Humphreys has obtained spectra for some bright stars in NGC 2403 and M101, the first spectra for stars outside the Local Group of galaxies [H9]. By comparing these stars with similar objects in our Galaxy, a spectroscopic parallax (see Sec. 2.5) to the galaxies can be estimated. For NGC 2403 the result is $26.9 < \mu_V < 28.2$, while for M101 comparison of the star IVb 59 with Cyg OB2 star 12 in our Galaxy yields $\mu_V \simeq 28.8$ ($\mu_{Hya} = 3.03$). The foreground extinction is estimated to be about $A_V \simeq 0.18$ to NGC 2403 (see Table 3.9) and to be negligible to M101, but there may be some internal extinction in the galaxies. Humphreys's estimates are consistent with the distances for these groups adopted in Table 4.13.

De Vaucouleurs [V6] has proposed the use of the brightest superassociations in spiral and irregular galaxies, of the 30 Doradus type, as a distance indicator and has shown that the resulting distances are reasonably well correlated with his distances based on the luminosity, size, and luminosity class of spirals.

4.4 *The HI 21-cm line width, or velocity dispersion*

In 1977 Brent Tully and Richard Fisher [T7] discovered that there is a correlation between the absolute magnitude of a spiral galaxy and the width of the radio emission line at 21 cm due to neutral atomic hydrogen, a quantity readily observable in external galaxies with large radio telescopes. Tremendous effort has gone into applying this correlation to the determination of galaxy distances in the past few years.

The motions of the atomic hydrogen in the line of sight broaden the observed spectral line due to the Doppler shift. The width of the 21-cm line in wavlength or frequency can be converted to a spread in velocities W_0.

The correlation found by Tully and Fisher took the form

$$M_{pg}^0 = -a \log \left(\frac{W_0}{\sin i} \right) - b \tag{4.14}$$

where M_{pg}^0 is the absolute photographic magnitude corrected for the effects of inclination and extinction, W_0 is the 21-cm line width in kilometers per second, i is the inclination angle between the normal to the plane of the galaxy and the line of sight, $a = 6.25$, and $b = 3.5 \pm 0.3$. They found a distance modulus to the Virgo cluster $\mu_0 = 30.6 \pm 0.2$ mag with this method. More recently Fisher and Tully [F1] have measured the 21-cm line widths for 1171 galaxies.

For $i = 90°$ (line of sight in the plane of the galaxy) W_0 will be an average of the rotation velocity of the galaxy at different locations in the galaxy along the line of sight, an average weighted by the amount of atomic hydrogen at the different locations. We may expect that W_0 will be on the order of $2V_m$, where V_m is the maximum velocity in the galaxy's rotation curve (see Fig. 2.44). For $i = 0°$, on the other hand, there will be no contribution to W_0 from the rotation of the galaxy, and the line will be broadened only by the random motions of the gas in the line of sight. For intermediate values of the inclination angle i we may expect that the observed line width will be reduced below that found at $i = 90°$ by a factor of sin i, hence the use of $W_0/\sin i$ as the independent variable in Eq. (4.14). Tully and Fisher restricted their analysis, from which Eq. (4.14) was derived, to galaxies with inclination angles $i \geq 45°$.

Sandage and Tammann have reanalyzed the Tully–Fisher relation, including in their study some galaxies with $i < 45°$ (M101, for instance, with $i = 18°$, is included), and used their distance moduli for the M81 and M101 groups ($\mu_0 = 27.56$ and 29.30 mag, respectively) to give a calibration [S12]

$$M^0_{pg} = -6.88(\pm 0.3) \log\left(\frac{W_0}{\sin i}\right) - 2.36. \tag{4.15}$$

Their value for the slope, $a = 6.88$, is steeper than that found by Tully and Fisher. This calibration is then used to deduce distance moduli to the M81 and M101 groups of 27.64 ± 0.09 and 29.08 ± 0.16, respectively, and to the Virgo cluster of 31.4 ± 0.2 (based on $\mu_{Hya} = 3.03$). The consistency of the figures for the M81 and M101 groups with those assumed in the calibration is hardly surprising.

L. Bottinelli and co-workers, on the other hand, used galaxies with distances determined mainly from HII regions and brightest stars by de Vaucouleurs [V4]. They attempted to correct the observed line width W_0 for the effects of noncircular and turbulent motions in the galaxy as well as for the effects of inclination. They obtained the very different calibration

$$M^0_B = -5.0(\pm 0.4) \log V_m - 8.40(\pm 0.15) \tag{4.16}$$

where M^0_B is the absolute blue magnitude of the galaxy, corrected for the effects of extinction and inclination, and

$$V_m = \frac{W_0/2 - V_t}{\sin i} \tag{4.17}$$

where V_t is the line-of-sight component of the velocity dispersion due to noncircular and turbulent motions [B8],[B9]. As a result of their different approach, Bottinelli et al. obtained a significantly flatter slope, $a = 5.0$, than Tully and Fisher or Sandage and Tammann. Bottinelli et al. estimated the distances of over 300 galaxies by this method. Bottinelli and L. Gouguenheim [B7] found that the use of galaxies with $i < 30°$ leads to a systematic overestimation of distance, and these have therefore been excluded.

An important correction, which is necessary when galaxies at different inclinations to the line of sight are used, is that due to the effect of internal obscuration in the galaxy being studied. Box 2.6 summarized several proposed forms for the correction for internal extinction. The correction is largest when the galaxy is edge on ($i = 90°$), whereas the correction to the line width for the effect of inclination is largest when i is small. The Holmberg correction used by Sandage and Tammann (see Box 2.6) results in a strong correlation of the absolute magnitude for Virgo galaxies with inclination, galaxies with higher inclination angle having brighter corrected absolute magnitudes. Since the corrected absolute magnitude should be independent of the angle from which the galaxy is viewed, this can be interpreted as evidence that the Holmberg formula overcorrects for internal extinction, and for this reason I have adopted the de Vaucouleurs formula in this book.

Sandage and Tammann differ from Tully and Fisher in a further respect. They corrected the galaxy magnitude for the total internal obscuration, that is, to an absorption-free magnitude, rather than simply to the magnitude the galaxy would have when observed face on, as they did in the other papers in their series on the distance scale (and as all other workers have done). Although they justified this on the grounds that the resulting scatter in the calibration relation is reduced, this seems a dubious procedure, and I do not follow it here.

I have reanalyzed the Sandage and Tammann data with the following modifications:

1 I restricted the sample of galaxies to those with inclinations in the range of $45° \leq i \leq 75°$ to avoid large corrections for internal extinction or to the velocity width.

2 I used the Fisher–Tully interstellar reddening law (see Box 2.5) and the de Vaucouleurs internal-extinction correction to face-on magnitudes (see Box 2.6).

3 I used the Local Group distances adopted in Table 3.13, which were corrected to a revised Hyades distance modulus of 3.30.

The slope of the calibration equation (4.14) was found from the M81 group to be 5.95, and the constant b was then determined from the Local Group galaxies M31 and M33 as $b = 4.52$. Applied to the M81 and M101 groups, this calibration yielded distance moduli of 27.24 and 29.27 mag, respectively. Using the final best-estimate distances for the M81 and M101 groups from Table 4.13, I obtained the following calibration based on a total of 10 galaxies (see Table 4.7 and Fig. 4.6),

$$M^0_{pg} = -5.90(\pm 0.34) \log \left(\frac{W_0}{\sin i} \right) - 4.39(\pm 0.08). \qquad (4.18)$$

Applied to seven Virgo galaxies with $45° \leq i \leq 75°$, I found an average distance modulus of $\mu_0 = 31.01$ mag, with an external root-mean-square error for a single galaxy of 0.50 mag. This agrees well with the root-mean-square scatter about the Tully–Fisher relation at blue wavelengths found by Marc Aaronson and Jeremy Mould [A4].

The slope of Eq. (4.18), $a = 5.90 \pm 0.34$, is now in much better agreement with that of Bottinelli et al., 5.0 ± 0.4. Most of the remaining difference is because

TABLE 4.7

Adopted calibration of Tully–Fisher relation at blue wavelengths.

*Galaxy**	ΔV^i_{21}† (*km s*$^{-1}$)	μ_0‡	A_B§	A_i‖	M^0_{pg} **	*Type*
M31	559	24.25	0.36	0.36	−20.73	Sab
M33	250	24.65	0.28	0.16	−18.91	Scd
M81	512	27.31	0.08	0.21	−20.14	Sab
NGC 2366	128	27.31	0.22	0.33	−16.46	Im
NGC 2403	323	27.31	0.24	0.17	−19.10	Scd
NGC 4236	196	27.31	0.09	0.34	−17.74	Sdm
IC 2574	127	27.31	0.05	0.29	−16.72	Sm
NGC 5204	156	29.16	0	0.18	−17.74	Sm
NGC 5585	218	29.16	0	0.14	−18.06	Sd
Ho IV	109	29.16	0	0.41	−16.52	Sm

* Restricted to galaxies with $45° \leq i \leq 75°$.
† From [S12].
‡ Local Group distance moduli from Table 3.13; M81 and M101 group moduli from Table 4.13.
§ From Table 3.9 for Local Group galaxies and NGC 2403; from Tully–Fisher formula (Box 2.5) for others.
‖ Inclination correction, from Table 3.9 for M31 and M33; from de Vaucouleurs formula (Box 2.6) for others.
** m_{pg} from [S12]

Bottinelli et al. applied an additional correction to the observed velocity width to allow for the random motions in the galaxy [Eq. (4.17)]. This has most effect on galaxies with lower values of $W_0/\sin i$ and so reduces the slope. However, Vera Rubin and co-workers have found much steeper slopes, ranging from 8 to 14, when they concentrated on galaxies of a particular Hubble type [R3],[R4],[R5],[B13].

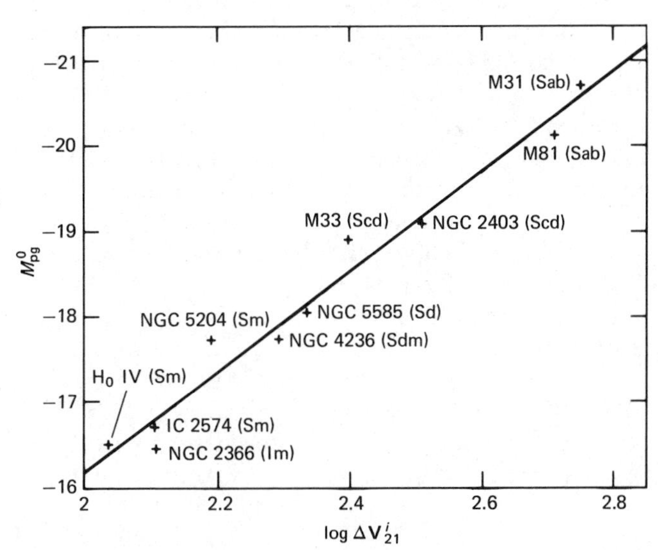

FIGURE 4.6

Revised calibration of optical Tully–Fisher method. Vertical axis is the absolute photographic magnitude of a galaxy M^0_{pg}, *corrected for interstellar and internal extinction; horizontal axis is logarithm of the width of the 21-cm line* ΔV^i_{21} *in kilometers per second, corrected for inclination. Straight line is Eq. (4.18).*

Mort Roberts has cast doubt on the 21-cm line-width method on the grounds that the Tully–Fisher relation seems to be dependent on the galaxy type (Fig. 4.7). The detailed studies by Rubin and co-workers of the optical rotation curves in over 60 spiral galaxies of different Hubble types confirmed this suggestion [R3]. They found that the slope of the Tully–Fisher relation in the blue is similar for galaxies of type Sa, Sb, or Sc (~ 10), but that at a given value of V_m, the maximum rotation velocity, the Sc galaxies are about 2 mag more luminous than the Sa galaxies. A similar dependence of the Tully–Fisher relation in the blue on the Hubble type, but less pronounced, has also been found by Aaronson and Mould [A4]. The lower slopes found above [Eqs. (4.14)–(4.16) and (4.18)] can be understood as arising because these are composite relations based on galaxies of a range of Hubble types.

The great range in corrections proposed for internal obscuration reflects the degree of our ignorance about the strength of this effect. For this reason Aaronson et al. have proposed that the 21-cm line width should be correlated not with the optical absolute magnitude, but with the infrared absolute magnitude at 1.6 μm, $M(1.6\ \mu\text{m})$ [A1]. The correction for internal obscuration at 1.6 μm is so small as to be negligible. Aaronson et al. found that the slope of the Tully–Fisher relation in the infrared is $a = 10.0 \pm 0.8$ using the galaxies of the M81 group. They then found the zero point using the galaxies M31 and M33 in the Local Group, for which they assumed distance moduli of 24.38 and 24.82 mag, respectively:

$$M(1.6\ \mu\text{m}) = -10.0 \log \left(\frac{W_0}{\sin i} \right) + 3.77(\pm 0.18). \tag{4.19}$$

This calibration can now be placed on a much securer basis using the recent study of M33 by Sandage and Carlson (see Sec. 3.1). Previously the distance to M33 was based entirely on observations made by Hubble in the 1930s.

Using their calibration, Eq. (4.19), Aaronson et al. deduced the distances of galaxies in the Sculptor, M81, and M101 groups to obtain mean distance moduli of 27.46 ± 0.27, 27.76 ± 0.20, and 29.16 ± 0.35 mag, respectively. Fig. 4.8 shows their calibrations compared with observations of the galaxies in these groups. With the revised M31 and M33 distance moduli adopted here (see Table 3.13), these distance moduli become 27.31, 27.61, and 29.01 mag, respectively. The data of Aaronson et al. have also been reanalyzed by Sandage and Tammann [S15], using a calibration based on galaxies of the Local, M81, M101, and Sculptor groups. They arrived at a distance modulus for the Virgo cluster of 31.43 ± 0.35 mag, consistent with the best estimate of 31.32 in Table 4.13.

Aaronson and co-workers claimed that there is no dependence of the infrared Tully–Fisher relation on the galaxy type [A1],[A4]. They gave a theoretical justification for Eq. (4.19) based on the fact that if the quantity $\Delta V = W_0/\sin i$ is, as assumed above, a measure of $2V_m$, where V_m is the maximum velocity in the galaxy rotation curve, then we expect the mass of the galaxy M to be on the order of $r(\Delta V)^2/G$, where r is the radius of the galaxy and G is the gravitational constant (see Sec. 2.9). Then if the luminosity-to-mass ratio is roughly the same in all types of galaxies (that is, $L \propto M$, where L is the luminosity of the galaxy), and if galaxies have

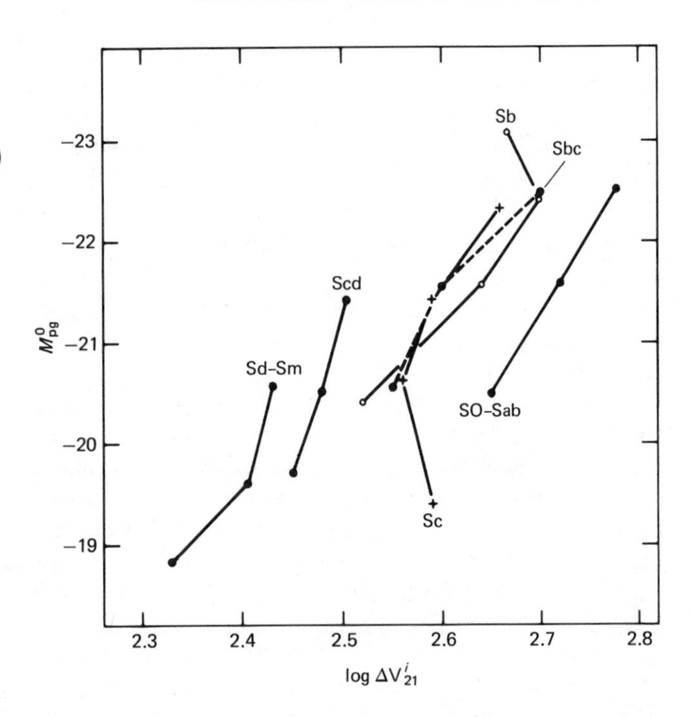

FIGURE 4.7

Dependence of the Tully–Fisher relation on galaxy type. (From [R2].)

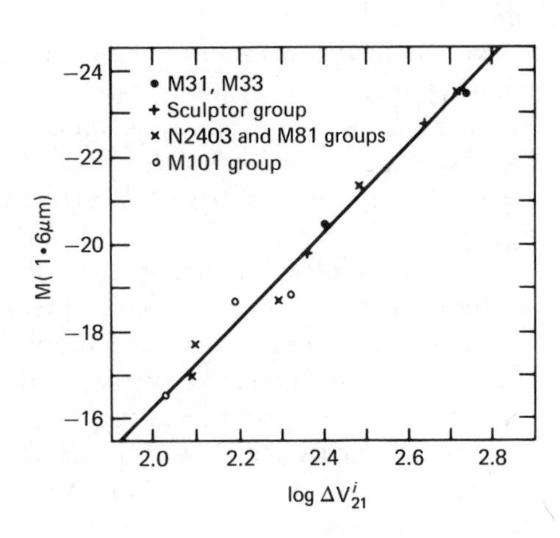

FIGURE 4.8

Calibration of infrared Tully–Fisher method. The absolute magnitude at 1.6 μm referred to a particular isophotal diameter M(1.6 μm) is plotted against the logarithm of the 21-cm velocity width ΔV^i_{21}, corrected for inclination, for 12 nearby galaxies. The zero point is determined by the Sandage and Tammann distances to M31 and M33. Straight line is Eq. (4.19). (From [A1].)

roughly the same surface brightness (that is, $L \propto r^2$), it follows that $L \propto (\Delta V)^4$, or $M_{\rm bol} = -10 \log \Delta V +$ constant. Aaronson et al. argued that it is in the infrared that the old stars responsible for most of the mass in a galaxy radiate, so that here the slope will correspond most closely to the theoretical value. This simple interpretation has been questioned by David Burstein [B12], who has analyzed how the Tully–Fisher relation depends on the galaxy type and other properties.

It is clear that the controversy about how the slope of the Tully–Fisher relation depends on galaxy type and wavelength needs to be resolved. On the whole there is a reasonable consensus that when galaxies of a particular type are used, the slope a is around 10, independent of wavelength. There is still disagreement about how much the intercept b depends on galaxy type and wavelength.

One area in which the infrared method requires special care is the correction of the observed infrared magnitudes to a standard galaxy diameter, to allow for the fact that galaxies at different distances have different angular sizes, whereas observations are usually made through a fixed aperture. At present this correction is done in terms of the galaxy's size at a standard intensity level in the visible band, but it would be desirable for this to be done entirely with infrared observations.

Aaronson and Mould have used Eq. (4.19) to derive the distances to 32 nearby groups and clusters of galaxies, including the Virgo cluster [A4]. Aaronson et al. have also applied the calibration (4.19) to galaxies in the more distant Pegasus I, Cancer, Zwicky 74-23, and Perseus clusters [A2]. Van den Bergh [B6] has pointed out a problem with this work; the mean infrared surface brightness in galaxies differs from cluster to cluster (see Sec. 4.7). The fact that only small numbers of galaxies are used in the solution for the distance means that chance superposition of one or two foreground galaxies can lead to a serious underestimate of the distance. Although the work of Aaronson and co-workers is highly significant in showing that the 21-cm line-width method can be applied to clusters of galaxies at distances greater than 50 Mpc, we have to be cautious about placing too much weight on the results.

To summarize this discussion, the Tully–Fisher relation has tremendous potential for the determination of galaxy distances, but there is evidence at both optical and infrared wavelengths that an additional parameter is required. This could be the Hubble type of the galaxy, its average surface brightness, or perhaps its color, since each of these three quantities is correlated with the other.

De Vaucouleurs and D.W. Olson have argued that a method similar to that of Tully and Fisher may be applied to elliptical and lenticular galaxies [V8]. However, since neutral hydrogen is rarely detected in these galaxies, the velocity dispersion in the central regions determined from optical line profiles has to be used in place of W_0. A plot of this velocity dispersion σ_v against the absolute magnitude M_B^0 for 161 galaxies with distances determined on the de Vaucouleurs scale is shown in Fig. 4.9. For 82 ellipticals the mean relation is

$$M_B^0 = -19.38 \pm 0.07 - (9.0 \pm 0.7)(\log \sigma_v - 2.3) \tag{4.20}$$

while for 71 lenticulars the relation is

$$M_B^0 = -19.65 \pm 0.08 - (8.4 \pm 0.8)(\log \sigma_v - 2.3). \tag{4.21}$$

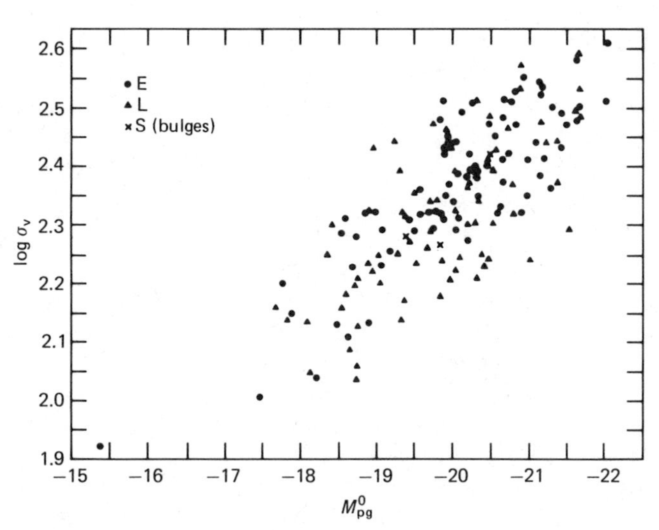

FIGURE 4.9
Correlation of logarithm of the velocity dispersion σ_v in kilometers per second with absolute photographic magnitude M^0_{pg}, corrected for interstellar extinction, in elliptical (E) and lenticular (L) galaxies. Data for the spheroidal "bulges" of 3 spiral galaxies are also plotted. (From [V8].)

In addition de Vaucouleurs and Olson found solutions in which the colors of the galaxies and their surface brightnesses are taken into account. For galaxies in groups, they compared the distances derived from these solutions with those derived by other methods and found reasonably good agreement. They estimated the uncertainty in the distance moduli derived in this way as ± 0.5. Unfortunately the absence of bright elliptical galaxies in the Local or M81 groups means that no calibration via primary indicators is yet available.

4.5 Color–luminosity relation for early-type galaxies

W.A. Baum pointed out in 1959 that there is a correlation between color and luminosity for early-type galaxies (ellipticals and lenticulars), the more luminous galaxies appearing redder, and this can be used to estimate distance. Unfortunately the absence of nearby luminous ellipticals means that the method cannot be calibrated via primary indicators. Sandage used a comparison of U-B versus V diagrams [Fig. 4.10(a)] for early-type galaxies in the Virgo and Coma clusters to try to establish the relative distance of the two clusters and obtained $\Delta\mu_0(\text{Coma}-\text{Virgo}) = 3.66 \pm 0.14$ mag., that is, Coma is 5.4 times further away than Virgo. The importance of this is that while the mean recession velocity of the Virgo cluster is liable to be affected by the motion of the Local Group with respect to Virgo (see Sec. 5.4), this should not be much of a factor at the distance of Coma. N. Visvanathan and Sandage applied this method to determine the relative distances of Virgo and several other clusters, with distances ranging from 0.8 to 6 times that of

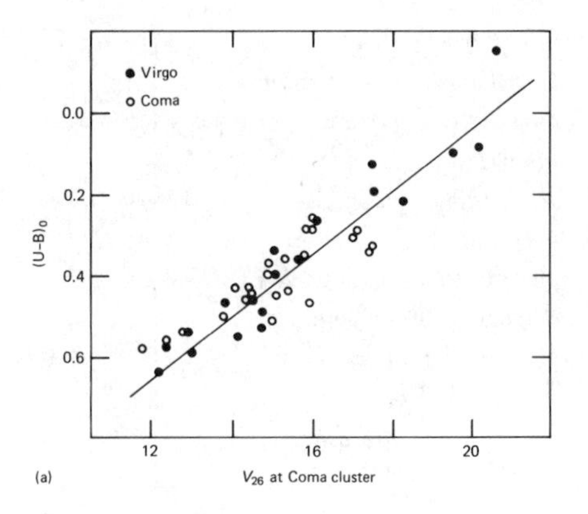

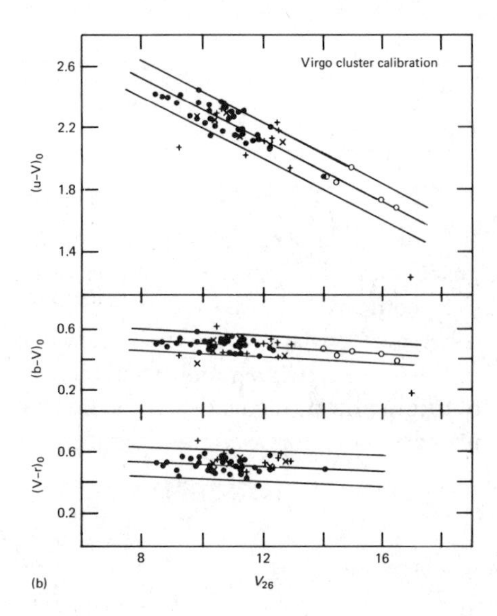

FIGURE 4.10

Color–luminosity relation for ellipticals. (a) U-B color, corrected for interstellar reddening, versus V magnitude, integrated to the 26 mag per square second of arc isophote V_{26}, for galaxies in the Coma and Virgo clusters, the latter having been shifted to the distance of Coma. (From [S1].) (b) u-V, b-V, and V-r colors, corrected for interstellar reddening, versus V_{26} for Virgo galaxies. ●—galaxies within 6° of cluster center; ×—galaxies more than 6° from cluster center; ○—colors transformed to ubV from UBV system; +—not used in calibration. Boundary lines are at $\pm 2\sigma$ of color residuals. (From [V11].)

Virgo. The results are given in Table 4.8. Other determinations of the relative distances of the Virgo and Coma clusters by several methods, including the use of infrared colors, are summarized in Table 4.9. These place Coma at 4.7–7.0 times the distance of Virgo. Internal obscuration is not a serious problem for early-type galaxies, but infrared wavelengths have the advantage that the infrared magnitudes of stars, and hence the integrated magnitudes of galaxies, are not much affected by the heavy-element abundance in the stars.

Several other versions of the color–luminosity relation have been proposed [V9],[V12],[M3],[T8]. Visvanathan and D. Griersmith made a study of groups of galaxies which contain both an early-type galaxy and at least one Sc galaxy. The distances of the Sc galaxies can be derived using the luminosity-class method (see Sec. 4.6), and hence the distance of the groups can be found. The relative distances of early-type galaxies (ellipticals, lenticulars, and early-type spirals) in the groups and

TABLE 4.8

Relative distances of Virgo and other clusters using the color–magnitude relation for E and SO galaxies.

Cluster	*Number of galaxies*	μ_0 *(cluster)* $- \mu_0$ *(Virgo)*	μ_0 *(adopted)*	*Reference*
Virgo	48	0	31.32	Table 4.13
Leo	8	-0.40 ± 0.43	30.92	[V11]
Dorado	8	-0.11 ± 0.30	31.21	[V11]
Fornax	24	0.23 ± 0.20	31.55	[G2]
Pegasus I	11	2.32 ± 0.37	33.64	[M4]
Centaurus	4	2.54 ± 0.36	33.86	[V11]
Hydra I	1	2.87 ± 0.5	34.19	[V11]
Sersic 129-1	11	3.29 ± 0.24	34.61	[G2]
0122 + 33	3	3.47 ± 0.41	34.79	[V11]
Coma	15	3.89 ± 0.24	35.21	[V11]

in the Virgo cluster are estimated using the U-V versus V luminosity–color relation. The distance of the Virgo cluster is therefore determined. With the calibration adopted in Table 4.10, line 2, for the absolute magnitude of an Sc galaxy as a function of luminosity class, I find that their method gives $\mu_0 = 30.8$ mag for the Virgo cluster. This method, though ingenious, seems to introduce several possible sources of error, through both uncertain group membership and the uncertainties in the distances to Sc galaxies.

R. Michard argued that the surface brightness of the galaxies is an additional parameter which needs to be taken into account in using the color–magnitude relation. Although plausible, the case for this does not seem clearcut yet. Finally Tully et al. [T8] have shown that there is a good correlation between the blue-to-infrared color and the infrared (1.65 μm) absolute magnitude for spiral galaxies and used it to determine the distance modulus to the Virgo cluster as 31.04 ± 0.16, consistent with my final best estimate in Table 4.13.

TABLE 4.9

Distance modulus difference between Virgo and Coma clusters.

Method	μ_0*(Coma)* $- \mu_0$*(Virgo)*	*Reference*
Brightest cluster member	3.37 ± 0.40	[S5]
Nuclear magnitude of 10 brightest cluster members	4.22 ± 0.28	[W1]
Color–luminosity relation for E and SO galaxies	3.89 ± 0.20	[V11]
Same, using U-K colors	3.66 ± 0.35	[T4], [P2]
Type I supernovae	3.89 ± 0.21	[T1]

SOURCE: Compiled from [T4].

4.6 *Luminosity class of spirals*

The idea that spiral galaxies could be classified into luminosity classes by the appearance of their spiral arms (see Sec. 2.10) was introduced by Sidney van den Bergh in 1960 [B3]. He also calculated the average absolute magnitude for each class, assuming that distance is proportional to recession velocity. The luminosity class of spiral galaxies played a large part in the distance scale put forward by Sandage and Tammann during the early 1970s. They believed that there was little variation in absolute magnitude for late-type spiral galaxies of a particular luminosity class, and hence that the luminosity class L_c could be used to map the distance scale out to distances of 300 Mpc [S11]. They calibrated the average magnitude as a function of luminosity class $M_{pg}(L_c)$, using galaxies whose distances had been determined through their brightest HII region diameters [S9]. Some problems with these distances were discussed in Sec. 4.1. A further difficulty was that the calibration of luminosity classes I to II depended entirely on the Sc I galaxy M101, whose distance Sandage and Tammann did not determine in a self-consistent manner. Bottinelli and Gouguenheim pointed out that the distances determined by Sandage and Tammann for galaxies in groups with lower values of L_c were greater on the average than the mean distance for galaxies in the same group [B7]. Part of this effect is due to systematic errors in Sandage and Tammann's HII region diameters (errors identified by de Vaucouleurs and Kennicutt, see Sec. 4.1), but a residual effect remains for Sc I and II galaxies, even when Kennicutt's more accurate HII region observations are used. Bottinelli and Gouguenheim suggested that M101 may be a foreground galaxy not associated with the rest of the M101 group, but we saw in Sec. 3.5 that this is not a very plausible suggestion.

Table 4.10 summarizes the calibration of $M_{pg}(L_c)$ by different authors and gives my calibrations based on (1) Local Group galaxies, (2) Local, M81, and M101 groups, and (3) Local, M81, and M101 groups and the Virgo cluster. A comparison of (2) and (3) shows that with my adopted distances M101 does not appear to be an exceptionally luminous object.

Van den Bergh [B5] has cautioned against reliance on L_c as a distance indicator, since an identical galaxy at different distances from us may not be given the same luminosity classification. The subjective nature of the classification makes it unlikely, a priori, that it could be a high-precision luminosity indicator. This is amply confirmed by more recent work. Tammann et al. studied the luminosity functions for galaxies of different Hubble types and luminosity classes and found a huge spread in M_B for each luminosity class [$\sigma(M_B) \gtrsim 1$ mag, see Fig. 4.11]. Sandage and Tammann estimated L_c for all the spiral galaxies in the revised Shapley–Ames (RSA) catalog and found that the standard deviation in absolute magnitude of a given luminosity class, $\sigma(M_B(L_c))$, is greater than 1 mag [S13],[T4]. Earlier Fisher and Tully [F1] had found this to be so for galaxies of luminosity classes IV–V and V. Kennicutt, however, has found that the standard deviation in absolute magnitude for galaxies in a particular luminosity class is reduced if the recession velocities, from which he estimates distances, are corrected for the effects of infall toward the Virgo

TABLE 4.10

Calibration of absolute photographic magnitudes of galaxies as a function of luminosity class $M_{pg}(L_c)$

	Luminosity class								
	I	*I–II*	*II*	*II–III*	*III*	*III–IV*	*IV*	*IV–V*	*V*
L_c	1	1.5	2	2.5	3	3.5	4	4.5	5
1 Calibration based on Local Group, N2403 ($\mu_0 = 27.24$), and M101 ($\mu_0 = 29.30$)									
Calibrating galaxies	M101	–	–	M33	N2403	LMC	–	SMC N6822	IC1613
$\langle -M^0_{pg} \rangle$	21.12	–	–	18.90	18.85	17.89	(16.94)	15.99	14.45
2 Calibration based on Local Group, M81 group ($\mu_0 = 27.31$), and M101 group ($\mu_0 = 29.16$)									
Calibrating galaxies	M101	–	–	M33	N2403	LMC	N4236 N5204 N5474 N5585	SMC N6822 N2366 IC2574 Ho II N5477 Ho IV	IC1613 Ho I
$\langle -M^0_{pg} \rangle$	20.98	–	–	18.90	18.92	17.89	17.86	16.13	14.33
3 As in 2, plus Virgo cluster ($\mu_0 = 31.32$)									
Number of calibrating galaxies	5	1	6	7	16	1	4	7	2
$\langle -M^0_{pg} \rangle$	**21.13**	**20.38**	**19.96**	**19.26**	**18.85**	**17.89**	**17.86**	**16.13**	**14.33**
Other calibrations									
Sandage and Tammann [S9]	21.25	20.74	20.23	19.72	19.21	18.70	17.79	16.33	14.46
Bottinelli and Gouguenheim [B7]	19.97	19.75	19.35	19.60	18.43	18.25			
Kennicutt [K2]	21.12	–	19.90	19.20	18.87	18.06	17.81	16.34	14.46
van den Bergh [B5]	–	–	–	18.24	18.56	18.27	17.54	15.78	14.24
Mould et al. [M5]*									
Maximum	21.21	–	20.10	19.46	18.99	18.30	17.88	16.52	14.72
Minimum	20.72	–	19.47	19.21	18.63	17.94	17.76	16.49	14.72

* $\mu_{Hya} = 3.29$; others $\mu_0 = 3.03$.

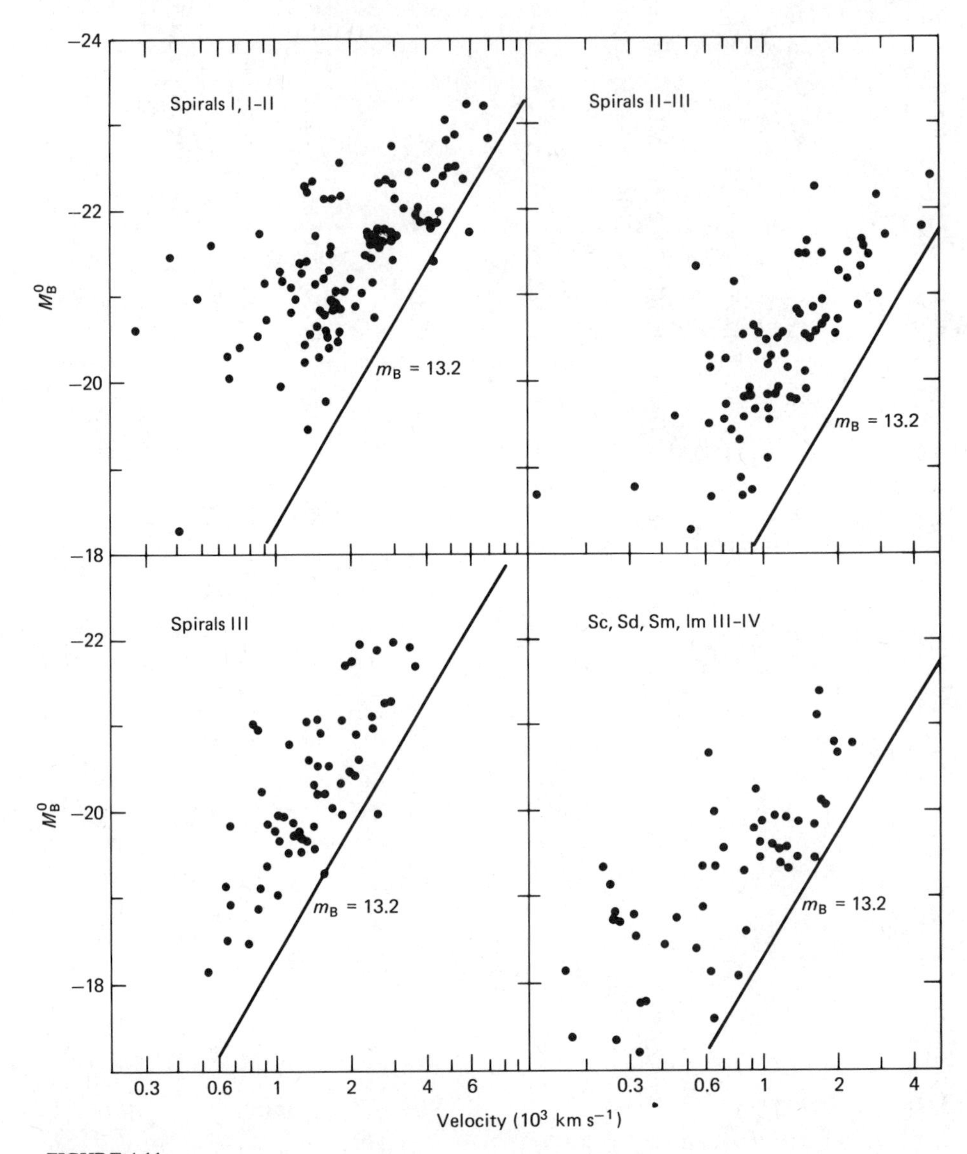

FIGURE 4.11

Variation of absolute blue magnitude M_B^0, corrected for interstellar and internal extinction, with recession velocity for spirals of luminosity classes I to III–IV in revised Shapley–Ames catalog at galactic latitude $|b| > 30°$ (excluding the Fornax cluster and the central 6° of the Virgo region). A limit line at $m_B = 13.2$ is shown. Note the very large spread in M_B^0 for each luminosity class. (From [T5].)

cluster (see Sec. 5.4), but only to $\sigma \sim 1$ mag. He has also tried to quantify the factors in spiral arm morphology which define the luminosity class, but found that these are only loosely correlated with the absolute magnitude of the galaxy, at the ± 1 mag level [K4],[K5]. This large standard deviation in the absolute magnitude of galaxies of a particular luminosity class, found by Sandage and Tammann, Kennicutt, and others, means that a magnitude-limited sample of galaxies would be subject to a bias known as the Malmquist effect (App. A.17), whereby at greater distances only outstandingly luminous galaxies are selected. Great care therefore has to be taken in the selection of galaxies for distance studies.

De Vaucouleurs [V5] has defined a slightly different parameter, the luminosity index, which is a linear combination of de Vaucouleurs's galaxy-type parameter T (see Sec. 2.10) and the luminosity class, corrected for the effects of inclination,

$$\Lambda_c = \frac{(T + l_c)}{10} \tag{4.22}$$

where T ranges from 2 to 10 for spirals and irregulars, and l_c ranges from 1 for luminosity class I to 9 for luminosity class V. De Vaucouleurs found a good correlation between corrected absolute magnitude M_B^0 and Λ_c for the 25 nearby galaxies whose distances he had obtained from primary and secondary indicators (Fig. 4.12). The correlation seems no worse than that using L_c alone, but in using this correlation, de Vaucouleurs was making the implicit assumption that an earlier-type galaxy with a larger value of L_c has the same luminosity as a later-type galaxy with a lower value of L_c. This has not been demonstrated, and there are no

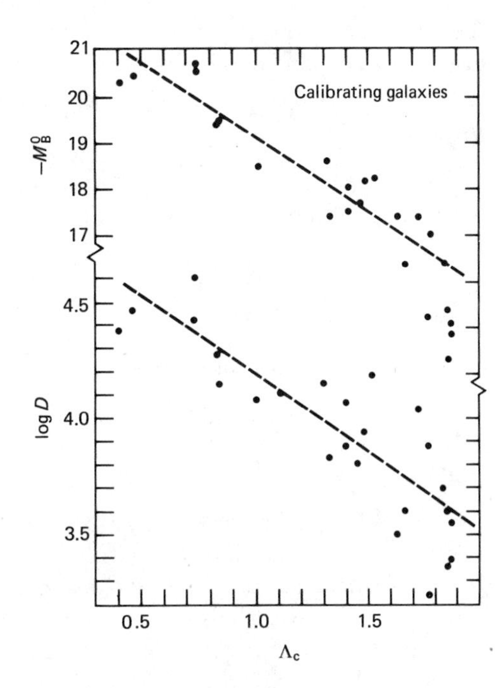

FIGURE 4.12

De Vaucouleurs' calibration of corrected absolute magnitude M_B^0 *and the logarithm of the linear diameter* D, *in parsecs, versus de Vaucouleurs luminosity index* Λ_c *for 25 calibrating galaxies. (From [V5].)*

theoretical grounds for believing that it should be true. The luminosity index must therefore be considered a rough, though useful, indicator of luminosity (and hence distance). In favor of Λ_c is the fact that it is partly based on the Hubble type, a far more objective and theoretically comprehensible parameter than L_c. Tammann and Sandage have investigated the standard deviation in absolute magnitude for spiral galaxies in Virgo with a particular value of Λ_c [T6]. They found $\sigma \sim 1.0$ mag, in agreement with the above estimates for $\sigma(M_B(L_c))$. As with L_c, I believe that it is incorrect to treat Λ_c as a continuous variable, as de Vaucouleurs does.

It should be noted that in estimating distances to nearby galaxies outside the Local Group, de Vaucouleurs did not stick to his chosen secondary indicators, but included ad hoc arguments based on star counts, on the four brightest and four largest galaxies in a group, and on the diameter of individual galaxies [V4]. The inclusion of the low galactic latitude galaxies NGC 1569 and 6946, and IC 10 and 342, for which the interstellar extinction correction is large and uncertain, must also reduce the reliability of his adopted tertiary calibrators. In using de Vaucouleurs's distance estimates I have, however, made no attempt to correct for these effects, but have corrected only for the mean difference between the Local Group distance moduli adopted in this book and those used by de Vaucouleurs, $\Delta\mu_0 = 0.29$ (see Table 3.13).

4.7 *Sizes of galaxies*

Galaxies do not have sharply defined edges, so any definition of the size of a galaxy is bound up with the light distribution across the galaxy. The characteristic distribution of intensity I with angular distance from the center of the galaxy, along the major axis θ, is shown in Fig. 4.13 for two galaxies. In the disks of spirals an exponential intensity distribution

$$I(\theta) = I_0 \exp\left(\frac{-\theta}{\theta_0}\right) \tag{4.23}$$

is usually a good fit, while for ellipticals either the Hubble intensity distribution

$$I(\theta) = I_0 \left[1 + \left(\frac{\theta}{\theta_0}\right)^2\right]^{-1} \tag{4.24}$$

or the de Vaucouleurs intensity distribution

$$I(\theta) = I_0 \exp\left[-\left(\frac{\theta}{\theta_0}\right)^{1/4}\right] \tag{4.25}$$

is in general a good overall fit. In each case θ_0 is a characteristic angular size of the galaxy. If the distance d of the galaxy is known, this can be converted to a linear ("metric") diameter D, since $D = \theta_0 d$. A model for the intensity profile in ellipticals due to Ivan King is also illustrated in Fig. 4.13.

In practice, when the size of a galaxy is measured on a photographic plate, we are measuring the size and shape of the galaxy down to some particular intensity level

FIGURE 4.13

Intensity profile of giant E1 galaxy NGC 3379 (dots and circles) compared with de Vaucouleurs law [Eq. (4.25)]. (From [V7].) (b) Intensity profile of giant E2 galaxy NGC 4472 compared with a model by King. (From [K7].)

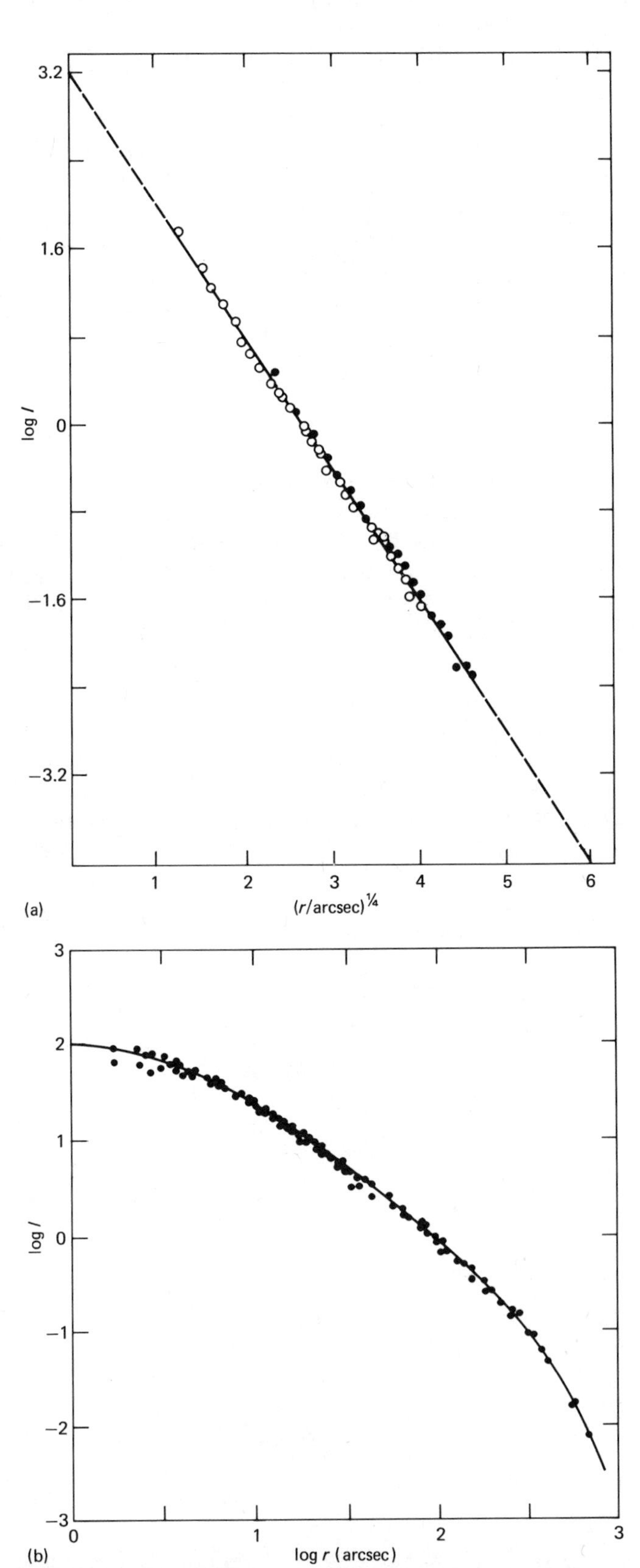

$I_{\rm lim}$, determined by the length of exposure, the sensitivity of the emulsion or detector, and the foreground intensity of the night sky (atmospheric emission + zodiacal light + light from Milky Way stars). Contours of constant intensity are called isophotes, so the size measured in this way is called the *isophotal diameter.* The measured size is usually corrected to a standard intensity level, for example, that of one 25th-magnitude star per square second of arc.

Erik Holmberg found in 1969 that the linear diameter D of different classes of galaxies (E-SO, Sa-b-c, Irr I) is correlated with the absolute magnitude of the galaxy $M_{\rm pg}$ by (Fig. 4.14)

$$\log D = a - b\, M_{\rm pg} \tag{4.26}$$

where a and b are constants. Measurements of the angular size and the apparent magnitude of a galaxy could therefore be used to determine its distance. The method has been used by G. Paturel, for example [P1]. However, Tammann and Sandage have pointed out that for galaxies of a fixed Hubble type, the surface brightness is almost independent of absolute magnitude, so this is not a very effective distance indicator [T6].

FIGURE 4.14

Correlation between absolute photographic magnitude M^0_{pg} and the logarithm of the intrinsic major diameter D, in parsecs, for galaxies of type E-SO (○), types Sa-Sb-Sc (●), and type Irr (×). The straight line has the form of eqn (4.26), with $a = 1.19$, $b = 0.167$. (From [H7].)

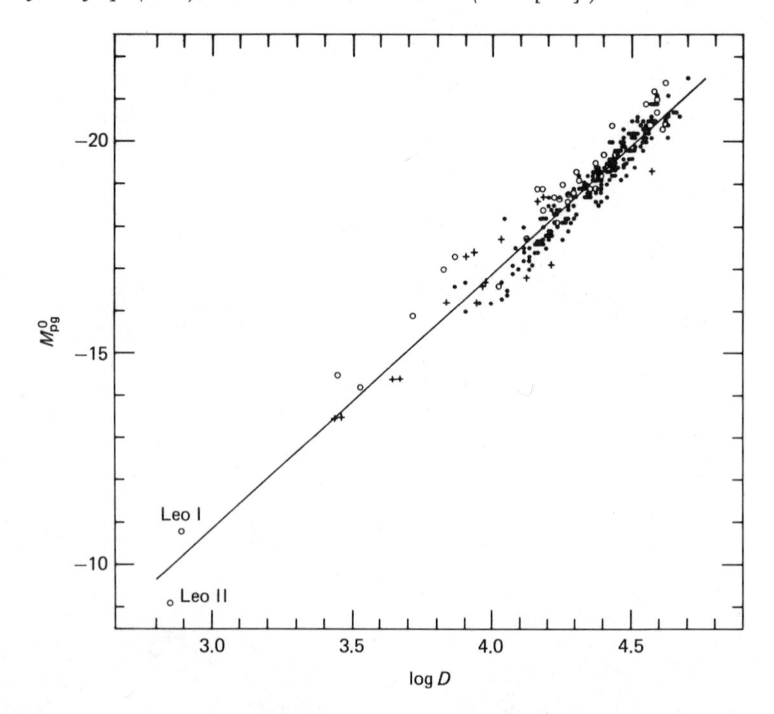

De Vaucouleurs has shown that there is a correlation between the linear isophotal diameter and his luminosity index Λ_c defined above (see Fig. 4.12), and has used this calibration to estimate distances to several hundred galaxies [V6]. He also defined a weighted mean of the distances estimated from the angular size and from the absolute magnitude for these galaxies. His average distances for groups and clusters of galaxies, corrected for the mean local group distance modulus difference $\Delta\mu_0 = 0.29$ (see Table 3.13), have been used in Table 4.13.

More recently R. Buta and de Vaucouleurs have proposed that the diameter of ring structures seen in spiral and lenticular galaxies can be used as a distance indicator, and that this is free of many of the problems associated with galaxy luminosities and isophotal diameters [B14]. They have applied this method to several hundred galaxies at distances of up to 100 Mpc. However, the validity of this distance indicator has been disputed by J. Kormendy [K8].

Van den Bergh [B6] has discussed the correlation of a galaxy's infrared (1.6 μm) surface brightness, measured by

$$\Sigma = m(1.6\ \mu\text{m}) + 5 \log D \tag{4.27}$$

where $m(1.6\ \mu\text{m})$ is the 1.6-μm infrared magnitude and D is the linear diameter, with absolute infrared magnitude. He found, using the data of Mould et al. [M5] for the Virgo cluster

$$\Sigma = 17.31 + (0.47 \pm 0.06)[M(1.6\ \mu\text{m}) + 20.60] \tag{4.28}$$

where $M(1.6\ \mu\text{m})$ is the 1.6-μm infrared absolute magnitude of the galaxy. The observations by Aaronson et al. [A2] of the Cancer and Pegasus clusters agree with this relation, but those of two other clusters (Zwicky 74-23 and Perseus) do not. If the differences between these clusters can be understood, the infrared surface brightness may have a considerable potential as a distance indicator.

4.8 Brightest cluster galaxies

As we look out to very large distances, further than 100 Mpc, say, few of the methods of measuring distances that we have discussed so far have contributed much to our knowledge of these remote regions. However, rich clusters of galaxies similar to Virgo and Coma can be recognized out to distances of several thousand Mpc, and if we assume that the properties of these clusters are similar, then the luminosity and size of the brightest galaxies in the cluster (or even the size of the whole cluster, see Sec. 6.2) can be used to estimate distance.

Sandage has pioneered the use of the first-ranked (most luminous) galaxy in rich clusters and found that the absolute magnitude of these galaxies has a surprisingly small dispersion from cluster to cluster [S3], assuming that relative distances can be deduced from the Hubble red-shift – distance law (see Sec. 6.1). Allowance has to be made for the fact that for galaxies at different distances, a fixed telescope aperture

TABLE 4.11

Calibration of absolute magnitudes of brightest galaxies in clusters.

Cluster	*Brightest galaxy*	μ_0 (*Table 4.13*)	W/Σ^{2}*	A_V†	$V_c^{BM,R}$‡	$M_V^{(1)}$§	V_0‖
Grus	IC 1459	31.18	7.95	0.06	9.49	−21.69	1581
Fornax	N1316	31.21	13.82	0.07	8.52	−22.69	1527
Dorado	N1553	31.21	3.9	0.11	8.50	−22.71	950
Virgo	N4472	31.32	34.40	0	8.23	−23.09	1019
Leo	N3368	31.50	16.98	0.02	8.26	−23.24	926
Cancer	N2563	33.33	5.85	0.19	11.62	−21.71	4758
Pegasus I	N7619	33.37	13.53	0.10	11.44	−21.93	3836
Centaurus	N4696	33.86	3.20	0.35	10.25	−23.61	3170
0122-33	N507	34.79	2.86	0.21	11.64	−23.15	5128
Coma	N4489	35.27	11.41	0	11.63	−23.64	6660
Weighted mean absolute magnitude						**−22.82**	
Standard deviation for single cluster						0.61	
(Zero-point) error in the mean						0.21	

* Weights are explained in text and in Table 4.12.
† Interstellar extinction, from Fisher–Tully formula (see Box 2.5).
‡ Visual magnitude of brightest galaxy in cluster, corrected for the Bautz–Morgan class and richness effects; From [S4].
§ Absolute visual magnitude of brightest galaxy in cluster, corrected for interstellar extinction.
‖ Recession velocity of cluster, corrected for the sun's motion around the Galaxy.

will be collecting light from areas with different linear extent in different galaxies. The measured brightness is therefore corrected to a standard linear size (the aperture correction [S2]).

Clusters have been classified by L. Bautz and W.W. Morgan according to the degree of concentration of the cluster [B2]. Sandage and E. Hardy [S5] found a slight correlation of the absolute magnitude of the first-ranked galaxy with the Bautz–Morgan class, for which they made a correction. There is also a correlation with richness, defined by the number of galaxies in the cluster within 2 mag of the third brightest galaxy, which are contained within a region of 1-Mpc diameter. The method can be calibrated via the Virgo, Coma, and other clusters, which can be reached by several of the methods discussed in this chapter (Table 4.11).

The small spread in absolute magnitude of the first-ranked galaxies in clusters $M^{(1)}$ can probably be understood as a statistical effect. The distribution of absolute magnitudes of the galaxies in a cluster, the galaxy luminosity function, is similar in different clusters. Fig. 4.15 shows the form of this luminosity function, based on a composite for several clusters. It is the steepness of this function at the bright end which leads to the small dispersion in $M^{(1)}$.

Dan Weedman [W1] has proposed a slightly different distance indicator, the average magnitude of the nucleus of the 10 brightest galaxies in a cluster. Sandage [S2] has argued that the isophotal diameter of the brightest galaxy in a cluster can be used as a distance indicator to great distances. Neither of these proposals appears to have attracted much further study.

In using the brightest galaxies in clusters as a distance indicator, two important effects have to be allowed for. First, as we look out to very great distances, we begin to look back in time, back through a significant fraction of the lifetime of the galaxies. We may therefore expect that the galaxies will look different when observed at great distances from how they look today, simply as a result of the evolution of the stellar populations of the galaxies. Second, dynamical studies of the gravitational interaction of galaxies in a cluster show that there is a tendency for large galaxies to gradually swallow up neighboring small ones. This is happening, for example, with our Galaxy and the Magellanic Clouds. The latter are losing material through tidal interaction, in the Magellanic Stream (see Sec. 3.6), and are slowly spiraling in toward our Galaxy. This galaxy "cannibalism" may be a very important factor in the evolution of first-ranked galaxies in clusters. These two effects are discussed further in Sec. 6.2.

FIGURE 4.15

Observed composite cluster galaxy luminosity distribution compared with simple analytical model. Horizontal axis is the absolute magnitude M *of a cluster galaxy; vertical axis is the number of galaxies found in each bin of absolute magnitude. Filled circles show the effect of including giant galaxies (Morgan class cD) in the composite. (From [S16].)*

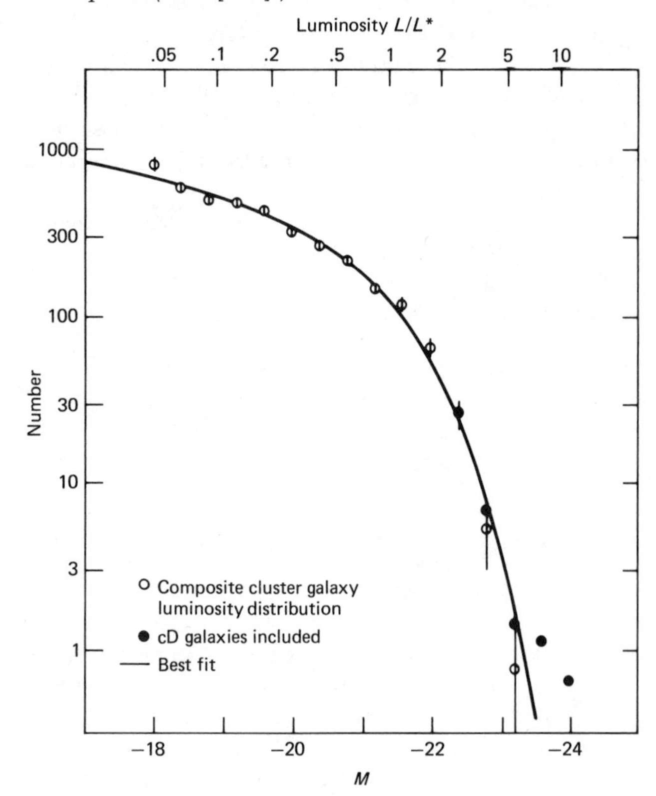

4.9 Distances to galaxies and groups of galaxies by secondary methods

In combining distance estimates by different methods to a single galaxy or to galaxies in groups and clusters, we need to weight the methods according to their reliability. This is to some extent a subjective matter, and it is important that the weighting used be made explicit. (It often takes the form that the weight is infinity for the observer's own work and zero for everyone else's!) In this book I adopt the philosophy advocated by van den Bergh in his review at the 1977 IAU Colloquium on "The Red Shift and the Expansion of the Universe", namely, that due weight should be given to all reliable methods. Given the strong possibility of systematic errors in any one method, it is inadvisable to place the whole weight of the distance ladder on a single method.

The methods of the present chapter are based on empirical correlations between observable properties of galaxies. Only a limited theoretical explanation can be provided for these correlations. The pattern of recent years has been that the more detail in which a method has been studied, and the more galaxies it has been studied in, the less precise it has been found to be as an indicator. The HII region and luminosity class methods are good examples of this. At present the accuracy of the methods based on the brightest red stars in galaxies and on the width of the 21-cm line of atomic hydrogen (the Tully–Fisher method) is claimed to be high. But my personal opinion is that the future of the distance scale lies with primary indicators whose use is justified by strong theoretical arguments, for example, Cepheids and supernovae.

The weights I use for combining distance estimates of galaxies in groups or clusters are W/Σ^2 (see Sec. 3.5 and App. A.17), where

$W = 2$ for methods with a clear theoretical basis

$W = 1$ for methods based on empirical correlations between galaxy properties

$\Sigma^2 = \sigma_{zp}^2 + \sigma^2/n$

σ_{zp} = total zero-point error in the calibration

σ = standard deviation about the mean calibration for a distance estimate to a single galaxy

n = number of galaxies used.

My estimates of W, σ_{zp}, and σ for each distance method are summarized in Table 4.12, with some comments on the sources for the estimates of σ_{zp} and σ. The results of applying the distance methods of Chaps. 3 and 4 to groups and clusters of galaxies are summarized in Table 4.13. Pairs of galaxies, and groups of galaxies for whose mean distances the total weight W/Σ^2 is less than 2.5, have been excluded.

TABLE 4.12

Adopted system of weights used in averaging distances by different methods.

The weight adopted for each method (see text) is W/Σ^2, where $\Sigma^2 = \sigma_{zp}^2 + \sigma^2/n$. ($\sigma_{zp}$—total zero-point error in calibration; σ—standard deviation about mean calibration for a distance estimate to a single galaxy; n—number of galaxies used; W—additional weight factor, which equals 1 or 2, depending on whether or not the method has a quantitative theoretical basis.)

Method	Error	Contributing error to σ	Contributing error to σ_{zp}	σ^2	σ_{zp}^2	Additional weight factor W
Cepheids	Intrinsic					
	LMC, SMC	0.05[a]	0.1[b]	0.031	0.010	2
	M31, N300	0.1[a]	0.1[b]	0.039		
	N2403	0.2[c]	0.1[b]	0.069		
	Error in A_B	0.12				
	Error in A_i	0.12				
RR Lyrae	Intrinsic, LMC, SMC	0.05[d]	0.2[e]	0.017	0.040	2
	Error in A_B	0.12				
Novae	Intrinsic					
	LMC, SMC	0.2[f]	0.2[g]	0.054	0.040	2
	M31	0.1[f]	0.2[g]	0.024		
	Error in A_B	0.12				
Type II supernovae (with model)	Intrinsic	0.3[h]	0.3[h]	0.119	0.090	2
	Error in A_B	0.12				
	Error in A_i	max(0.12, $A_i/2$)				
Type I supernovae	Intrinsic	0.57[i]	0.57[i]	0.354	0.324	2
	Error in A_B	0.12				
	Error in A_i	0.12				

[a] Based on estimates for a single Cepheid of $\sigma = 0.26$ [B4], 0.25 [V2], and 0.14 [M1].
[b] Compromise between estimates of 0.2 ([V2], period–luminosity method) and 0.07 [M1].
[c] From data in [T3].
[d] Based on estimates for a single RR Lyrae variable of $\sigma = 0.3$ [B4],[V2].
[e] From [V2].
[f] Based on estimates for a single nova of $\sigma = 0.4$ [B4] and 0.3–0.5 [V2].
[g] From [V2].
[h] From [S17].
[i] From [B10]. Tammann [T2] estimates $\sigma = 0.43$ for the best observed type I supernovae in ellipticals and lenticulars, for which internal extinction should be negligible.

Table 4.12 continues next page

Table 4.12 continued

Method	Error	Contributing error to σ	Contributing error to σ_{zp}	σ^2	σ^2_{zp}	Additional weight factor W
Tully–Fisher 21-cm line width						
1 Optical	Intrinsic	0.51[j]	0.39[k]	0.289	0.192	1
	Local Group error		0.20[l]			
	Error in A_B	0.12				
	Error in A_i	0.12				
2 Infrared	Intrinsic	0.45[j]	0.35[k]	0.203	0.163	1
	Local Group error		0.20[l]			
Globular clusters	Intrinsic	0.67[m]	0.47[m]	0.463	0.261	1
	RR Lyrae calibration	0.20				
	Error in A_B	0.12				
$\langle D(\Lambda_c), M_B(\Lambda_c)\rangle$	Intrinsic	0.7[n]	0.7[n]	0.518	0.580	1
	Error in A_B	0.12				
	Error in A_i	0.12				
	Calibrating galaxies		0.3[l]			
HII region diameters	Intrinsic	1.0[o]	1.0[o]	1.0	1.04	1
	Local Group error		0.2[l]			
HII region fluxes	Intrinsic	1.0[o]	1.0[p]	1.028	1.04	1
	Local Group error		0.2[l]			

[j] From [A4]. Somewhat larger values for σ have been found recently by Sidney van den Bergh (1984, Quarterly Journal R. astr. Soc. 25, pp. 137–146).
[k] For application to M81 and M101 groups [$n = 2$, $m = 10$ in Eq. (27) in App. A.17]. For application to more distant groups the intrinsic contribution to σ_{zp} is 0.23 for optical method and 0.20 for infrared method ($n = m = 10$), and the contribution of the uncertainty in the distances of the calibrating galaxies is 0.3; so $\sigma^2_{zp} = 0.143$ for the optical method and 0.130 for the infrared method.
[l] From Table 6.3.
[m] Assuming that Hanes's uncertainty estimate of ± 0.3 means $\sigma/\sqrt{5}$ [H3]. The corresponding zero-point error is then $\sigma/\sqrt{2}$, since the calibration is based on two galaxies, our own and M31.
[n] Assuming $\sigma \simeq \sigma_{zp} \simeq 1.0$ for each method separately (see Sec. 4.6).
[o] See Sec. 4.1.
[p] For application to M81, M101 groups, and Virgo cluster. For calibration 3 of Table 4.10 and 5 of Table 4.1, $\sigma_{zp} = 0.5$.

Table 4.12 continues next page

Table 4.12 continued

Method	Error	Contributing error to σ	Contributing error to σ_{zp}	σ^2	σ^2_{zp}	Additional weight factor W
$M_{pg}(L_c)$	Intrinsic	1.0[q]	1.0[p]	1.028	1.04	1
	Local Group error		0.2[l]			
	Error in A_B	0.12				
	Error in A_i	0.12				
Brightest red stars	Intrinsic	0.3[r]	0.2[s]	0.119	0.08	1
	Local Group error		0.2[l]			
	Error in A_B	0.12				
	Error in A_i	0.12				
Color–magnitude for E and SO galaxies	Intrinsic					
	u-V	0.66[t]	0.2[t]	0.448	0.20	1
	u-K	0.69	0.3	0.488	0.25	1
	Virgo cluster error		0.4[l]			
	Error in A_B	0.12				
Color–magnitude for Es + $M_{pg}(L_c)$ for Sc	Intrinsic	1.0[u]	1.0[u]	1.028	1.04	1
	Local Group error	0.2[l]				
	Error in A_B	0.12				
	Error in A_i	0.12				
Brightest cluster galaxies	Intrinsic	0.61[v]	0.210[v]	0.386	0.29	1
	Calibrating galaxies		0.5			
	Error in A_V	0.09				

[q] See Sec. 4.6.
[r] De Vaucouleurs [V3] gives $\sigma = 0.2$, Humphreys [H10] gives $\sigma = 0.16$, and Sandage and Tammann [S14] give $\sigma = 0.17$, but the latter is increased to 0.24 if three galaxies of the M81 group are included, while for three galaxies in the Canes Venatici I cluster $\sigma \approx 0.4$.
[s] Humphreys [H10] gives $\sigma_{zp} = 0.07$ and Sandage and Tammann [S14] give $\sigma_{zp} = 0.06$, but the estimate $\sigma_{zp} = 0.2$ of de Vaucouleurs [V3] seems more realistic, given that the dependence of $M_V(3)$ on the absolute magnitude of the galaxy has to be demonstrated from the same data as the calibration.
[t] From [A3].
[u] Adopted; see Sec. 4.6.
[v] From Table 4.11.

TABLE 4.13

Weighted mean distance moduli to groups and clusters of galaxies.

For each method the distance modulus, the values of σ^2 and σ^2_{zp} (from Table 4.12), the number of galaxies n from which the distance modulus was determined, and the corresponding values of $\Sigma^2(=\sigma^2_{zp}+\sigma^2/n)$ and W/Σ^2 are given. For each group the weighted mean distance modulus (see App. A.17), the total weight $\Sigma W/\Sigma^2$, and the internal and external errors of the weighted mean σ_{int} and σ_{ext} are given.

Group	*Method*	μ_0	σ^2	σ^2_{zp}	n	Σ^2	W/Σ^2
M81	Cepheids	27.24	0.069	0.010	1	0.079	25.3
	21 cm, infrared	27.61	0.203	0.163	5	0.204	4.9
	21 cm, optical	27.74	0.289	0.192	5	0.250	4.0
	HII diameters	26.59	1.0	1.04	5	1.24	0.8
	HII fluxes	26.46	1.0	1.04	5	1.24	0.8
	$M_{pg}(L_c)$	26.88	1.03	1.04	5	1.25	0.8
	Weighted mean $\mu_0 =$	**27.31;**	$\Sigma W/\Sigma^2 = 36.6$; $\sigma_{int} = 0.22$; $\sigma_{ext} = 0.12$				
M101	Type II supernova (M101)	29.30	0.119	0.090	1	0.209	9.6
	21 cm, infrared	29.01	0.203	0.163	3	0.231	4.3
	21 cm, optical	29.27	0.289	0.192	3	0.268	3.7
	HII diameters	28.55	1.0	1.04	5	1.24	0.8
	HII fluxes	28.92	1.0	1.04	5	1.24	0.8
	$M_{pg}(L_c)$	28.70	1.03	1.04	5	1.25	0.8
	Weighted mean $\mu_0 =$	**29.16;**	$\Sigma W/\Sigma^2 = 20.0$; $\sigma_{int} = 0.27$; $\sigma_{ext} = 0.09$				
Virgo	21 cm, infrared	30.93	0.203	0.130	16	0.143	7.0
	21 cm, optical	31.03	0.289	0.143	7	0.184	5.4
	Type I supernovae	31.73	0.354	0.324	7	0.375	5.4
	Type II supernova (M100)	31.80	0.119	0.090	1	0.209	9.6
	HII fluxes	31.50	1.0	1.04	17	1.10	0.9
	$M_{pg}(L_c)$	31.24	1.03	1.04	25	1.08	0.9
	$\langle D(\Lambda_c), M_B(\Lambda_c)\rangle$	30.82	0.518	0.580	6	0.666	1.5
	Globular clusters	30.80	0.463	0.261	5	0.354	2.8
	Color–magnitude $+ M_{pg}(L_c)$	30.80	1.03	1.04	20	1.09	0.9
	Weighted mean $\mu_0 =$	**31.32;**	$\Sigma W/\Sigma^2 = 34.4$; $\sigma_{int} = 0.20$; $\sigma_{ext} = 0.15$				
Sculptor (S. polar)	Cepheids	26.15	0.039	0.010	1	0.049	20.4
	21 cm, infrared	27.49	0.203	0.163	3	0.231	4.3
	$M_{pg}(L_c)$	27.34	1.03	0.29	3	0.63	1.6
	Weighted mean $\mu_0 =$	**26.44;**	$\Sigma W/\Sigma^2 = 26.3$; $\sigma_{int} = 0.26$; $\sigma_{ext} = 0.38$				

Group	*Method*	μ_0	σ^2	σ^2_{zp}	n	Σ^2	W/Σ^2
N1023	21 cm, infrared	29.83	0.203	0.130	5	0.171	5.8
	HII diameters	29.84	1.0	1.04	1	2.04	0.49
	HII fluxes	29.68	1.0	0.29	1	1.29	0.78
	$M_{pg}(L_c)$	29.20	1.03	0.29	2	0.80	1.25
	Type I supernova (N1003)	31.72	0.354	0.324	1	0.68	2.94
	Type II supernova (N1058)	30.70	0.119	0.090	1	0.21	9.6
	$\langle D(\Lambda_c), M_B(\Lambda_c)\rangle$	29.82	0.518	0.580	2	0.84	1.2
	Weighted mean $\mu_0 =$	**30.42;**	$\Sigma W/\Sigma^2 = 22.1$; $\sigma_{int} = 0.27$; $\sigma_{ext} = 0.28$				
N1068	HII fluxes	31.48	1.0	0.29	1	1.29	0.78
	$M_{pg}(L_c)$	30.85	1.03	0.29	3	0.63	1.59
	Type I supernova (N1084)	33.13	0.354	0.324	1	0.68	2.94
	$\langle D(\Lambda_c), M_B(\Lambda_c)\rangle$	31.12	0.518	0.580	3	.75	1.33
	Weighted mean $\mu_0 =$	**31.99;**	$\Sigma W/\Sigma^2 = 6.6$; $\sigma_{int} = 0.47$; $\sigma_{ext} = 0.60$				
Eridanus	21 cm, infrared	30.97	0.203	0.130	8	0.155	6.54
	HII diameters	31.81	1.0	1.04	1	2.04	0.49
	HII fluxes	31.70	1.0	0.29	1	1.29	0.78
	$M_{pg}(L_c)$	31.45	1.03	0.29	5	0.50	2.02
	$\langle D(\Lambda_c), M_B(\Lambda_c)\rangle$	31.15	0.518	0.580	4	0.71	1.41
	Weighted mean $\mu_0 =$	**31.17;**	$\Sigma W/\Sigma^2 = 11.1$; $\sigma_{int} = 0.30$; $\sigma_{ext} = 0.14$				
N2841	21 cm, infrared	30.81	0.203	0.130	3	0.198	5.05
	HII diameters	29.32	1.0	1.04	1	2.04	0.49
	HII fluxes	29.88	1.0	0.29	1	1.29	0.78
	$M_{pg}(L_c)$	29.25	1.03	0.29	3	0.63	1.59
	$\langle D(\Lambda_c), M_B(\Lambda_c)\rangle$	30.24	0.518	0.580	4	0.71	1.41
	Weighted mean $\mu_0 =$	**30.30;**	$\Sigma W/\Sigma^2 = 9.3$; $\sigma_{int} = 0.33$; $\sigma_{ext} = 0.31$				
N3184	21 cm, infrared	30.50	0.203	0.130	3	0.198	5.05
	HII fluxes	30.68	1.0	0.29	1	1.29	0.78
	$M_{pg}(L_c)$	30.72	1.03	0.29	3	0.63	1.59
	$\langle D(\Lambda_c), M_B(\Lambda_c)\rangle$	30.28	0.518	0.580	3	0.75	1.33
	Weighted mean $\mu_0 =$	**30.52;**	$\Sigma W/\Sigma^2 = 8.7$; $\sigma_{int} = 0.34$; $\sigma_{ext} = 0.08$				
Ursa Major I(N) (N3631)	HII diameters	31.89	1.0	1.04	1	2.04	0.49
	HII fluxes	32.03	1.0	0.29	1	1.29	0.78
	$M_{pg}(L_c)$	32.00	1.03	0.29	1	1.32	0.76
(entry continues)							

Table 4.13 continues next page

Table 4.13 continued

Group	Method	μ_0	σ^2	σ^2_{zp}	n	Σ^2	W/Σ^2
	$\langle D(\Lambda_c), M_B(\Lambda_c)\rangle$	31.45	0.518	0.580	4	0.71	1.41
	Type I supernovae (N3992, N3913)	32.68	0.354	0.324	2	0.50	4.00
	Weighted mean μ_0 = **32.26;**	$\Sigma W/\Sigma^2 = 7.4$; $\sigma_{int} = 0.46$; $\sigma_{ext} = 0.25$					
Leo	21 cm, infrared	31.17	0.230	0.130	5	0.171	5.85
	HII diameters	31.60	1.0	1.04	1	2.04	0.49
	HII fluxes	31.73	1.0	0.29	1	1.29	0.78
	$M_{pg}(L_c)$	31.75	1.03	0.29	4	0.55	1.82
	Color–magnitude for Es	30.92	0.448	0.20	8	0.256	3.90
	$\langle D(\Lambda_c), M_B(\Lambda_c)\rangle$	31.81	0.518	0.580	2	0.84	1.20
	Type I supernova (N3389)	32.54	0.354	0.324	1	0.68	2.94
	Weighted mean μ_0 = **31.50;**	$\Sigma W/\Sigma^2 = 17.0$; $\sigma_{int} = 0.26$; $\sigma_{ext} = 0.23$					
N3938	HII diameters	32.04	1.0	1.04	1	2.04	0.49
	HII fluxes	32.15	1.0	0.29	1	1.29	0.78
	$M_{pg}(L_c)$	31.93	1.03	0.29	2	0.80	1.25
	Type I supernova (N3938)	33.04	0.354	0.324	1	0.68	2.94
	$\langle D(\Lambda_c), M_B(\Lambda_c)\rangle$	31.23	0.518	0.580	4	0.71	1.41
	Weighted mean μ_0 = **32.29;**	$\Sigma W/\Sigma^2 = 6.9$; $\sigma_{int} = 0.46$; $\sigma_{ext} = 0.35$					
Canes Venatici I cloud	21 cm, infrared	28.30	0.203	0.130	3	0.198	5.05
	HII diameters	27.89	1.0	1.04	3	1.37	0.72
	HII fluxes	27.82	1.0	0.29	3	0.62	1.61
	$M_{pg}(L_c)$	28.59	1.03	0.29	5	0.50	2.02
	Type I supernovae (N4214, I4182)	28.81	0.354	0.324	2	0.50	4.00
	$\langle D(\Lambda_c), M_B(\Lambda_c)\rangle$	28.74	0.518	0.580	6	0.67	1.50
	Brightest red stars	28.65	0.119	0.08	3	0.12	8.33
	Weighted mean μ_0 = **28.54;**	$\Sigma W/\Sigma^2 = 23.2$; $\sigma_{int} = 0.23$; $\sigma_{ext} = 0.12$					
Coma I cloud	21 cm, infrared	30.27	0.203	0.130	6	0.164	6.10
	$M_{pg}(L_c)$	30.02	1.03	0.29	2	0.80	1.25
	Type I supernova (N4414)	31.77	0.354	0.324	1	0.68	2.94
	$\langle D(\Lambda_c), M_B(\Lambda_c)\rangle$	29.0	0.518	0.580	6	0.67	1.50
	Weighted mean μ_0 = **30.57;**	$\Sigma W/\Sigma^2 = 11.8$; $\sigma_{int} = 0.33$; $\sigma_{ext} = 0.40$					
Canes Venatici II cloud	$M_{pg}(L_c)$	29.32	1.03	0.29	4	0.55	1.82
	$\langle D(\Lambda_c), M_B(\Lambda_c)\rangle$	29.99	0.518	0.580	7	0.65	1.54
	Weighted mean μ_0 = **29.63;**	$\Sigma W/\Sigma^2 = 3.4$; $\sigma_{int} = 0.55$; $\sigma_{ext} = 0.33$					

Table 4.13 continues next page

Table 4.13 continued

Group	Method	μ_0	σ^2	σ^2_{zp}	n	Σ^2	W/Σ^2
N5128	HII diameters	28.76	1.0	1.04	1	2.04	0.49
	HII fluxes	29.50	1.0	0.29	1	1.29	0.78
	$M_{pg}(L_c)$	28.50	1.03	0.29	2	0.80	1.25
	Type I supernova (N5253)	27.39	0.354	0.324	1	0.68	2.94
	$\langle D(\Lambda_c), M_B(\Lambda_c)\rangle$	28.71	0.518	0.580	2	0.84	1.20
	Weighted mean $\mu_0 =$	**28.18;**	$\Sigma W/\Sigma^2 = 6.7$; $\sigma_{int} = 0.47$; $\sigma_{ext} = 0.38$				
N6643	$M_{pg}(L_c)$	31.85	1.03	0.29	3	0.63	1.59
	$\langle D(\Lambda_c), M_B(\Lambda_c)\rangle$	31.83	0.518	0.580	3	0.75	1.33
	Weighted mean $\mu_0 =$	**31.84;**	$\Sigma W/\Sigma^2 = 2.9$; $\sigma_{int} = 0.59$; $\sigma_{ext} = 0.01$				
Virgo X	21 cm, infrared	30.73	0.203	0.130	7	0.159	6.29
	Type I supernovae (N4636, N4496)	31.67	0.354	0.324	2	0.50	4.00
	$\langle D(\Lambda_c), M_B(\Lambda_c)\rangle$	30.83	0.518	0.580	2	0.84	1.20
	Weighted mean $\mu_0 =$	**31.07;**	$\Sigma W/\Sigma^2 = 11.5$; $\sigma_{int} = 0.36$; $\sigma_{ext} = 0.31$				
Virgo III	21 cm, infrared	31.56	0.203	0.130	9	0.153	6.54
	Type I supernova (N5668)	31.97	0.354	0.324	1	0.68	2.94
	$\langle D(\Lambda_c), M_B(\Lambda_c)\rangle$	31.39	0.518	0.580	2	0.84	1.20
	Weighted mean $\mu_0 =$	**31.65;**	$\Sigma W/\Sigma^2 = 10.6$; $\sigma_{int} = 0.35$; $\sigma_{ext} = 0.14$				
N2207	Type I supernova (N2207)	33.35	0.354	0.324	1	0.68	2.94
	$\langle D(\Lambda_c), M_B(\Lambda_c)\rangle$	31.22	0.518	0.580	2	0.84	1.20
	Weighted mean $\mu_0 =$	**32.73;**	$\Sigma W/\Sigma^2 = 4.1$; $\sigma_{int} = 0.66$; $\sigma_{ext} = 0.97$				
Pegasus I	Color–magnitude for Es	33.64	0.448	0.20	11	0.24	4.13
	21 cm, infrared excl. N7537, N7541	33.16	0.203	0.130	8	0.155	6.48
	Type I supernova (N7634)	33.47	0.354	0.324	1	0.68	2.94
	Weighted mean $\mu_0 =$	**33.37;**	$\Sigma W/\Sigma^2 = 13.51$; $\sigma_{int} = 0.30$; $\sigma_{ext} = 0.15$				
Coma	Color–magnitude for Es	35.21	0.448	0.20	15	0.23	4.35
	U-K color–magnitude for Es	34.98	0.488	0.25	13	0.287	3.48
	Type I supernovae (A1304+28, A1255+28)	35.59	0.354	0.324	2	0.50	4.00
	Weighted mean $\mu_0 =$	**35.27;**	$\Sigma W/\Sigma^2 = 11.8$; $\sigma_{int} = 0$; $\sigma_{ext} = 0$				

Table 4.13 continues next page

Table 4.13 continued

Group	Method	μ_0	σ^2	σ^2_{zp}	n	Σ^2	W/Σ^2
Cancer	21 cm, infrared (excl. U4334, U4354)	33.33	0.203	0.130	4	0.171	5.85
Dorado	Color–magnitude for Es	31.21	0.448	0.20	8	0.256	3.90
Fornax	21 cm, infrared	30.73	0.203	0.130	7	0.159	6.29
	Color–magnitude for Es	31.55	0.448	0.20	24	0.218	4.59
	Type I supernova in N1316	31.72	0.354	0.324	1	0.68	2.94
	Weighted mean $\mu_0 =$ **31.21;**	$\Sigma W/\Sigma^2 = 13.8$; $\sigma_{int} = 0.30$; $\sigma_{ext} = 0.32$					
Centaurus	Color–magnitude for Es	33.86	0.448	0.20	4	0.312	3.2
0122+33	Color–magnitude for Es	34.79	0.448	0.20	3	0.35	2.86
Sersic 129-1	Color–magnitude for Es	34.61	0.448	0.20	11	0.241	4.15
N134	21 cm, infrared	30.85	0.203	0.130	2	0.231	4.33
N2336	21 cm, infrared	32.09	0.203	0.130	3	0.198	5.05
Leo triplet	21 cm, infrared	29.69	0.203	0.130	3	0.198	5.05
N3521	21 cm, infrared	30.79	0.203	0.130	3	0.198	5.05
M51 group	21 cm, infrared	29.54	0.203	0.130	2	0.231	4.33
	HII diameters	30.36	1.0	1.04	1	2.04	0.49
	HII fluxes	30.58	1.0	0.29	1	1.29	0.78
	$M_{pg}(L_c)$	29.89	1.03	0.29	1	1.32	0.76
	Weighted mean $\mu_0 =$ **29.77;**	$\Sigma W/\Sigma^2 = 6.4$; $\sigma_{int} = 0.40$; $\sigma_{ext} = 0.22$					
N5371	21 cm, infrared	32.14	0.203	0.130	3	0.198	5.05
N5364	21 cm, infrared	30.95	0.203	0.130	2	0.231	4.33
N5676	21 cm, infrared	32.45	0.203	0.130	3	0.198	5.05
N5866	21 cm, infrared	30.33	0.203	0.130	2	0.231	4.33
	$\langle D(\Lambda_c), M_B(\Lambda_c)\rangle$	31.63	0.518	0.580	3	0.75	1.33
	Weighted mean $\mu_0 =$ **30.64;**	$\Sigma W/\Sigma^2 = 5.7$; $\sigma_{int} = 0.42$; $\sigma_{ext} = 0.55$					
N6070	21 cm, infrared	32.17	0.203	0.130	3	0.198	5.05
Grus	21 cm, infrared	31.05	0.203	0.130	8	0.155	6.45
	$\langle D(\Lambda_c), M_B(\Lambda_c)\rangle$	31.73	0.518	0.580	6	0.666	1.50
	Weighted mean $\mu_0 =$ **31.18;**	$\Sigma W/\Sigma^2 = 7.95$; $\sigma_{int} = 0.36$; $\sigma_{ext} = 0.27$					

To test whether systematic errors are present in these distance estimates, we can compare the internal error σ_{int} for the weighted mean distance moduli of the groups in Table 4.13, that is, the weighted mean error of all the methods used, with the external error σ_{ext}, the weighted estimate of the standard deviation of the different distance modulus estimates from the mean (see App. A.17 for a detailed definition of these two types of error). The ratios of the two types of error $\sigma_{\mathrm{ext}}/\sigma_{\mathrm{int}}$ differ from unity by no more than a few multiples of the expected standard deviation of this ratio [which is $(2(m-1))^{-1/2}$, where m is the number of independent distance estimates] and so there is no clear evidence for systematic errors.

The methods which contribute the most weight to distance estimates beyond 20 Mpc are the type I supernova method and the infrared Tully–Fisher method. It can be seen that the type I supernova method gives a distance to Virgo higher than most other methods and that the infrared Tully–Fisher method gives a distance on the low side. The same trend was found in most clusters where these methods have been applied, so that there is a possibility that there is a systematic error in either or both of these methods [see Fig. 4.16(a) and (b)]. The discrepancy between the methods is highlighted in Fig. 4.16(c), in which distance estimates by the two methods are compared directly. The mean difference in the moduli is 1.03 mag, suggesting that either method may be in error by, on the average, ~1 mag, or that both methods may be in error (in opposite senses) by ~0.5 mag. Recent work by Visvanathan, using the Tully-Fisher method at wavelengths of 0.55, 0.67 and 1.05 μm, appears to confirm that the distance moduli determined by Aaronson et al are underestimated by about 0.5 mag [V10].

This possibility of systematic error means that the calculated uncertainty of ± 0.18 mag for the Virgo cluster distance modulus given in Table 4.13 is almost certainly an underestimate, and in Table 4.12 I have assigned an uncertainty of ± 0.4 mag to this modulus for methods which depend on the Virgo distance. Table 4.14, adapted from the 1981 review by Paul Hodge [H6], is a summary of recent estimates of the distance modulus to the Virgo cluster, including some methods which I did not include in the weighted means given in Table 4.13, due to doubts about their reliability.

The mean distance moduli for groups and clusters of galaxies adopted in Table 4.13 are compared in Table 4.15 with those given by Sandage and Tammann and by de Vaucouleurs. A straight mean of the differences shows that my distance scale is 0.10 mag (5%) shorter than that of Tammann and Sandage. (If their distances were corrected to $\mu_{\mathrm{Hya}} = 3.30$ and for the effect of line blanketing, this difference would by 0.23 mag, or 11%.) It is 0.33 mag (16%) longer than that of de Vaucouleurs. It appears that these differences arise mainly from disagreement about the distances of Local Group galaxies, for which my mean distance moduli were 0.03 mag less than those of Tammann and Sandage and 0.29 mag greater than those of de Vaucouleurs (Table 3.13). Fig. 4.17 shows a comparison of my weighted mean distances to groups and clusters of galaxies with those of de Vaucouleurs and of Sandage and Tammann, after having corrected the latter for the mean differences between the Local Group distance moduli I have adopted and those used by de Vaucouleurs and by Sandage and Tammann. It can be seen that apart from the disagreement about

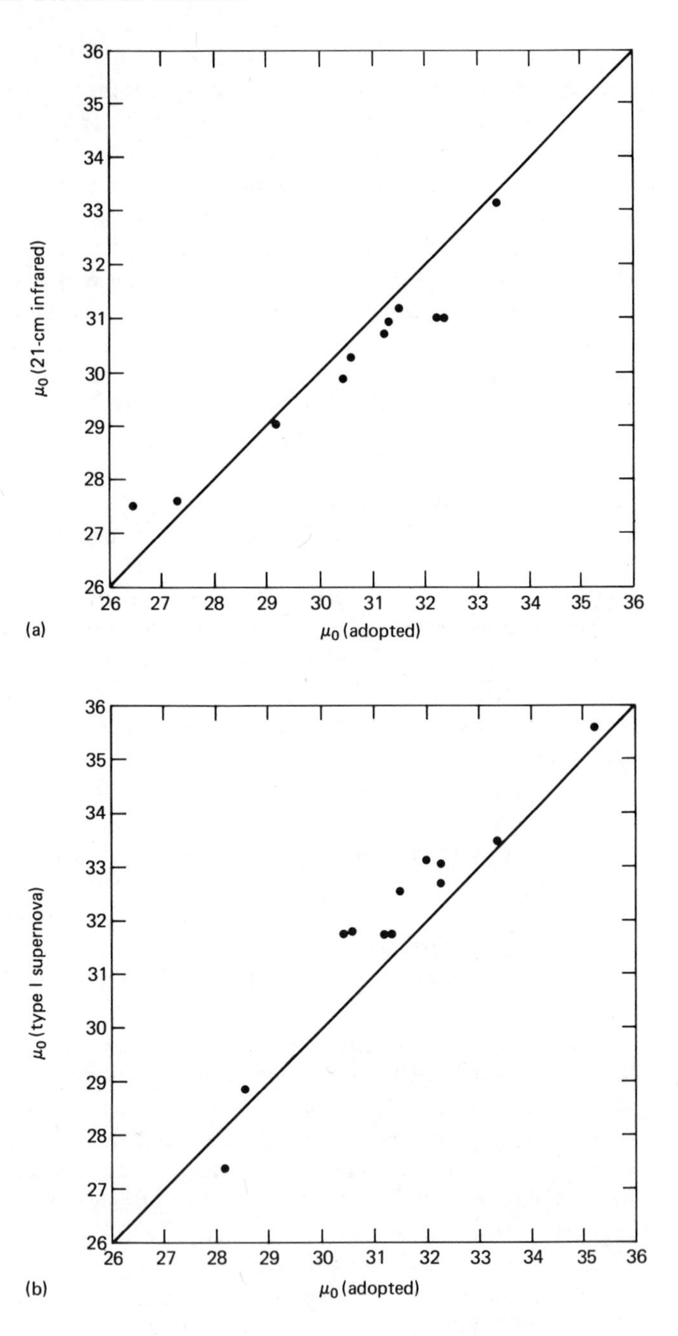

36
35
34
33
32
31
30
29
28
27
26
μ_0 (21-cm infrared)
26 27 28 29 30 31 32 33 34 35 36
μ_0 (adopted)
(a)
μ_0 (type I supernova)
μ_0 (adopted)
(b)

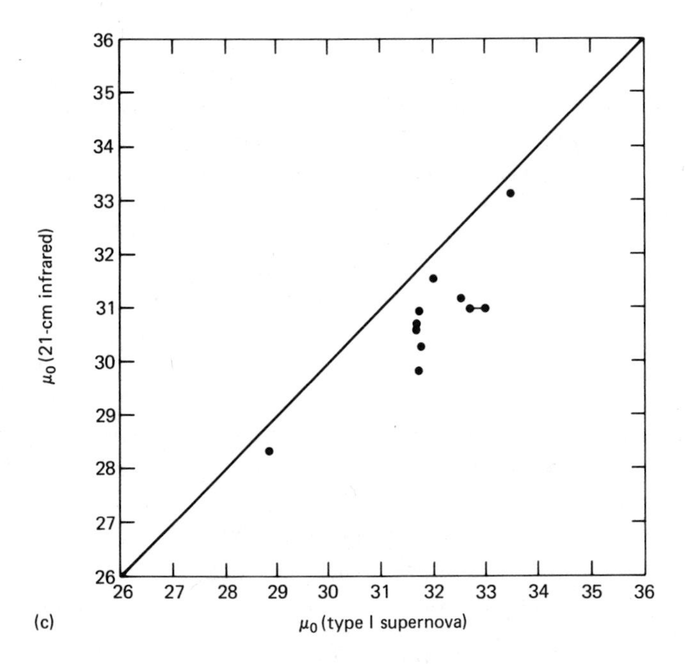

FIGURE 4.16

(a) Comparison of distance moduli for groups and clusters derived by infrared Tully–Fisher method with weighted mean distance moduli adopted in Table 4.13. (b) Similar comparison of distance moduli derived by type I supernova method with distance moduli adopted in Table 4.13. (c) Comparison of distance moduli for groups and clusters derived by infrared Tully–Fisher method with those derived by type I supernova method. On the average the type I supernova distance moduli are 1.0 mag greater than the infrared Tully–Fisher moduli (corresponding to distances 20% greater).

Local Group distances, my distance scale is consistent with both that of de Vaucouleurs and that of Sandage and Tammann.

The controversy over the distance scale has arisen from the following:

1 de Vaucouleurs's probably excessive correction for interstellar extinction and underestimate of Cepheid luminosities, leading to rather low Local Group distances

2 Sandage and Tammann's probable underestimation of interstellar and internal extinction, systematic errors in their use of HII region diameters, and a lack of self-consistency in their treatment of the M101 group, resulting in distances on the high side

3 a general tendency to overestimate the accuracy of extragalactic distance indicators

The secondary and tertiary methods on which the distance scales of de Vaucouleurs and Sandage and Tammann are based (HII regions, brightest stars, luminosity classes) do not carry a very high weight in my weighted mean distances to groups and clusters of galaxies, summarized in Table 4.13.

The controversy has had a very beneficial effect, though, in stimulating interest in the distance scale and inspiring many younger astronomers to develop new distance methods. In the next few years I expect to see the disagreement between the infrared Tully–Fisher and supernova methods resolved and the controversy about interstellar and internal dust extinction settled, principally by new observations.

TABLE 4.14

Comparison of recent estimates of the distance modulus to the Virgo cluster.

Method	μ_0	*Authors*
Resolution difference	31.56 ± 0.50	Sandage and Tammann (1976) [S12]
	31.15 ± 0.50	Mould et al. (1980) [M5]
Luminosity class	31.70 ± 0.11	Sandage and Tammann (1976) [S12]
	30.91–31.35	Kennicutt (1979) [K2], Mould et al. (1980) [M5]
Luminosity index	31.25 ± 0.14	De Vaucouleurs (1979) [V6]
HII region size	31.99 ± 0.30	Sandage and Tammann (1976) [S12]
	30.93–31.47	Kennicutt (1979) [K1], Mould et al. (1980) [M3]
HII region luminosities	31.4 ± 0.25	Kennicutt (1981) [K3]
Globular clusters	31.45 ± 0.50	Sandage and Tammann (1976) [S12]
	31.70 ± 0.30	Hanes (1979) [H3]
	30.90 ± 0.30	Harris and Racine (1979) [H5]
Velocity ratio	31.75 ± 0.42	Sandage and Tammann (1976) [S12]
	31.17 ± 0.35	Mould et al. (1980) [M5]
Color–magnitude diagram	30.73 ± 0.39	Visvanathan (1978) [V9]
Color–magnitude–L_c method	31.52 ± 0.16	Visvanathan and Griersmith (1979) [V12]
Supernovae		
Type I, compared with M101	31.81 ± 0.85	Sandage and Tammann (1976) [S12]
Type I, Baade–Wesselink	32.91 ± 0.80	Sandage and Tammann (1976) [S12], Branch and Pachett (1973) [B11]
Type I, Baade–Wesselink	31.73 ± 0.62	Branch (1979) [B10], Mould et al. (1980) [M5]
Type II	31.35 ± 0.80	Sandage and Tammann (1976) [S12]
	30.53 ± 0.80	Carney (1980) [C1]
Magnitude–surface brightness relation	31.04 ± 0.56	Holmberg (1969) [H7], Mould et al. (1980) [M5]
21-cm infrared method	30.98 ± 0.09	Mould et al. (1980) [M5]

SOURCE: Compiled from [H6], assuming $\mu_{Hya} = 3.29$.

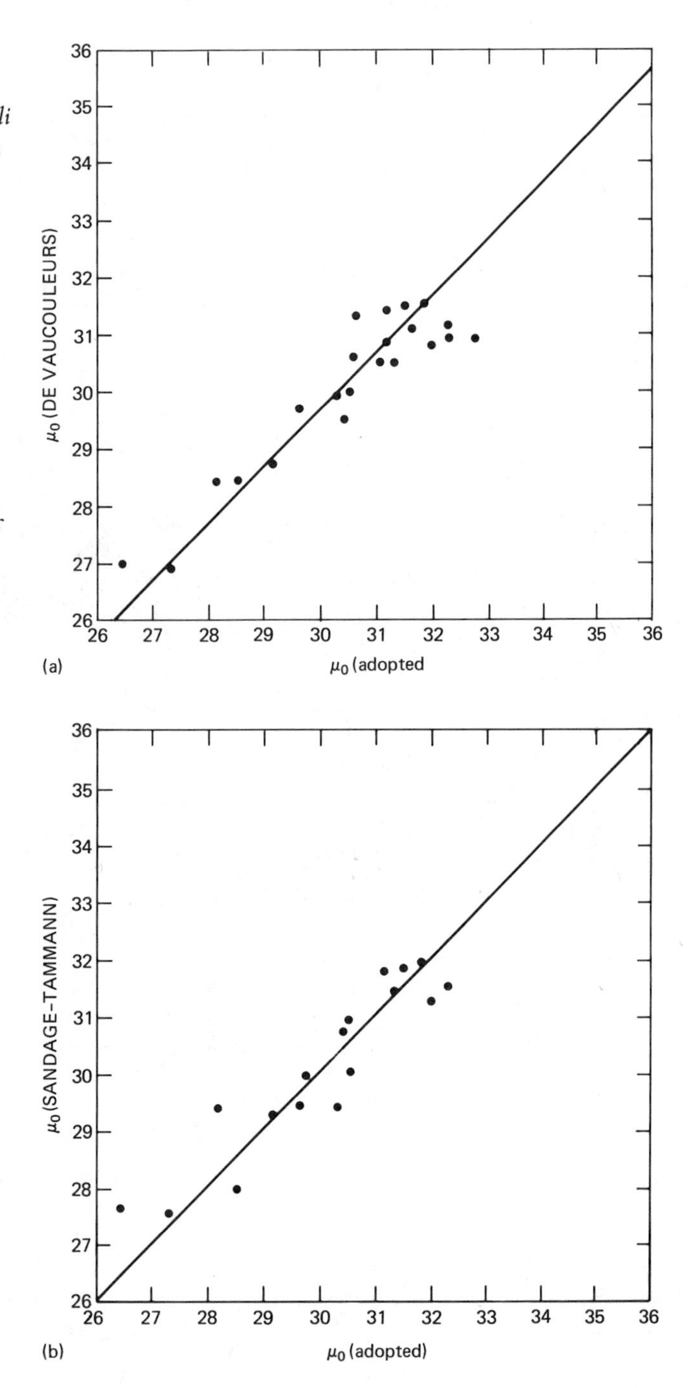

FIGURE 4.17

(a) Comparison of de Vaucouleurs distance moduli for groups and clusters with those adopted in Table 4.13. Solid line shows the effect of applying a mean correction of +0.29 mag (the mean difference between the Local Group distance moduli of de Vaucouleurs and those adopted in this book, see Table 3.13) to the de Vaucouleurs distance moduli. (b) Comparison of Sandage and Tammann distance moduli for groups and clusters with those adopted in Table 4.13. Solid line shows the effect of applying a mean correction of −0.03 mag (see Table 3.13) to their moduli.

TABLE 4.15

Comparison of adopted distance moduli for groups and clusters of galaxies with those given by Sandage and Tammann [S10] and de Vaucouleurs [V5].

*Group*** *number*	*Name**	μ_0 (*adopted*)	$\mu_0(ST)$	$\mu_0(V)$	V_0†	V_c‡	d§	d_V‖	H_c#
G1	Sculptor	26.44	27.65	26.99	281	−59	1.94	20.2	−31
2	M81	27.31	27.56	27.42	226	298	2.90	17.2	103
4	N5128	28.18	29.43	28.42	317	695	4.34	16.3	160
3	Canes Venatici I	28.54	28.50	28.45	339	574	5.1	14.5	113
5	M101	29.16	29.30	28.73	402	498	6.8	14.5	73
10	Canes Venatici II	29.63	29.40	29.70	698	974	8.4	11.7	116
	Leo Triplet	29.69			683	1165	8.7	10.4	134
	M51	29.77	29.94		543	721	9.0	12.5	80
6	N2841	30.30	29.40	29.95	601	820	11.5		71
7	N1023	30.42	30.75	29.53	721	432	12.1	26.7	36
12	N3184	30.52	30.93	29.99	673	987	12.7	12.2	78
13	Coma I	30.57	30.04	30.64	922	1280	13.0	6.8	98
30	N5866	30.64		31.34	920	926	13.4	15.2	69
	N3521	30.79			808	1328	14.4	7.8	92
39	N134	30.85			1581	1241	14.8	32.6	84
	N5364	30.95			1268	1779	15.5	5.9	115
26	Virgo X	31.07		30.54	1272	1730	16.4	3.4	106
	N5033	31.09			935	1198	16.5	8.2	73
31	Eridanus	31.17	31.78	30.86	1520	1435	17.1	32.7	84
27	Grus	31.18		31.44	1581	1225	17.2	34.1	71
53	Fornax	31.21			1527	1478	17.5	32.7	85
	Dorado	31.21			935	1002	17.5	30.8	57
18,19	Virgo	31.32	31.45	30.53	1019	1455	18.3	0.0	79
49	Leo	31.50	31.85	31.91	926	1418	19.9	8.7	71
	(N6384)	31.65			1801	1672	21.4	23.9	78

29	Virgo III	31.65		31.10	1729	2010	21.4	11.1	94
51	N6643	31.84	31.94	31.54	1842	1667	23.3	26.3	71
15	N1068	31.99	31.28	30.83	1332	1068	25.0	41.4	43
	N2336	32.09			2349	2285	26.2	28.2	87
	N5371	32.14			2664	2854	26.8	15.6	107
	N6070	32.17			2040	2110	27.2	22.5	78
34	Ursa Major I (N)	32.26		31.16	1074	1274	28.3	19.5	45
32	Ursa Major I (S)	32.29	31.53	30.94	873	1155	28.8	16.7	40
37	N5676	32.45			2343	2428	30.9	22.4	79
	(N6835)	32.46			1715	1357	31.1	41.6	44
36	N2207	32.73		30.93	1827	2060	35.2	42.2	59
	(N3811)	32.97			3121	3371	39.3	27.3	86
	Cancer	33.33			4758	5117	46.3	40.8	110
	Pegasus I	33.37			3836	3299	47.3	64.2	70
	(A1308+03)	33.49			2904	3318	50	32.3	66
	(A0232+37)	33.49			3851	3352	50	62	71
	Centaurus	33.86			3170	3589	59	50	61
	Sersic 129-1	34.61			4230	4227	84	89	51
	(N7495)	34.67			5109	4556	86	102	53
	(A0316+41)	34.68			8700	8323	86	96	96
	0122+33	34.79			5128	4721	91	104	52
	Coma	35.27			6890	7222	11	94	65
	(1248+28)	35.97			7420	7760	156	139	50
	(1255+27)	36.54			12050	12386	203	186	61

** Group numbers are taken from ref [V13].
* Only groups for which $\Sigma W/\Sigma^2 \geq 2.5$ are included. Those in parentheses are based on a type I supernova distance to a single galaxy.
† Recession velocity corrected for galactic rotation [300 km s^{-1} to $(l, b) = (90°, 0°)$].
‡ Recession velocity corrected for the Galaxy's motion in space [546 km s^{-1} to $(L, B) = (122, =35)$].
§ Distance from our Galaxy, in megaparsecs.
‖ Distance from Virgo.
$H_c = V_c/d$.

BIBLIOGRAPHY

[A1] Aaronson, M., Mould, J., and Huchra, J., "A Distance Scale from the Infrared Magnitude/HI Velocity Width Relation. I—The Calibration," *Astrophys. J.*, 1980, **237,** pp. 655–665.

[A2] Aaronson, M., Mould, J., Huchra, J., Sullivan, W.T. III, Schomer, R.A., and Bothun, G.D., "A Distance Scale from the Infrared Magnitude/HI Velocity Width Relation. III—The Expansion Rate Outside the Local Supercluster," *Astrophys. J.*, 1980, **239,** pp. 12–37.

[A3] Aaronson, M., Persson, S.E., and Frogel, J.A., "The Infrared Colour-Magnitude Relation for Early-Type Galaxies in Virgo and Coma," *Astrophys. J.*, 1981, **245,** pp. 18–24.

[A4] Aaronson, M., and Mould, J., "A Distance Scale from the Infrared Magnitude/HI Velocity Width Relation. IV—The Morphological Type Dependence and Scatter in the Relation: The Distances to Nearby Groups," *Astrophys. J.*, 1983, **265,** pp. 1–17.

[B1] Baum, W.A., "Analysis of Composite Radiation. IV—Population Inferences from Star Counts, Surface Brightnesses, and Colors," *Publ. Astron. Soc. Pacific,* 1959, **71,** pp. 106–117.

[B2] Bautz, L.P., and Morgan, W.W., "On the Classification of the Forms of Clusters of Galaxies," *Astrophys. J.*, 1970, **162,** pp. L149–L153.

[B3] van den Bergh, S., "A Preliminary Luminosity Classification of Late-Type Galaxies," *Astrophys. J.*, 1960, **131,** pp. 215–223.

[B4] van den Bergh, S., "The Distance Scale within the Local Group," in *Décalages vers le rouge et expansion de l'univers—L'évolution des galaxies et ses implications cosmologiques* (IAU Colloq. 37), Balkowski, C., and Westerlund, B.E., Eds., Editions du CNRS, Paris, 1977, pp. 13–22.

[B5] van den Bergh, S., "Can the Hubble Constant Be Determined from HII Region Diameters?" *Astrophys. J.*, 1980, **235,** pp. 1–3.

[B6] van den Bergh, S., "Infrared Surface Brightness and Absolute Magnitude of Spiral Galaxies," *Astrophys. J.*, 1981, **248,** pp. L9–L11.

[B7] Bottinelli, L., and Gouguenheim, L., "The Luminosity Classes of Galaxies and the Hubble Constant," *Astr. Astrophys.*, 1976, **51,** pp. 275–282.

[B8] Bottinelli, L., Gouguenheim, L., Paturel, G., and de Vaucouleurs, G., "The 21 Centimeter Line Width as an Extragalactic Distance Indicator," *Astrophys. J.*, 1980, **242,** pp. L153–L156.

[B9] Bottinelli, L., Gouguenheim, L., and Paturel, G.,"Study of spiral galaxies from 392 new measurements of 21-cm line data," *Astr. Astrophys.*, 1982, **113**, pp. 61–72.

[B10] Branch, D., "On the Use of Type I Supernovae to Determine the Hubble Constant," *Month. Not. R. Astron. Soc.*, 1979, **186,** pp. 609–616.

[B11] Branch, D., and Pachett, B., "Type I Supernovae," *Month. Not. R. Astron. Soc.*, 1973, **161,** pp. 71–83.

[B12] Burstein, D., "Mass and Luminosity in Spiral Galaxies and the Tully–Fisher Relation," *Astrophys. J.*, 1982, **253,** pp. 539–551.

[B13] Burstein, D., Rubin, V.C., Thonnard, N., and Ford, W.K., Jr., "The Distribution of Mass in Sc Galaxies," *Astrophys. J.*, 1982, **253,** pp. 70–85.

[B14] Buta, R., and de Vaucouleurs, G., "Inner Ring Structures in Galaxies as Distance Indicators. II—Calibration of Inner Ring Diameters as Quaternary Indicators," *Astrophys. J.,* suppl. ser., 1982, **48,** pp. 219-237; "Inner Ring Structures in Galaxies as Distance Indicators. IV—Distances to Several Groups, Clusters, the Hercules Supercluster, and the Value of the Hubble Constant," *Astrophys. J.,* 1983, **266,** pp. 1-17.

[C1] Carney, B.W., "Spectrophotometry of Supernova 1979c in M101," *Publ. Astron. Soc. Pacific,* 1980, **92,** pp. 56-59.

[D1] Dawe, J.A., and Dickens, R.J., "Suspected Globular Clusters in the Fornax I Cluster of Galaxies," *Nature,* 1976, **263,** pp. 395-396.

[E1] Elias, J.H., Frogel, J.A., Humphreys, R.M., and Persson, S.E., "Infrared Luminosities of M Supergiants and Their Use as Distance Indicators," *Astrophys. J.,* 1981, **249,** pp. L55-L59.

[F1] Fisher, J.R., and Tully, R.B., "Neutral Hydrogen Observations of DDO Dwarf Galaxies," *Astr. Astrophys.,* 1975, **44,** pp. 151-171.

[F2] Ford, H.C., and Jenner, D.C., "Planetary Nebulae in the Nuclear Bulge of M81: A New Distance Determination," *Bull. Am. Astron. Soc.,* 1978, **10,** p. 665.

[G1] Graham, J.A., "Distance Indicators in the Magellanic Clouds," in *Highlights in Astronomy,* vol. 6, West, R.M., Ed., Reidel, 1983, pp. 209-216.

[G2] Griersmith, D., "The CM Relation at UBV Wavelengths in the Fornax, Virgo, and Sersic 129-1 Clusters," *Astr. J.,* 1982, **87,** pp. 462-476.

[H1] Habing, H. J., and Israel, F. P., "Compact HII Regions and OB Star Formation," *Ann. Rev. Astron. Astrophys.,* 1979, **17,** pp. 345-385.

[H2] Hanes, D.A., "Globular Clusters and the Virgo Cluster Distance Modulus," *Month. Not. R. Astron. Soc.,* 1977, **180,** pp. 309-321.

[H3] Hanes, D.A., "A New Determination of the Hubble Constant," *Month. Not. R. Astron. Soc.,* 1979, **188,** pp. 901-909.

[H4] Hanes, D.A., "A Re-examination of the Sandage-Tammann Extragalactic Distance Scale," *Month. Not. R. Astron. Soc.,* 1982, **201,** pp. 145-170.

[H5] Harris, W.E., and Racine, R., "Globular Clusters in Galaxies," *Ann. Rev. Astron. Astrophys.,* 1979, **17,** pp. 241-274.

[H6] Hodge, P.W., "The Extragalactic Distance Scale," *Ann. Rev. Astron. Astrophys.,* 1981, **19,** pp. 357-372.

[H7] Holmberg, E., "A Study of Physical Groups of Galaxies," *Arkiv Astronomi,* 1969, **5,** pp. 305-343; "Magnitudes, Colors, Surface Brightness, Intensity Distributions, Absolute Luminosities, and Diameters of Galaxies," *Stars and Stellar Systems,* vol. 9; "Galaxies and the Universe," Sandage, A., Sandage, M., and Kristian, J., Eds., University of Chicago Press, 1975, pp. 123-157.

[H8] Humphreys, R.M., "M Supergiants and the Low Metal Abundances in the Small Magellanic Cloud," *Astrophys. J.,* 1979, **231,** pp. 384-387; "Studies of Luminous Stars in Nearby Galaxies. II—M Supergiants in the Large Magellanic Cloud," *Astrophys. J.,* suppl. ser., 1979, **39,** pp. 389-403; "IV—Baade's Field IV in M31," *Astrophys. J.,* 1979, **234,** pp. 854-860; "V—The Local Group of Irregulars NGC 6822 and IC 1613," *Astrophys. J.,* 1980, **238,** pp. 65-78; "VI—The Brightest Supergiants and the Distance to M33," *Astrophys. J.,* 1980, **241,** pp. 587-597; "VIII—The Small Magellanic Cloud," *Astrophys. J.,* 1983, **265,** pp. 176-193.

[H9] Humphreys, R.M., "Studies of Luminous Stars in nearby Galaxies. VII—The Brightest Blue Stars in the Spiral Galaxies M101 and NGC 2403," *Astrophys. J.*, 1980, **241,** pp. 598-611.

[H10] Humphreys, R.M., "The Brightest Stars as Extragalactic Distance Indicators," *Astrophys. J.*, 1983, **269,** pp. 335-351.

[H11] Humphreys, R.M., and Davidson, K., "Studies of Luminous Stars in nearby Galaxies. III—Comments on the Evolution of the Most Massive Stars in the Milky Way and the Large Magellanic Cloud," *Astrophys. J.*, 1979, **232,** pp. 409-420.

[H12] Humphreys, R.M., and Sandage, A., "On the Stellar Content and Structure of the Spiral Galaxy M33," *Astrophys. J.*, suppl. ser., 1980, **44,** pp. 319-381.

[H13] Humphreys, R.M., and Strom, S.E., "M Supergiants and the Distance to the Sc I Galaxy M101," *Bull. Am. Astron. Soc.*, 1981, **13,** p. 892.

[J1] Jacoby, G.H., "The Luminosity Function for Planetary Nebulae and the Number of Planetary Nebulae in the Local Group Galaxies," *Astrophys. J.*, suppl. ser., 1980, **42,** pp. 1-18.

[K1] Kennicutt, R.C., Jr., "HII Regions as Extragalactic Distance Indicators. II—Applications of Isophotal Diameters," *Astrophys. J.*, 1979, **228,** pp. 696-703.

[K2] Kennicutt, R.C., Jr., "HII Regions as Extragalactic Distance Indicators. III—Application of the HII Region Fluxes and Galaxy Diameters." *Astrophys. J.*, 1979, **228,** pp. 704-711.

[K3] Kennicutt, R.C., Jr., "HII Regions as Extragalactic Distance Indicators. IV—The Virgo Cluster," *Astrophys. J.* 1981, **247,** pp. 9-16.

[K4] Kennicutt, R.C., Jr., "The Local Velocity Field and the Calibration of the Luminosity Classification of Spiral Galaxies," *Astrophys. J.*, 1982, **259,** pp. 530-534.

[K5] Kennicutt, R.C., Jr., "The Luminosity Dependence of Spiral Arm Morphology and the Luminosity Classification," *Astr. J.*, 1982, **87,** pp. 255-263.

[K6] Kennicutt, R.C., Jr., and Hodge, P.W., "HII Regions in NGC 628. III—Hα luminosities and the IR Luminosity Function," *Astrophys. J.*, 1980, **241,** pp. 573-586.

[K7] King, I.R., "Surface Photometry of Elliptical Galaxies," *Astrophys. J.*, 1978, **222,** pp. 1-13.

[K8] Kormendy, J., "A Morphological Survey of Bar, Lens, and Ring Components in Galaxies: Secular Evolution in Galaxy Structure," *Astrophys. J.*, 1979, **227,** pp. 714-728.

[L1] Laurie, D.G., and Kwitter, K.B., "Giant ringlike HII regions and the distance to M101," *Astrophys. J.*, 1982, **255,** pp. L29-31.

[M1] Martin, W.L., Warren, P.R., and Feast, M.W., "Multicolour Photoelectric Photometry of Magellanic Cloud Cepheids. II—An Analysis of BVI Observations in the LMC," *Month. Not. R. Astron. Soc.*, 1979, **188,** pp. 139-157.

[M2] Melnick, J., "Velocity Dispersions in Giant HII Regions: Relation with Their Linear Diameters," *Astrophys. J.*, 1977, **213,** pp. 15-17.

[M3] Michard, R., "Sequences and Populations of Early-Type Galaxies," *Astr. Astrophys.*, 1979, **74,** pp. 206-217.

[M4] Mould, J.R., "The Infrared Color-Magnitude Relation for Early-Type Galaxies in the Pegasus I Cluster," *Publ. Astron. Soc. Pacific*, 1981, **93,** pp. 25-28.

[M5] Mould, J., Aaronson, M., and Huchra, J., "A Distance Scale from the Infrared Magnitude/HI Velocity-Width Relation. II — The Virgo Cluster," *Astrophys. J.*, 1980, **238,** pp. 458–470.

[P1] Paturel, G., "Relation diamètre Luminosité. IV-Application à la détermination de la constante de Hubble," Astr. Astrophys., 1979, **71,** pp. 19–28.

[P2] Persson, S.E., Frogal, J.A., and Aaronson, M., "Photometric Studies of Composite Stellar Systems. III — UBVR and JHK Observations of E and SO Galaxies," *Astrophys. J.*, suppl. ser., 1979, **39,** pp. 61–87.

[R1] Racine, R., "Globular Cluster Systems in Galaxies" in *Star Clusters* (IAU Symp. 85), Hesser, J.E., Ed., Reidel, 1980, pp. 369–383.

[R2] Roberts, M.S., "Twenty-One Centimeter Line Widths of Galaxies," *Astr. J.*, 1978, **83,** pp. 1026–1035.

[R3] Rubin, V.C., "Systematics of HII rotation curves," *Internal Kinematics and Dynamics of Galaxies* (IAU Symp. 100), Athanassoula, E., Ed., Reidel, 1983, pp. 3–10.

[R4] Rubin, V.C., Kent Ford, W., Jr., and Thonnard, N., "Rotational Properties of 21 Sc Galaxies with a Large Range of Luminosities and Radii, from NGC 4605 ($R = 4$ kpc) to UGC 2885 ($R = 122$ kpc)." *Astrophys. J.*, 1980, **238,** pp. 471–487.

[R5] Rubin, V.C., Burstein, D., and Thonnard, N., "A New Relation for Estimating the Intrinsic Luminosities of Spiral Galaxies," *Astrophys. J.*, 1980, **242,** pp. L149–L159.

[S1] Sandage, A., "Absolute Magnitudes of E and SO Galaxies in the Virgo and Coma Clusters as a Function of U-B Color," *Astrophys. J.*, 1972, **176,** pp. 21–30.

[S2] Sandage, A., "The Redshift-Distance Relation. I — Angular Diameter of First-Ranked Cluster Galaxies as a Function of Redshift: The Aperture Correction to Magnitudes," *Astrophys. J.*, 1972, **173,** pp. 485–499.

[S3] Sandage, A., "The Redshift-Distance Relation. II — The Hubble Diagram and Its Scatter for First-Ranked Cluster Galaxies: A Formal Value for q_0," *Astrophys. J.*, 1972, **178,** pp. 1–24.

[S4] Sandage, A., "The Redshift-Distance Relation. VIII — Magnitudes and Redshifts of Southern Galaxies in Groups: A Further Mapping of the Local Velocity Field and an Estimate of q_0," *Astrophys. J.*, 1975, **202,** pp. 563–582.

[S5] Sandage, A., and Hardy, E., "The Redshift-Distance Relation. VII — Absolute Magnitudes of the First Three Ranked Cluster Galaxies as a Function of Cluster Richness and Bautz–Morgan Cluster Type: The Effect on q_0," *Astrophys. J.*, 1973, **183,** pp. 743–757.

[S6] Sandage, A., and Tammann, G.A., "Steps toward the Hubble Constant. I — Calibration of the Linear Sizes of Extragalactic HII Regions," *Astrophys. J.*, 1974, **190,** pp. 525–538.

[S7] Sandage, A., and Tammann, G.A., "Steps toward the Hubble Constant. II — The Brightest Stars in Late-Type Spiral Galaxies," *Astrophys. J.*, 1974, **191,** pp. 603–621.

[S8] Sandage, A., and Tammann, G.A., "Steps toward the Hubble Constant. III — The Distance and Stellar Content of the M101 Group of Galaxies," *Astrophys. J.*, 1974, **194,** pp. 223–243.

[S9] Sandage, A., and Tammann, G.A., "Steps toward the Hubble Constant. IV — Distances to 39 Galaxies in the General Field Leading to a Calibration of the Galaxy Luminosity Classes and a First Hint of the Value of q_0," *Astrophys. J.*, 1974, **194,** pp. 559–568.

[S10] Sandage, A., and Tammann, G.A., "Steps toward the Hubble Constant. V—The Hubble Constant from nearby Galaxies and the Regularity of the Local Velocity Field," *Astrophys. J.,* 1975, **196,** pp. 313–328.

[S11] Sandage, A., and Tammann, G.A., "Steps toward the Hubble Constant. VI—The Hubble Constant Determined from Redshifts and Magnitudes of Remote Sc I Galaxies: The Value of q_0," *Astrophys. J.,* 1975, **197,** pp. 265–280.

[S12] Sandage, A., and Tammann, G.A., "Steps toward the Hubble Constant. VII—Distances to NGC 2403, M101, and the Virgo Cluster Using 21 Centimeter Line Widths Compared with Optical Methods: The Global Value of H_0," *Astrophys. J.,* 1976, **210,** pp. 7–24.

[S13] Sandage, A., and Tammann, G.A., *Revised Shapley–Ames Catalog of Bright Galaxies,* Carnegie Institution of Washington, 1981.

[S14] Sandage, A., and Tammann, G.A., "Steps toward the Hubble Constant. VIII—The Global Value," *Astrophys. J.,* 1982, **256,** pp. 339–345.

[S15] Sandage, A. and Tammann, G.A., "The Hubble Constant as Derived From 21 cm Line Widths," *Nature,* 1984, **307,** pp. 326–329.

[S16] Schecter, P., "An Analytic Expression for the Luminosity Function for Galaxies," *Astrophys. J.,* **203,** pp. 297–306.

[S17] Schurmann, S.R., Arnett, W.D., and Falk, S.W., "Type II Supernovae: Nonstandard Candles as Extragalactic Distance Indicators," *Astrophys. J.,* 1979, **230,** pp. 11–25.

[S18] Seaton, M.J., "Distances of Planetary Nebulae," *Astrophys. Lett.,* 1968, **2,** pp. 55–58.

[S19] Sersic, J.L., "The HII Regions in Galaxies," *Z. Astrophys.,* 1960, **50,** pp. 168–177.

[T1] Tammann, G.A., "Some Statistical Properties of Supernovae," *Mem. Soc. Astronomia Ital.,* 1978, **49,** pp. 315–329.

[T2] Tammann, G.A., "Supernovae Statistics and Related Problems," in *Supernovae: A Survey of Current Research,* Rees, M.J., and Stoneham, R.J., Eds., Reidel, 1982, pp. 371–403.

[T3] Tammann, G.A., and Sandage, A., "The Stellar Content and Distance of the Galaxy NGC 2403 in the M101 Group," *Astrophys. J.,* 1968, **151,** pp. 825–860.

[T4] Tammann, G.A., Sandage, A., and Yahil, A., "The Determination of Cosmological Parameters," in *Physical Cosmology,* Balian, R., Audouze, J., and Schramm, D.N., Eds., North-Holland, 1979, pp. 53–125.

[T5] Tammann, G.A., Yahil, A., and Sandage, A., "The Velocity Field of Bright Nearby Galaxies. II—Luminosity Functions for Various Hubble Types and Luminosity Classes: The Peculiar Motion of the Local Group Relative to the Virgo Cluster," *Astrophys. J.,* 1979, **234,** pp. 775–785.

[T6] Tammann, G.A., and Sandage, A., "The value of H_0," *Highlights in Astronomy,* vol. 6, West, R.M., Ed., Reidel, 1983, pp. 301–313.

[T7] Tully, R.B., and Fisher, J.R., "A New Method of Determining Distances to Galaxies," *Astr. Astrophys.,* 1977, **54,** pp. 661–673.

[T8] Tully, Mould, J., and Aaronson, M., "A Color-Magnitude Relation for Spiral Galaxies," *Astrophys. J.,* 1982, **257,** pp. 527–537.

[V1] Vaucouleurs, G. de, "Distances of the Virgo, Fornax and Hydra Clusters of Galaxies and the Local Value of the Hubble Ratio," *Nature,* 1977, **266,** pp. 126–129.

[V2] Vaucouleurs, G. de, "The Extragalactic Distance Scale. I—A Review of Distance Indicators: Zero Points and Errors of Primary Indicators," *Astrophys. J.*, 1978, **223**, pp. 351–363.

[V3] Vaucouleurs, G. de, "The Extragalactic Distance Scale. III—Secondary Distance Indicators," *Astrophys. J.*, 1978, **224**, pp. 14–21.

[V4] Vaucouleurs, G. de, "The Extragalactic Distance Scale. IV—Distances of Nearest Groups and Field Galaxies from Secondary Indicators," *Astrophys. J.*, 1978, **224**, pp. 710–717.

[V5] Vaucouleurs, G. de, "The Extragalactic Distance Scale. V—Tertiary Distance Indicators," *Astrophys. J.*, 1979, **227**, pp. 380–390.

[V6] Vaucouleurs, G. de, "The Extragalactic Distance Scale. VI—Distances of 458 Spiral Galaxies from Tertiary Indicators," *Astrophys. J.*, 1979, **227**, pp. 729–755.

[V7] Vaucouleurs, G de, and Cappacioli, M., "Luminosity Distribution in Galaxies. I—The Elliptical Galaxy NGC 3379 as a Luminosity Distribution Standard," *Astrophys. J.*, suppl. ser., 1979, **40**, pp. 699–731.

[V8] Vaucouleurs, G. de, and Olsen, D.W., "The Central Velocity Dispersion in Elliptical and Lenticular Galaxies as an Extragalactic Distance Indicator, *Astrophys. J.*, 1982, **256**, pp. 346–369.

[V9] Visvanathan, N., "Distance to Virgo I Cluster via CM Data of M31 Group and Virgo I Cluster," *Astr. Astrophys.*, 1978, **67**, pp. L17–L19.

[V10] Visvanathan, N., "A Global Value of the Hubble Constant," *Astrophys. J.*, 1983, **275**, 430–444.

[V11] Visvanathan, N., and Sandage, A., "The Color-Absolute Magnitude Relation for E and SO Galaxies. I—Calibration and Tests for Universality Using Virgo and Eight Other nearby Clusters," *Astrophys. J.*, 1977, **216**, pp. 214–226.

[V12] Visvanathan, N., and Griersmith, D., "Distance to the Virgo I cluster and the value of the Hubble constant," *Astrophys. J.*, 1979, **230**, pp. 1–10.

[V13] Vaucouleurs, G. de, "Nearby Groups of Galaxies," *Stars and Stellar Systems,* vol. 9; "Galaxies and the Universe," Sandage, A., Sandage, M., and Kristian, J., Eds., University of Chicago Press, 1975, pp. 557–600.

[W1] Weedman, D.W., "The Hubble Diagram for Nuclear Magnitudes of Cluster Galaxies," *Astrophys. J.*, 1976, **203**, pp. 6–13.

CHAPTER FIVE

THE COSMOLOGICAL MODELS

We now try to extend our distance ladder to the largest scale possible, the cosmological scale. To understand distances on the cosmological scale, we need to develop models for the universe. These have to be rooted both in physical theory, in this case the general theory of relativity, and in observation. Two phenomena in particular allow us to probe the large-scale structure of the universe, the redshift and the microwave background radiation.

By 1925 Vesto Slipher had shown that the lines in the spectra of the majority of distant galaxies are shifted toward the red end of the spectrum, and interpreting this as a Doppler shift, he concluded that the galaxies are receding from us. In 1929 Edwin Hubble showed that the velocity of the galaxies, as measured by the redshift of their spectral lines, increases linearly with distance. This redshift–distance relation, or Hubble law, which has been strongly confirmed by subsequent work, had been predicted by Albert Einstein and Willem de Sitter as a consequence of simple models for an expanding universe. Controversy has surrounded the cosmological interpretation of the redshifts of quasars, but for most cosmologists this controversy has been largely resolved (Box 5.1). Efforts to explain this redshift as due to causes other than motion have been unconvincing.

The microwave background radiation was discovered by Arno Penzias and Robert Wilson in 1965, working at Bell Laboratories [P8]. It was soon established that this radiation is extremely isotropic (the same in every direction on the sky) and that its spectrum is close to that of a 3-K blackbody. Both these properties point to a cosmological origin, and the simplest explanation is that it is the relic of the fireball phase of a hot big-bang universe. In these models, first put forward by George Gamov in 1948, the universe explodes out from a state of infinite density about 10–20 billion years ago. For the first million years or so, the energy density of the universe is dominated by radiation, and it is the relic of this radiation that we see today at microwave wavelengths. The properties of the hot big-bang models are outlined in Secs. 5.1 and 5.2, and some alternative approaches to cosmology are discussed in Sec. 5.3.

One of the crucial preliminaries to any discussion of the cosmological distance

BOX 5.1 THE QUASAR REDSHIFT CONTROVERSY

The enormous energies implied by the cosmological interpretation of quasar redshifts, up to 1000 times the luminosity of our Galaxy being generated from a region not much larger than the solar system, led several astronomers to doubt whether these redshifts were of cosmological origin. Halton Arp found examples of quasars which appeared to be associated with foreground galaxies and Geoffrey Burbidge claimed that there were significant peaks and periodicities in the quasar redshift distribution. The issue of whether quasar redshifts are cosmological was debated, somewhat inconclusively, between Arp and John Bahcall at the American Association for the Advancement of Science in 1972 [F2]. Since then Alan Stockton has shown that many low-redshift quasars are associated with small groups of galaxies with the same redshift as the quasar [S12]. For many low-redshift quasars, faint optical emission has been detected from the underlying galaxy in which the quasar event has occurred [W4],[H5]. These together with other arguments have convinced most astronomers that the redshifts of quasars are indeed cosmological and that the effects reported by Arp and Burbidge are therefore due to chance. However, Arp still believes that the associations he finds demonstrate that quasar redshifts are non-cosmological [A3],[A4],[A5]. For further reference see [R2],[N1].

scale is the establishment of the correct local frame of reference. This has been done fairly convincingly by measuring the velocity of our Galaxy with respect to the microwave background radiation (Sec. 5.4). Sec. 5.5 discusses some of the observable consequences of the density inhomogeneities, which must have been present in the early universe in order that galaxies can exist today.

The less mathematically minded reader might want to skip this chapter at a first reading.

5.1 Introduction to big-bang cosmology

SPECIAL AND GENERAL RELATIVITY

As soon as we try to talk about the large-scale structure of the universe, we come up against two problems. First we have to discuss more carefully what distance means on the large scale. Second we have to take account of the dynamics of the universe, in particular the mutual gravitational attraction of galaxies. Both problems are solved in Einstein's general theory of relativity. For a deep understanding of cosmology it is essential to study general relativity (see, for example, Misner et al. [M5], Narlikar [N1] Weinberg [W1]), but in this book we give only a brief sketch of how the cosmological models are derived.

In relativity theory the idea of distance has to be extended to take account of the fourth dimension, time. This is done by introducing the concept of an *event,* that is, a particular location at a particular moment in time. The three dimensions of space and the dimension of time are bound together into a four-dimensional *space-time continuum.* If (x^1, x^2, x^3) are the spatial coordinates of a point and $t = x^4$ is the time coordinate, the event can be characterized by the coordinates (x^1, x^2, x^3, x^4). Now consider two events which are near each other in space and time, characterized by the coordinates (x^1, x^2, x^3, x^4) and $(x^1 + dx^1, x^2 + dx^2, x^3 + dx^3, x^4 + dx^4)$. The spatial distance between these two events in a normal Euclidean space would then be dl, where

$$dl^2 = (dx^1)^2 + (dx^2)^2 + (dx^3)^2. \tag{5.1}$$

In Einstein's special theory of relativity he recognized that the distances measured, the length of a rod, for example, depended on the relative motion of the observer and the rod, although the dependence is very slight unless the relative velocities are a significant fraction of the speed of light. In order to find a quantity that did not depend on this relative motion, Einstein introduced the four-dimensional *interval* between two events ds, defined by

$$\begin{aligned} ds^2 &= dt^2 - \frac{1}{c^2}(dx^1)^2 + (dx^2)^2 + (dx^3)^2 \\ &= dt^2 - \frac{dl^2}{c^2} \end{aligned} \tag{5.2}$$

where c is the velocity of light. Then ds is an *invariant* for all observers in an inertial frame, that is, it is the same for all observers who are not being acted upon by any forces, regardless of their relative motions. ds measures the time interval for an observer at rest in the spatial frame of reference (x^1, x^2, x^3), since if $dx^1 = dx^2 = dx^3 = 0$, $dx = dt$. The time measured in a frame at rest with respect to an observer is called the *proper time* for the observer. If the two events considered above are connected by a light signal such that one is the event of the light signal setting off and the other is the event of it arriving, then the velocity $dl/dt = c$. Thus in this case the interval $ds = 0$ for all inertial observers (a *null* interval).

To deal with noninertial observers, for example, those experiencing a gravitational field, Einstein in his general theory of relativity generalized Eq. (5.2) to

$$ds^2 = \sum_{\mu,\nu=1}^{4} g_{\mu\nu}\, dx^\mu\, dx^\nu \tag{5.3}$$

where the quantities $g_{\mu\nu}$ are functions of position and time. Eq. (5.3) is called the *metric* of space-time, and $g_{\mu\nu}$ is the *metric tensor.* The meaning of this general expression for ds is that we are no longer restricting ourselves to a Euclidean space and are now allowing the four-dimensional space-time continuum to be curved. The metric of special relativity (obtained by taking $g_{11} = 1$, $g_{22} = g_{33} = g_{44} =$

$-1/c^2$, $g_{\mu\nu} = 0$ if $\mu \neq \nu$) is called the Minkowski metric, after one of the early pioneers of relativity theory, Hermann Minkowski.

If we make a change of coordinates from x^μ, $\mu = 1, \ldots, 4$, to x'^μ, then the metric tensor will be changed. It is possible to find a change of coordinate system which not only changes $g_{\mu\nu}$ to the Minkowski form at a given event, but also makes the derivatives of the metric tensor 0 there. This means that in a small region of space-time about this event the metric is Minkowskian and special relativity holds. This is an expression of one of the fundamental ideas of general relativity, the *principle of equivalence:* in a freely falling (and nonrotating) frame, gravity vanishes locally and special relativity holds.

That the effects of gravity are eliminated in free fall is a natural consequence of the fact discovered by Galileo, namely, that all bodies experience the same acceleration under gravity. In a frame of reference with just this acceleration, the effects of gravity disappear. When the astronauts orbit the earth without firing their rocket motors, they are in effect in free fall, and they become weightless even though the earth's gravity 100 miles up is almost the same as at ground level. But the gravitational field can be eliminated only locally, over a region of space where the gravity is roughly uniform and over a time span during which the gravity is roughly steady. The frame where the effects of gravity (and other noninertial forces like the centrifugal force) have been eliminated locally is called the *local inertial frame.*

We now examine how distances can be measured on the large scale, taking into account the effects of gravitation. Let us return to the general expression (5.3) and consider an observer at rest in a general frame of reference x^μ, so that $dx^1 = dx^2 = dx^3 = 0$. Then $ds^2 = g_{44}\, dt^2 = d\tau^2$, where $d\tau$ is the proper time interval. We see that unless $g_{44} = 1$, t is no longer a measure of proper time, though they are related by $\tau = \int g_{44}\, dt$.

We have more of a problem when we want to measure distances, though. In an inertial frame, where $ds^2 = dt^2 - dl^2/c^2$, we can measure the distance from A to B by sending a light signal and measuring the light travel time $\Delta t = t_B - t_A$. Then $c\,\Delta t$ is the radar distance*, or *proper distance,* from A to B. Now since $ds = 0$ for a light signal, $\Delta l = c\,\Delta t$, so the proper distance is the same as the normal Euclidean distance,

$$\Delta l = [(\Delta x^1)^2 + (\Delta x^2)^2 + (\Delta x^3)^2]^{1/2}.$$

For a general frame of reference, we can still define the distance between two nearby events in terms of the proper distance, that is, we solve the equation

$$ds^2 = \sum_{\mu,\nu=1}^{4} g_{\mu\nu}\, \Delta x^\mu\, \Delta x^\nu = 0$$

for Δx^4 ($= \Delta t$) and then calculate $c\,\Delta\tau$, using the fact that $\Delta\tau = g_{44}\,\Delta t$.

The problem is that this expression is a function of time in general, so there is no

* Strictly speaking, a radar measurement involves sending a signal from A to B and back again and then halving the elapsed time.

unique way of adding up distance elements over the large scale. Only if the $g_{\mu\nu}$ are independent of time or if, for some reason, there is a universal time coordinate, on which all observers can agree and by which they can synchronize their distance measurements, can proper distance be defined uniquely over the large scale. Fortunately there is a universal cosmic time in the simple cosmological models we will describe.

In general relativity the metric tensor $g_{\mu\nu}$ essentially describes the gravitational field. The inverse-square law of gravity is then replaced by differential equations, Einstein's field equations, which relate the functions $g_{\mu\nu}$ to the distribution of matter. (These equations are beyond the scope of this book.) When the gravitational field is weak and varies only slowly with time, the theory reduces to Newtonian gravity.

HOMOGENEOUS AND ISOTROPIC COSMOLOGIES

How can we make models for the large-scale structure of the universe within the framework of general relativity? Shortly after introducing his general theory, Einstein made the daring speculation that on the large scale the structure of the universe is *homogeneous,* that is, every region is the same as every other region, and *isotropic,* that is, the universe looks the same in every direction. This is now called the *cosmological principle.* It is almost the strongest assumption that can be made about the simplicity of the universe.* Models based on the cosmological principle have proved to be surprisingly successful in describing the large-scale structure of the universe. The evidence for departures from homogeneity and isotropy will be discussed in Sec. 5.4.

We can now outline how the properties of homogeneous and isotropic cosmological models are derived in general relativity (but without going into the mathematical details). It can be shown that the metric for homogeneous and isotropic models of the universe takes the form

$$ds^2 = dt^2 - \frac{R^2(t)}{c^2}\, d\sigma^2 \tag{5.4}$$

where $R(t)$ is a function of time, called the *scale factor* (see below) and $d\sigma^2$ is the metric for a three-dimensional space with constant curvature. This metric can be written in terms of spherical polar coordinates (r, θ, ϕ) as

$$d\sigma^2 = \frac{dr^2}{1 - kr^2} + r^2(d\theta^2 + \sin^2\theta\, d\phi^2) \tag{5.5}$$

where k is the curvature constant, which describes the geometry of space at a particular instant of time. $k > 0$ corresponds to a space of positive curvature, $k = 0$

* A stronger assumption is that the universe does not vary in time either. This is called the perfect cosmological principle and forms the basis of the steady-state theory (see Sec. 5.3).

FIGURE 5.1

Two-dimensional curved surfaces. (a) Sphere, surface of positive curvature. The sum of the angles in a triangle upon a sphere is greater than π. (b) Hyperboloid, surface of negative curvature. The sum of the angles in a triangle upon a hyperboloid is less than π

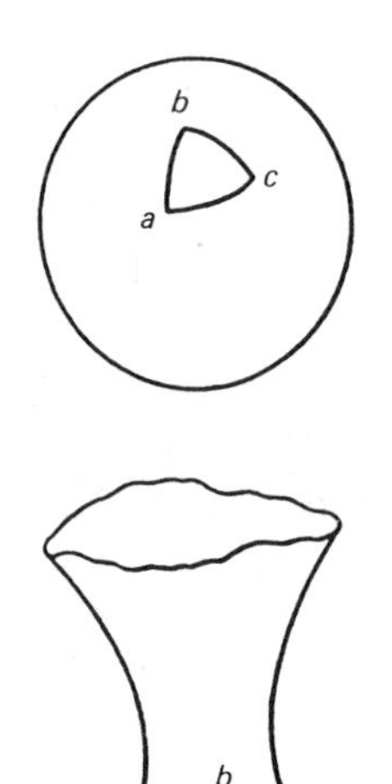

to normal flat space, and $k < 0$ to a space of negative curvature (Fig. 5.1). k can be taken as $+1$, 0, or -1 by a suitable rescaling of the radial coordinate r. In determining this metric, which is called the Robertson–Walker metric, it was assumed that r is a *comoving* coordinate, that is, it is a label for a particular galaxy and does not alter as the universe expands or contracts. Since the universe is isotropic, θ and ϕ also remain fixed for a particular galaxy. The time coordinate t has a special meaning too, since if we consider a particular galaxy at (r, θ, ϕ), so that $dr = d\theta = d\phi = 0$, then $ds = dt$, and hence t measures the proper time for the galaxy. There is therefore a universal cosmic time t, which is the same for all observers at rest with respect to the local matter. In practice, galaxies are moving around with their own random motions, so the local inertial frame has to be defined in terms of the average motion of galaxies over a sufficiently large region, or with respect to the microwave background radiation (see Sec. 5.4).

If we determine the proper distance $c\,\Delta t$ between two galaxies from the equation $ds = 0$, we find $c\,dt = R(t)\,d\sigma$, and so

$$c\,\Delta t = R(t)\,\Delta\sigma \tag{5.6}$$

where all measurements are understood to be made at the same epoch t. Since r is a comoving label and remains fixed for each galaxy, and θ and ϕ remain fixed for isotropic motion, then $\Delta\sigma$ is fixed, and the proper distance between two galaxies is just scaled by the function $R(t)$ as t varies. For this reason $R(t)$ is called the *scale-factor:* the universe simply expands or contracts isotropically as R increases or decreases with time.

The way in which the scale-factor R varies with time has to be determined by substituting the metric (5.4) into the field equations. For a homogeneous gas of density $\rho(t)$ and pressure $p(t)$, two equations result:

$$\ddot{R} = -4\pi G\left(\rho + \frac{3p}{c^2}\right)\frac{R}{3} + \Lambda\frac{R}{3} \tag{5.7}$$

$$\dot{R}^2 = 8\pi G\rho\frac{R^2}{3} - kc^2 + \Lambda\frac{R^2}{3} \tag{5.8}$$

where the dot means d/dt.

Here G is the usual gravitational constant, but Λ is a new constant, the *cosmological constant,* which appears in the most general form of the general theory of relativity. Einstein regretted introducing the Λ term, and most relativists today prefer to take $\Lambda = 0$. We follow this line for the moment, but will consider the effects of $\Lambda \neq 0$ in Sec. 5.3. The equations for R can then be written

$$\dot{R}^2 = 8\pi G\rho\frac{R^2}{3} - kc^2 \tag{5.9}$$

and

$$\frac{d}{dt}(\rho R^3) + \frac{p}{c^2}\frac{d}{dt}(R^3) = 0 \tag{5.10}$$

[differentiating Eq. (5.8) and substituting for $\ddot{R}$ from Eq. (5.7)]. If the pressure can be neglected (valid for the matter in the universe at the present epoch), then

$$\begin{aligned} \rho R^3 &= \text{constant} = \rho_0 R_0^3 \\ \dot{R}^2 &= 8\pi G\rho_0\frac{R_0^3}{3R} - kc^2. \end{aligned} \tag{5.11}$$

The solutions of Eq. (5.11) are illustrated in Fig. 5.2. For $k = 0$ they have the simple

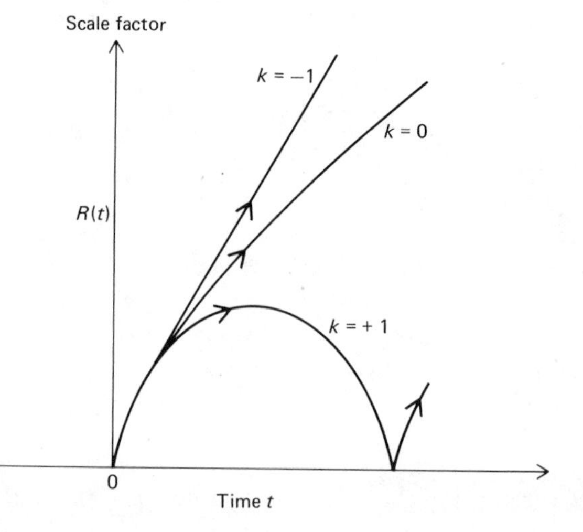

FIGURE 5.2
Scale factor R(t) *for* $\Lambda = 0$ *cosmological models for the three possible cases.* $k = +1$*—oscillating model;* $k = 0$*—Einstein–de Sitter model,* $R(t) \propto t^{2/3}$, $k = -1$*—monotonically expanding model.*

form $R \propto t^{2/3}$, which is the Einstein–de Sitter model. Since the universe is observed to be expanding, we can reject the corresponding contracting models. The models all have the property that if we start from any set of values for R, $\dot{R}$, and t and extrapolate backward in time, then $R \rightarrow 0$, a finite time in the past. We denote this epoch when $R = 0$ by $t = 0$. From Eq. (5.6) the proper distance between any pair of galaxies was zero at this epoch and the universe had infinite density. The models are therefore called the *big-bang* models, and $t = 0$ is referred to as the *initial singularity.* Since general relativity breaks down under these extreme conditions, the models are valid only for t greater than some minimum value. The earliest time for which the big-bang models can be valid is the Planck era, $t_p \sim 10^{-43}$ s (see Sec. 1.6).

There are only two possible futures for an expanding universe with $\Lambda = 0$. Either it keeps on expanding forever (if $k \leq 0$), or the expansion is eventually reversed and the universe collapses into a second singularity (if $k > 0$). The latter models are often called "oscillating" models, though there is no reason to suppose that this cycle is performed more than once.

We can define some useful parameters in terms of the scale factor R and its derivates, the density ρ, the pressure p, the gravitational constant G, the cosmological constant Λ, and the velocity of light c.

1 HUBBLE PARAMETER H Defined by $H = \dot{R}/R$, it has the dimensions of $(\text{time})^{-1}$ and characterizes the rate of expansion of the universe at any epoch. The value at the present epoch H_0 appears as the constant of proportionality in the Hubble velocity–distance law (see sec. 5.2).

2 HUBBLE TIME τ Defined by $\tau = 1/H$, it has the dimensions of a time, and characterizes the time scale for the expansion of the universe at any epoch. At the present epoch the Hubble time almost certainly lies in the range of 10–20 billion years.

3 DECELERATION PARAMETER q Defined by $q = -R\ddot{R}/\dot{R}^2$, it is dimensionless. It characterizes the extent to which the self-gravitation of the universe is slowing down the expansion at any epoch.

4 DENSITY PARAMETER Ω Defined by $\Omega = 8\pi G\rho H^2/3$, this is a dimensionless measure of the density of the universe. If $\Lambda = 0$, then $\Omega = 2q$.

5 PRESSURE PARAMETER Ψ Defined as $\Psi = 3p/\rho c^2$, it is a dimensionless measure of the extent to which the pressure of the matter and radiation in the universe makes a dynamical contribution to the evolution of the universe.

From Eq. (5.9),

$$kc^2 = R^2H^2(\Omega - 1) \qquad \text{if } \Lambda = 0.$$

Hence $k = +1$, 0, and -1 corresponds to $\Omega > 1$, 1, and <1, respectively. Thus whether the universe keeps on expanding forever or will eventually collapse again depends on whether $\Omega < 1$ or ≥ 1 if $\Lambda = 0$. The values of these parameters at the present epoch $t = t_0$ are denoted by H_0, τ_0, q_0, Ω_0, and Ψ_0.

TABLE 5.1

Age of the universe for $\Lambda = 0$ (units of Hubble time).

Ω_0	t_0/τ_0
0	1
0.01	0.980
0.03	0.955
0.1	0.896
0.2	0.847
0.3	0.809
0.4	0.783
0.5	0.755
0.75	0.688
1	0.667
1.5	0.613
2	0.570
3	0.499
4	0.471
5	0.440
10	0.314

[If $\Lambda \neq 0$, $p \neq 0$, we note that Eqs. (5.7) and (5.8) can be written as

$$\frac{\Lambda}{3} = H^2\left(\Omega\frac{1+\Psi}{2} - q\right)$$
$$kc^2 = R^2H^2\left(\Omega - q - 1 + \Omega\frac{1+\Psi}{2}\right). \tag{5.12}$$

Since the expansion of the universe is always being decelerated if $\Lambda = 0$, the expansion time scale τ_0 will always be longer than the age of the universe t_0. (In the physically unrealistic case of an empty universe, with $\rho_0 = 0$ and $\Lambda = 0$, then $\tau_0 = t_0$.) We can determine the exact relationship between τ_0 and t_0 from

$$t_0 = \int_0^{t_0} dt = \int_0^{R_0} \frac{dR}{\dot{R}} \tag{5.13}$$

where $R_0 = R(t_0)$. Using Eq. (5.11), valid if $\Lambda = 0$ and $p = 0$,

$$t_0 = \tau_0 \int_0^1 \frac{dx}{(\Omega_0/x + 1 - \Omega_0)^{1/2}} \tag{5.14}$$

where we have written $x = R/R_0$. If $\Omega_0 = 0$, $t_0 = \tau_0$; if $\Omega_0 = 1$, $t_0 = 2\tau_0/3$. For $0 < \Omega_0 < 1$ and $\Omega_0 > 1$ the substitutions $x = [\Omega_0/(1 - \Omega_0)] \sinh^2\theta$ and $x = [\Omega_0/(\Omega_0 - 1)] \sin^2\theta$ allow Eq. (5.14) to be evaluated. Table 5.1 shows how t_0/τ_0 varies with Ω_0, and we see that if $\Lambda = 0$, the Hubble time τ_0 gives an upper limit to the age of the universe.

THE HISTORY OF THE EARLY UNIVERSE

Let us now consider a universe containing both matter and radiation. Thus the mass density ρ becomes the sum of contributions due to the matter ρ_{m} and the radiation ρ_{r}, that is,

$$\rho = \rho_{\mathrm{m}} + \rho_{\mathrm{r}}.$$

We continue to neglect the contribution of matter to the pressure, but for the radiation the pressure is given by the usual expression for a relativistic fluid,

$$p_{\mathrm{r}} = \tfrac{1}{3}\rho_{\mathrm{r}}c^2. \tag{5.15}$$

Eq. (5.10) becomes

$$\frac{d}{dt}(\rho_{\mathrm{m}}R^3) + \frac{1}{R}\frac{d}{dt}(\rho_{\mathrm{r}}R^4) = 0. \tag{5.16}$$

If there is strict conservation of matter, that is, we neglect any conversion of matter to radiation, then each of the two terms in Eq. (5.15) will separately be zero, so

$$\begin{aligned} \rho_{\mathrm{m}} &= \rho_{\mathrm{m},0}\left(\frac{R_0}{R}\right)^3 \\ \rho_{\mathrm{r}} &= \rho_{\mathrm{r},0}\left(\frac{R_0}{R}\right)^4. \end{aligned} \tag{5.17}$$

This means that as the universe expands, the matter density ρ_{m} falls off inversely as the cube of the scale factor, while the mass density of radiation ρ_{r} falls off at a faster rate, inversely as the fourth power of the scale factor. At the present epoch the mass density of the universe is dominated by the contribution of the matter, and $\rho_{\mathrm{r},0}/\rho_{\mathrm{m},0} \sim 10^{-3} \sim \Psi_0$. But at epochs such that $R < R_0\Psi_0$ the universe would have been radiation dominated $(\rho_{\mathrm{r}} > \rho_{\mathrm{m}})$. When $\rho_{\mathrm{r}} \gg \rho_{\mathrm{m}}$, Eq. (5.9) has the solution

$$R \propto t^{1/2}.$$

Thus the early universe can be divided into two phases:

$$\begin{aligned} &\frac{R}{R_0} \ll \Psi_0 \qquad \text{radiation dominated } R \propto t^{1/2} \\ &\Psi_0 \ll \frac{R}{R_0} \qquad \text{Einstein–de Sitter (matter dominated), } R \propto t^{2/3}. \end{aligned} \tag{5.18}$$

If $\Omega_0 \simeq 1$ this second phase continues to the present day. If $\Omega_0 \neq 1$, it is valid for $R/R_0 \ll \Omega_0/|\Omega_0 - 1|$. If $\Omega_0 \ll 1$, then $R \propto t$ for $R/R_0 \gg \Omega_0$.

The temperature of the radiation T_{r} can be defined by the blackbody relation $\rho_{\mathrm{r}} = aT_r^4/c^2$. Thus from Eq. (5.17),

$$T_r(t) \propto \frac{1}{R(t)}. \tag{5.19}$$

As $R \rightarrow 0$, it is clear that T_r becomes infinite. The universe must also become infinitely optically thick. At the present epoch radiation traverses the universe freely, with only a small probability of being scattered or absorbed by dust or gas. But when the radiation density was high enough ($T_r \gtrsim 1500$ K, in fact), the universe was ionized, and the optical depth due to electron scattering was high. In the time it took the universe to double its size, each photon was scattered many times. In these circumstances the radiation may be expected to have had a perfectly blackbody spectrum during the radiation-dominated era, since matter and radiation were in perfect thermal equilibrium. Somewhat unexpectedly, this blackbody spectrum is maintained to the present day. The era of recombination, when the pregalactic gas became neutral, was so short that there was little opportunity for interaction of matter and radiation to distort the spectrum of the radiation from a blackbody form. And the subsequent isotropic expansion of the universe did not alter the blackbody character of the spectrum, though it shifted the temperature of the radiation (and the peak frequency, since $h\nu_{peak} \sim 0.3kT$ for a blackbody, where h is Planck's constant and k is the Boltzmann constant) to progressively lower values according to Eq. (5.19).

In 1965 Robert Dicke, James Peebles, and co-workers at Princeton identified the microwave background radiation discovered by Penzias and Wilson as this relic blackbody radiation. A recent compilation of the best available observations of the spectrum of this radiation is illustrated in Fig. 5.3. It is interesting that as early as 1950 Ralph Alpher and Robert Hermann calculated that the temperature of the relic radiation should be about 5 K, but they do not seem to have considered that this radiation could be observable.

Prior to the epoch of recombination, matter and radiation were locked together in thermal equilibrium, and the temperatures of matter and radiation were identical, $T_m(t) = T_r(t)$. After recombination the matter cooled off rapidly until the formation of galaxies, or, possibly, until pregalactic objects heated it up again (Fig. 5.4).

According to Eq. (5.19), for any particular elementary particle of mass m there must be an epoch at which the temperature is such that the energy of a typical photon is comparable to the rest-mass energy of the particle, that is, $kT \sim mc^2$. Prior to this epoch a collision between two photons can result in the creation of a particle pair, the particle and its antiparticle. The thermal equilibrium between matter and radiation ensures that at these earlier epochs, when $kT > mc^2$, there are roughly as many of these particles as photons. When the temperature drops below mc^2/k, particle pairs can no longer be created, and the particles and antiparticles rapidly annihilate each other, provided they are abundant enough to collide frequently. The history of the early universe can be roughly outlined as follows:

$t \sim 10^{-43}$, Planck era General relativity does not hold at this era because quantum effects cannot be neglected, and a quantum theory of gravity is required (see Sec. 1.6).

$10^{-43} < t < 10^{-23}$ s, Compton era the idea of a metric (see earlier in this section) is

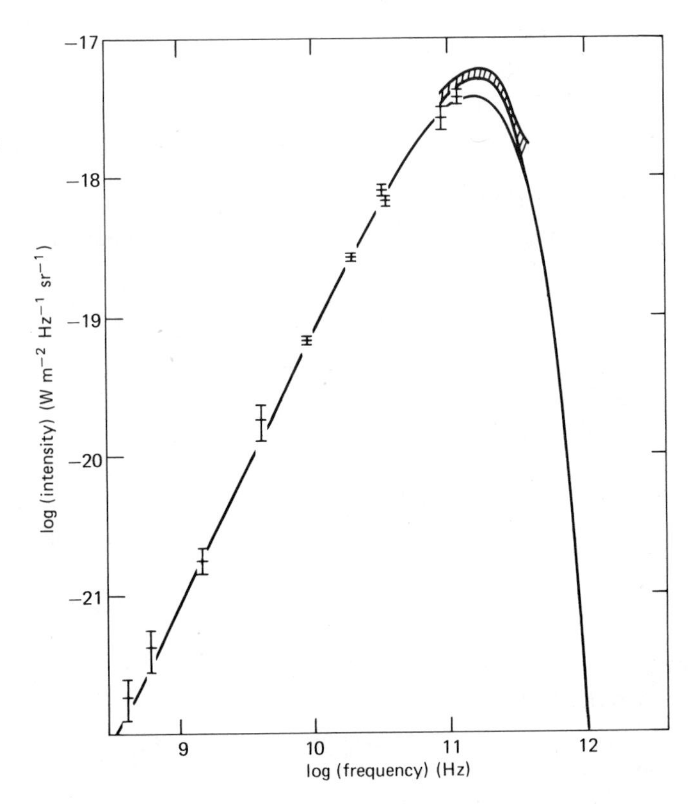

FIGURE 5.3

Summary of best observations of the intensity of the microwave background radiation as a function of frequency, compared with a 2.7 K blackbody. (From [R3].)

valid at this stage, but there are still some quantum gravitational effects, for example, the spontaneous creation of particle pairs and radiation at the expense of gravitational energy.

$t \sim 10^{-35}$ *s* The Grand Unified Theories predict that a phase transition occurs and

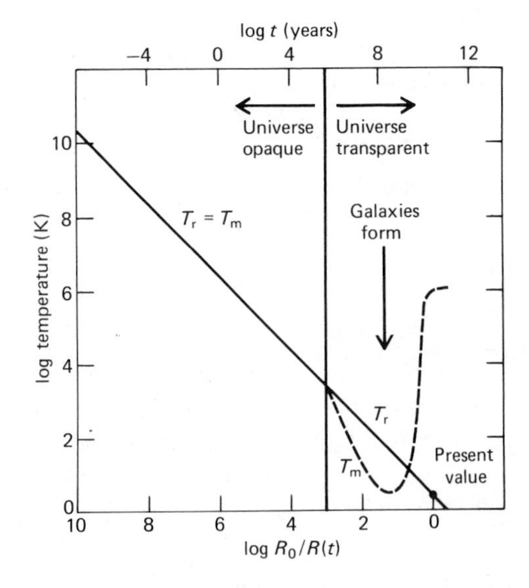

FIGURE 5.4

Variation of the temperature of matter and radiation with epoch. T_r—*radiation temperature;* T_m—*matter temperature. These remain equal until the epoch of recombination, indicated by the vertical line. After this,* T_m *cools off rapidly until galaxies form, when any intergalactic material remaining will be heated to a high temperature. (From [R3].)*

the grand unified force is broken down into the strong nuclear and the electroweak forces (see Sec. 1.6). Baryon nonconserving processes at this epoch would have resulted in the net excess of baryons over antibaryons that we see today, $\sim 10^{-9}$ baryons per photon.

$10^{-35} < t < 10^{-6}$ s, hadron era During this era the universe consists of a soup of quarks and leptons and their antiparticles in thermodynamic equilibrium.

$t \sim 10^{-12}$ s ($T \sim 10^{16}$ K) The unified electroweak force breaks down into the electromagnetic and the weak nuclear forces, leaving us with the four separate forces of physics (strong and weak nuclear, electromagnetic, and gravitational) that we see today (see Sec. 1.6). Because nothing much happens over the period from $t \sim 10^{-35}$ to $t \sim 10^{-12}$ s, this era is known to particle physicists as the "desert."

$t \sim 10^{-6}$ s ($T \sim 10^{13}$ K) Quarks and their antiparticles annihilate each other, and the residues combine to form protons and neutrons in equal numbers.

$t \sim 10^{-4}$ s ($T \sim 10^{12}$ K) Muons and antimuons annihilate each other. Muon neutrinos and antineutrinos decouple from everything else.

$t > 0.01$ s ($T < 10^{11}$ K) The neutron–proton mass difference (1.3 MeV/c^2) begins to shift the neutron–proton ratio through equilibrium of the weak interaction processes

$$n + \nu_e \leftrightarrow p + e^-$$
$$n + e^+ \leftrightarrow p + \bar{\nu}_e$$
$$n \leftrightarrow p + e^- + \bar{\nu}_e$$

where n, p, e^-, e^+, ν_e, and $\bar{\nu}_e$ denote the neutron, proton, electron, positron, and electron neutrino and antineutrino, respectively. The ratio of the number of neutrons to the number of protons depends on the temperature according to

$$\frac{N_n}{N_p} = \exp\left(-\frac{1.5 \times 10^{10}}{T}\right). \tag{5.20}$$

$t \sim 1$ s ($T \sim 10^{10}$ K) Electron neutrinos and antineutrinos start to decouple from everything else.

$t \sim 4$ s ($T \sim 5 \times 10^9$ K) Electrons and positrons annihilate each other, leaving a small excess of electrons. This, together with the cooling of the neutrinos by the expansion of the universe, puts an end to the weak interaction processes, with the exception of β-decay ($n \rightarrow p + e^- + \bar{\nu}_e$). The resulting ratio of neutrons to all nucleons

$$\frac{N_n}{N_n + N_p} = X_n \sim 0.16 \exp\left(-\frac{t}{1013}\,\mathrm{s}\right) \tag{5.21}$$

depends only on how long β-decay continues.

$t \sim 3$ min ($T \sim 10^9$ K) Nucleosynthesis begins through sequences of 2-body reactions like

$$n + p \leftrightarrow {}^2\mathrm{H} + \gamma$$
$${}^2\mathrm{H} + {}^2\mathrm{H} \leftrightarrow {}^3\mathrm{He} + n$$
$${}^3\mathrm{He} + n \leftrightarrow {}^3\mathrm{H} + p$$
$${}^3\mathrm{H} + {}^2\mathrm{H} \leftrightarrow {}^4\mathrm{He} + n.$$

The crucial step is the formation of deuterium (^{2}H), which, because of the low binding energy of the proton and neutron in its nucleus (2.2 MeV), is destroyed by collisions as soon as it is made if the temperature is higher than about 10^9 K. The absence of stable nuclei with atomic numbers $A = 5$ and 8 means that very little production of elements heavier than helium takes place. Almost all neutrons end up in ^{4}He nuclei, which have by far the highest binding energy of all nuclei with $A < 5$. In addition traces of ^{2}H, ^{3}He, and ^{7}Li are produced (Fig. 5.5).

Nucleosynthesis turns off the decay of free neutrons and fixes the neutron–proton ratio at the value just before the onset of nucleosynthesis. After nucleosynthesis is over, all the neutrons have been converted to helium nuclei, so the mass fraction in the form of helium is just twice the mass fraction in the form of neutrons before the onset of nucleosynthesis. This depends weakly on the density of matter in the universe (Fig. 5.5), since the latter affects the time that neutrons decay according to Eq. (5.21). Deuterium on the other hand is very sensitive to the density of matter in the universe because its final abundance is controlled by the efficiency of the processes which destroy it, and the cross-section for these reactions is proportional to the square of the matter density.

$t \sim 10^6$ years ($T \sim 1500$ K), epoch of recombination The temperature of the matter becomes too low to keep hydrogen ionized. Protons and electrons combine to form neutral atoms of hydrogen. The matter therefore becomes transparent to radiation, and matter and radiation decouple. In any region with above average density of matter, the self-gravity will now start to slow down the expansion of the region. The existence of galaxies and clusters today shows that there must have been irregularities in density present at the epoch of recombination. However, there is still controversy about the character of these density irregularities, or fluctuations. If the density fluctuations are *adiabatic,* that is, the densities of both matter and radiation are enhanced together, then during the radiation-dominated fireball era these fluctuations would have undergone oscillations in density. It is found that the lower mass fluctuations ($M < 10^{13} M_\odot$) would have been damped out, and so the only surviving ones are on the scale of galaxy clusters or larger. On the other hand, if the fluctuations are *isothermal,* that is, only the matter density is enhanced and the

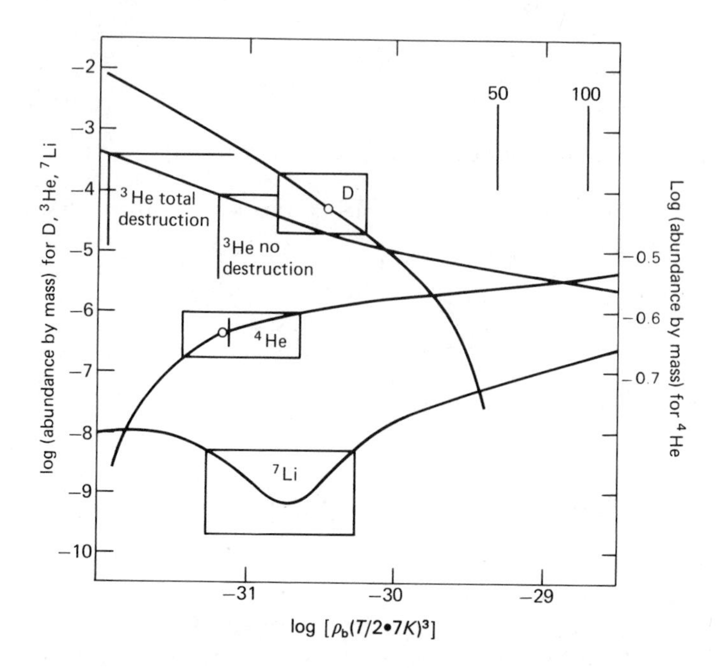

FIGURE 5.5

Comparison of primordial abundance estimates with predictions of the three-neutrino hot big-bang model. The logarithm of abundance is plotted against the logarithm of average baryon density in grams per cubic centimeter for an assumed microwave background temperature of 2.7 K. The ^{4}He scale (on the right) is expanded by a factor of 10 for clarity. Open circles indicate preferred values and the corresponding densitites, with error boxes. Vertical bars in the top right-hand corner indicate the densities corresponding to $\Omega_0 = 1$ when $H_0 = 50$ or 100 km s^{-1} Mpc^{-1}, while the vertical mark on the ^{4}He curve indicates the effect of 1 neutrino flavor more or less. Upper limits for the primordial ^{3}He abundance are shown depending on whether or not ^{3}He is destroyed in stars. (Adapted from [P1].)

radiation density remains homogeneous, the fluctuations would have survived the fireball era unscathed. Fluctuations could exist on arbitrarily small mass scales. However, the lowest mass regions in which gravity is sufficiently strong to overcome the resistance of pressure forces and cause eventual collapse are those with $M \gtrsim 10^6 M_\odot$ at recombination.

An important parameter for characterizing the density fluctuations is the density contrast $\Delta\rho/\rho = (\rho' - \rho)/\rho$, where ρ is the average density and ρ' is the density within the fluctuation. $\Delta\rho/\rho$ will be different on different mass scales. By studying clustering of galaxies it can be shown that $\Delta\rho/\rho$ at recombination must have increased as mass decreased. If this relation is extrapolated down to $10^6 M_\odot$, such fluctuations would have been far denser than their surroundings and would have immediately started to collapse. The time scale for collapse is proportional to $(\Delta\rho/\rho)^{-3/2}$, so the smallest mass fluctuations collapse first. Thus with isothermal fluctuations there is the possibility that a pregalactic population of objects of mass $\sim 10^6 M_\odot$ may have formed; and these objects might have fragmented into stars (perhaps with masses $\gg 100 M_\odot$). Such a generation of "stars" has been called

Population III, as an extension of the normal Populations I and II. With John Negroponte and Joe Silk, I have suggested that population III could explain an apparent distortion of the microwave background spectrum at wavelengths of 0.5–2 mm [R4]. If they existed, Population III stars would probably have completed their evolution by $t \sim 10^7$ or 10^8 years and would exist today only as dark remnants, for example, black holes or neutron stars.

$t \sim 10^8$–10^9 years Galaxies form either through the fragmentation of clusters (if fluctuations were adiabatic) or directly from isothermal fluctuations, or possibly through aggregation of lower mass objects. Since most of the oldest stars in the disk of our Galaxy already have a heavy-element abundance of about 1% by mass, the first 10^9 years in the life of a spiral galaxy may be a period of exceptionally active massive-star formation and high total luminosity. Ellipticals too are likely to be initially luminous since all their stars form at once, and massive stars will dominate the galaxy light output while they live.

5.2 Observable consequences of cosmological models

REDSHIFT AND THE PROPER DISTANCE

We now look at some of the observable consequences of the models. First we show that the light from distant sources is redshifted, and that the redshift increases with distance. To derive these results, we have to study how light propagates to us from a distant source, using the fact that according to general relativity, light travels along a null geodesic (see Sec. 5.1).

Consider a photon emitted by a galaxy Q at t_e, so that the event of emission is (r_0, θ_0, ϕ_0, t_e). If this photon is received at earth O at a later time t_0, the event of observation is (0, θ_0, ϕ_0, t_0) (Fig. 5.6). The relationship between t_e, t_0, and r_0 can be found from Eq. (5.4) by considering an element ds of the radial light ray from Q to O,

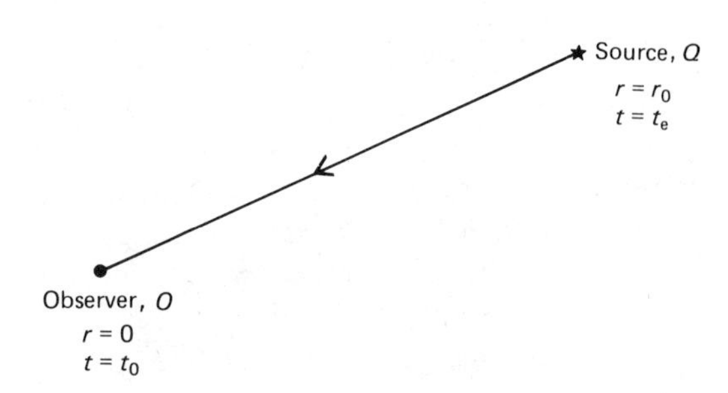

FIGURE 5.6

Radial null geodesic linking the events of emission and reception of a light signal sent from Q to O.

$$ds^2 = dt^2 - \frac{R^2(t)\,dr^2}{c^2(1 - kr^2)}$$

$$= 0$$

since light travels on a null geodesic. Thus

$$\frac{dr}{(1 - kr^2)^{1/2}} = -\frac{c\,dt}{R(t)} \tag{5.22}$$

for any element of the ray from Q to O (the minus sign has been chosen to correspond to an incoming signal). Summing this over the ray,

$$\int_0^{r_0} \frac{dr}{(1 - kr^2)^{1/2}} = \int_{t_e}^{t_0} \frac{c\,dt}{R(t)}. \tag{5.23}$$

Now r_0 was chosen to be a comoving coordinate. Thus it depends only on the particular galaxy Q, not on t_e or t_0. If we now consider a later photon emitted at $t_e + dt_e$ and observed at $t_0 + dt_0$,

$$\int_0^{r_0} \frac{dr}{(1 - kr^2)^{1/2}} = \int_{t_e+dt_e}^{t_0+dt_0} \frac{c\,dt}{R(t)} \tag{5.24}$$

or for small dt_0 and dt_e,

$$\frac{dt_e}{R(t_e)} = \frac{dt_0}{R(t_0)}. \tag{5.25}$$

Now suppose that these two events of emission correspond to successive wave crests. Then the frequencies at which the light is emitted and observed are given by $\nu_e = 1/dt_e$, $\nu_0 = 1/dt_0$, and so

$$\frac{\nu_e}{\nu_0} = \frac{dt_0}{dt_e} = \frac{R(t_0)}{R(t_e)} = 1 + z. \tag{5.26}$$

The quantity

$$z = \frac{\nu_e - \nu_0}{\nu_0} = \frac{\Delta\nu}{\nu_0} \tag{5.27}$$

is called the *redshift*. If the universe has expanded between t_e and t_0, then $z > 0$, corresponding to a reduction in frequency or an increase in wavelength. For visible light this means a shift toward the red end of the spectrum, but of course light of all frequencies from radio to γ-ray will be affected in the same way.

If we could observe light which was emitted from a source at $t_e = 0$, so $R(t_e) = 0$, it would be redshifted to infinite wavelength. However, as the universe is opaque for

$R(t)/R(t_0) \lesssim 10^{-3}$, we know that we cannot observe sources with redshift greater than that of the recombination epoch $z_{\rm rec} \sim 10^3$.

We now want to evaluate the relationship between the redshift and the radial proper distance to a source. First we have to find the relationship between the radial coordinate of a galaxy r_0 and its redshift. This can be calculated from Eq. (5.23) and (5.9) by (assuming $\Lambda = 0$, $p = 0$)

$$\int_0^{r_0} \frac{dr}{(1-kr^2)^{1/2}} = \int_{t_e}^{t_0} \frac{c\,dR}{R\dot{R}} = \frac{c\tau_0}{R_0} \int_{1/(1+z)}^{1} \frac{dx}{x(\Omega_0/x + 1 - \Omega_0)^{1/2}}$$
$$= \chi(z), \text{ say} \tag{5.28}$$

where $x = R/R_0$, and from Eq. (5.12),

$$\frac{c\tau_0}{R_0} = \begin{cases} |\Omega_0 - 1|^{1/2} & \text{if } \Omega_0 \neq 1,\ k \neq 0 \\ 1 & \text{if } \Omega_0 = 1,\ k = 0. \end{cases}$$

Then for $k = +1, 0, -1$, Eq. (5.28) implies $r_0 = \sin\chi(z)$, $\chi(z)$, and $\sinh\chi(z)$, respectively.

The radial proper distance from O to Q at epoch t_0 is then

$$d_{\rm pr}(r_0) = R_0 \int_0^{r_0} \frac{dr}{(1-kr^2)^{1/2}} = c\tau_0 \int_{1/(1+z)}^{1} \frac{dx}{[\Omega_0 x + (1-\Omega_0)x^2]^{1/2}}$$
$$= c\tau_0 \left(\frac{R_0}{c\tau_0}\right) \chi(z). \tag{5.29}$$

For example, if $\Omega_0 = 1$, then $d_{\rm pr} = 2c\tau_0[1 - (1+z)^{-1/2}]$.

At small distances ($d_{\rm pr} \ll c\tau_0$) and redshifts ($z \ll 1$) the relationship between proper distance and redshift takes the simple form

$$d_{\rm pr} \simeq c\tau_0 z = \frac{V}{H_0}$$

where V is the inferred recession velocity and H_0 is the Hubble constant. This is the famous Hubble law, that the distance is directly proportional to the redshift, or to the velocity of recession. Enormous effort has gone into measuring the constant of proportionality H_0, the Hubble constant, and this work will be reviewed in Sec. 6.1.

LUMINOSITY DISTANCE

We now want to see how the brightness of a distant source changes with distance from us. To calculate the flux of radiant energy observed at O from a distant source Q, consider a spherical surface $r = r_0$ centered on Q passing through O. The element of area at O defined by the four directions (θ, ϕ), $(\theta + d\theta, \phi)$, $(\theta, \phi + d\phi)$, and $(\theta + d\theta, \phi + d\phi)$ will subtend a solid angle (see Fig. 5.7),

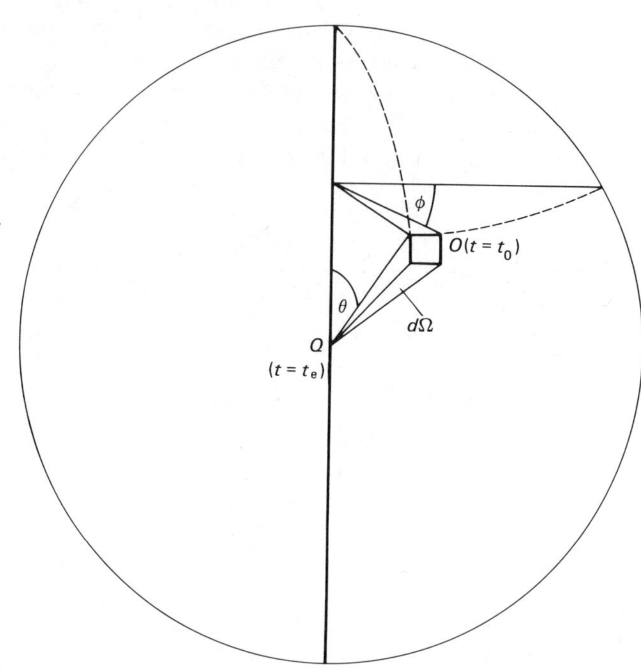

FIGURE 5.7
Light emitted by a source Q at $t = t_e$ into a small element of solid angle $d\Omega$ is received by O at $t = t_0$ on the small rectangular element of area, which forms part of the sphere $r = r_0$ centered on Q.

$$d\Omega = \sin\theta \, d\theta \, d\phi \qquad \text{at Q.}$$

To calculate the area A of this element, we must use the proper distance $(-c^2\,ds^2)^{1/2}|_{dt=0}$, so

$$\begin{aligned} A &= R(t_0)r_0\,d\theta \cdot R(t_0)r_0 \sin\theta\,d\phi \\ &= R_0^2 r_0^2\,d\Omega. \end{aligned}$$

For a unit area,

$$d\Omega = (R_0^2 r_0^2)^{-1}.$$

The total energy emitted per second into $d\Omega$ is $L\,d\Omega/4\pi$, where L is the total energy emitted by the source per second, that is, the luminosity. The flux received by O per unit area* is therefore

$$S = \frac{L}{4\pi R_0^2 r_0^2}(1+z)^{-2} \tag{5.30}$$

where one factor $(1+z)^{-1}$ is needed because the photons arrive with less energy than they set out with (since $E = h\nu$ for a photon where h is Planck's constant) and

* Neglecting interstellar extinction.

the second factor $(1+z)^{-1}$ because the photons arrive less frequently than they set off, by Eq. (5.26).

We define the *luminosity distance* d_{lum} as the distance the source would be assigned if the inverse-square law held, that is,

$$S = \frac{L}{4\pi d_{\text{lum}}^2} \tag{5.31}$$

so

$$d_{\text{lum}} = R_0 r_0 (1+z) \tag{5.32}$$

which can be evaluated using Eq. (5.28).

If $k=0$, $d_{\text{lum}} = (1+z)d_{\text{pr}}$, but otherwise the relationship between luminosity and proper distances is more complex. Eq. (5.32) can be expanded in a power series in z to give

$$d_{\text{lum}} = c\tau_0\{z - \tfrac{1}{2}(q_0 - 1)z^2 + \cdots\}. \tag{5.33}$$

To first order in z we have the Hubble redshift–distance law. The higher order terms depend on the cosmological parameters q_0 and Ω_0. This power series expansion is not very accurate, however, and it is usually better to use the exact formula. In terms of bolometric magnitudes, Eq. (5.31) to (5.33) become

$$\begin{aligned} m_{\text{bol}} &= M_{\text{bol}} + 5\log\left[\frac{d_{\text{lum}}(z)}{10\ \text{pc}}\right] \\ &= M_{\text{bol}} + 5\log\left[\frac{c\tau_0}{10\ \text{pc}}\right] + 5\log z - \frac{2.5(q_0-1)z}{\ln 10} + \cdots. \end{aligned} \tag{5.34}$$

In the above discussion L and S represent the total energy emitted by the source and the total energy per second per unit area received by the observer. In practice we usually observe in some relatively narrow band of wavelengths. Let $L(\nu_e)\,d\nu_e$ be the energy emitted by the source per second in the frequency range $(\nu_e, \nu_e + d\nu_e)$, and suppose that the corresponding energy received per second per unit area at the top of the earth's atmosphere is $S(\nu_0)\,d\nu_0$, where $\nu_0 = \nu_e(1+z)$. Then by Eq. (5.31),

$$S(\nu_0)\,d\nu_0 = \frac{P(\nu_e)\,d\nu_e}{d_{\text{lum}}^2}$$

so

$$S(\nu_0) = P[\nu_0(1+z)]\,\frac{1+z}{d_{\text{lum}}^2}. \tag{5.35}$$

Suppose the earth's atmosphere,* the telescope, and the detecting system result in a

* If there is interstellar extinction this will also contribute to $\phi(\nu_0)$.

fraction $\phi(\nu_0)$ of the energy incident on the atmosphere at frequency ν_0 being recorded. The measured flux in band B, say, is then

$$\begin{aligned} S_{\mathrm{B}} &= \int_{\mathrm{B}} \phi(\nu_0) S(\nu_0)\, d\nu_0 \\ &= \frac{1+z}{d_{\mathrm{lum}}^2} \int_{\mathrm{B}} \phi(\nu_0) P[\nu_0(1+z)]\, d\nu_0 . \end{aligned} \tag{5.36}$$

In terms of magnitudes,

$$m_{\mathrm{B}} = M_{\mathrm{B}} + 5 \log \left[\frac{d_{\mathrm{lum}}(z)}{10 \text{ pc}} \right] + K_{\mathrm{B}}(z) \tag{5.37}$$

where

$$K_{\mathrm{B}}(z) = -2.5 \log \left\{ (1+z) \frac{\int_{\mathrm{B}} \phi(\nu_0) P[\nu_0(1+z)]\, d\nu_0}{\int_{\mathrm{B}} \phi(\nu_0) P(\nu_0)\, d\nu_0} \right\}. \tag{5.38}$$

This quantity, called the K-correction, has been calculated for the standard photometric bands (such as U, B, V) for different types of galaxies from visible and ultraviolet observations of nearby galaxies (see, for example, [P7],[E1],[C9]).

DIAMETER DISTANCE

We now calculate how the apparent size of an object varies with distance from us. Consider an object of linear size D with radial coordinate $r = r_0$, subtending a small angle $\delta\theta$ at the observer O (Fig. 5.8). From the metric [Eq. (5.5)] the proper distance at time t_e between the ends of the object is $R(t_e) r_0 \delta\theta$, and this, by definition, is D. Thus

$$\delta\theta = \frac{D(1+z)}{R_0 r_0}. \tag{5.39}$$

We define the *diameter distance* as the distance that would be assigned to the object in a Euclidean geometry, that is,

$$d_{\mathrm{diam}} = \frac{D}{\delta\theta} = R_0 r_0 (1+z)^{-1} = d_{\mathrm{lum}} (1+z)^{-2}. \tag{5.40}$$

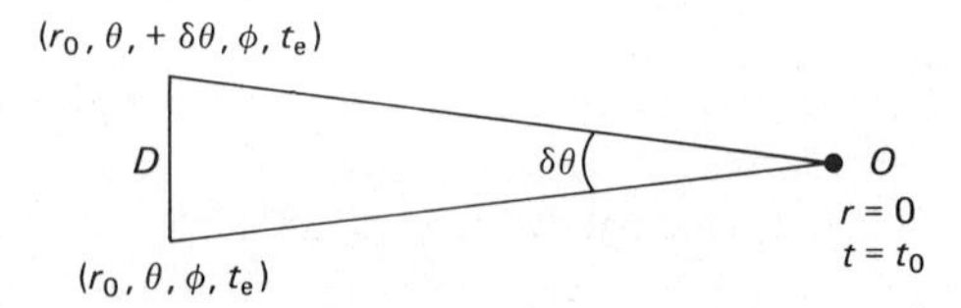

FIGURE 5.8

Source of size D, *emitting radiation at* $t = t_e$, *subtends an angle* $\delta\theta$ *at the observer* O, *observing at* $t = t_0$.

This refers to metric diameters (see Sec. 4.7). Isophotal diameters behave differently. The specific intensity of a source $I(r)$, or flux per unit solid angle as a function of radial metric distance r, has the form $S/(\delta\theta)^2$ and thus depends on redshift as

$$I_z(r) = I_0(r)(1+z)^{-4}. \tag{5.41}$$

Thus the metric diameter defined by a particular intensity level I_{lim} behaves like

$$D_z = d_{\text{diam}}(z)\theta_z$$

where θ_z is the isophotal diameter corresponding to the intensity level I_{lim}. Then from Eq. (5.41),

$$\begin{aligned} I_{\text{lim}} &= I_z\left(\frac{D_z}{2}\right) \\ &= (1+z)^{-4}\, I_0\left(\frac{D_z}{2}\right) \\ &= (1+z)^{-4}\, I_0\left[d_{\text{diam}}(z)\,\frac{\theta_z}{2}\right]. \end{aligned}$$

The isophotal diameter therefore depends on redshift as

$$\theta_z = \frac{2I_0^{-1}[(1+z)^{-4}I_{\text{lim}}]}{d_{\text{diam}}(z)}. \tag{5.42}$$

For example, if $I_0(r) \propto r^{-p}$, then

$$\theta_z = \frac{2\,r_1(1+z)^{-4/p}}{d_{\text{diam}}(z)} \tag{5.43}$$

where r_1 is such that $I_0(r_1) = I_{\text{lim}}$, and the isophotal diameter distance in this case is $(1+z)^{4/p}\, d_{\text{diam}}(z)$.

5.3 *Other cosmological models*

GENERAL RELATIVITY WITH THE Λ TERM

The introduction of a nonzero cosmological constant allows some interesting new cosmological models. For example, cosmological repulsion can exactly balance the self-gravity of the universe, giving a static universe. Of more physical interest are the Lemaître models, which start with a big bang and then have a long "quasi-stationary" phase with gravity and cosmological repulsion almost in balance, before cosmological repulsion wins and the expansion continues.

From Eq. (5.8) we see that the Λ term behaves like a constant energy density, present even when $\rho = 0$. $\Lambda/4\pi G$ can therefore be thought of as the energy density of the vacuum.

To derive the properties of models with $\Lambda \neq 0$, we write Eq. (5.8) as

$$\dot{R}^2 = 8\pi G\rho_0 \frac{R_0^3}{3R} - kc^2 + \Lambda \frac{R^2}{3} = G(R) \tag{5.44}$$

assuming $p = 0$, so $\rho \propto R^{-3}$. The properties of the models depend on the actual value of Λ.

☐ *If* $\Lambda < 0$, there exists a critical value of R, R_c, such that $G(R_c) = 0$ and $G(R) < 0$ for $R > R_c$. This is physically impossible since we cannot have $\dot{R}^2 < 0$. The scale factor R is therefore restricted to the range $0 \leq R \leq R_c$ for all t, and we have "oscillating" models for all $\Lambda < 0$.

☐ *If* $\Lambda > 0$ *and* $k \leq 0$, then $\dot{R}^2 > 0$ for all R, so we have a monotonically expanding universe, the only difference from that with $\Lambda = 0$ being that at large t, $\dot{R}^2 \simeq R^2/3$, so

$$R \propto \exp\left[\left(\frac{\Lambda}{3}\right)^{1/3} t\right]. \tag{5.45}$$

(If $k = 0$ and $\Omega_0 = 0$, then this is true for all t, and we have the de Sitter model.)

☐ *If* $\Lambda > 0$ *and* $k = 1$, there is a critical value of Λ, Λ_c, such that $R = 0$ and $\dot{R} = 0$ can both be satisfied at once, given by $\Lambda_c = (kc^2)^3/(4\pi G\rho_0 R_0^3)^2$. This means that there is the possibility of a static model with $R = R_c$ for all t, and $\Lambda_c = 4\pi G\rho_c = kc^2/R_c^2$. This is the Einstein static model and was the first cosmological model derived from general relativity. Indeed, Einstein introduced the Λ term in order to be able to make a static model for the universe, a move that he later regretted.

☐ *If* $\Lambda > \Lambda_c$, $G(R) > 0$ for all R, so we again have a monotonically expanding universe.

☐ *If* $\Lambda = \Lambda_c$, then apart from the Einstein static model, there are two models that approach this asymptotically. One expands out gradually from the Einstein static state at $t = -\infty$, and then turns into an exponential expansion; the other expands out from a big bang, and then tends asymptotically to the Einstein model as $t \rightarrow \infty$. These two models are called the Eddington–Lemaître models (Fig. 5.9). If Λ is close to Λ_c but slightly larger, that is, $\Lambda = \Lambda_c(1 + \epsilon)$, with $\epsilon \ll 1$, we have the Eddington–Lemaître models: for a long period of time R is close to R_c (the quasi-stationary epoch) with gravitation and cosmological repulsion almost in balance. Finally repulsion wins and the expansion continues. We might expect to see a big concentration of galaxies with redshifts given by $1 + z \simeq R_0/R_c$.

☐ *If* $0 < \Lambda < \Lambda_c$, finally, we have two possibilities: either the familiar oscillating model or a bounce model, in which the universe contracts from an infinitely extended state until the cosmological repulsion halts the contraction and turns it into an expansion out to infinity again (Fig. 5.9b).

The cosmological term has a large effect on the age of the universe, extending it indefinitely in some $\Lambda > 0$ models. For $\Lambda \neq 0$, the expression for the age of the universe becomes [cf Eq. (5.14)]

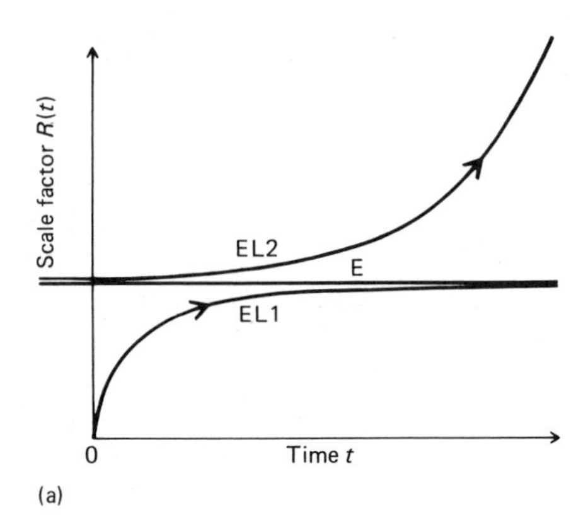

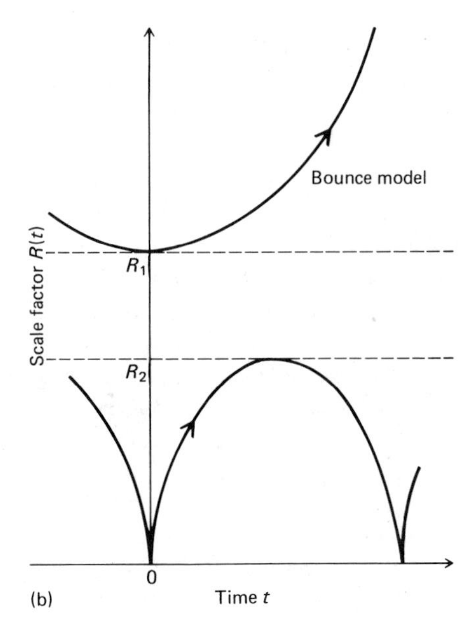

FIGURE 5.9

Dependence of scale factor R(t) *on time* t *for models with cosmological constant* $\Lambda > 0$. *(a)* $k = +1$, $\Lambda = \Lambda_c$. *E—Einstein static model;* EL_1, EL_2*—Eddington–Lemaître models. (b)* $k = +1$, $0 < \Lambda < \Lambda_c$, *showing bounce and oscillating models.*

$$t_0 = \tau_0 \int_0^1 \frac{dx}{[\Omega_0/x - 3\Omega_0/2 + q_0 + 1 + (\Omega_0/2 - q_0)x^2]^{1/2}} \tag{5.46}$$

and Eq. (5.28) becomes

$$\chi(z) = \frac{c\tau_0}{R_0} \int_{1/(1+z)}^1 \frac{dx}{x[\Omega_0/x - 3\Omega_0/2 + q_0 + 1 + (\Omega_0/2 - q_0)x^2]^{1/2}} \tag{5.47}$$

where

$$c\tau_0/R_0 = \begin{cases} |3\Omega_0/2 - q_0 - 1|^{1/2} & \text{if } k \neq 0 \\ 1 & \text{if } k = 0. \end{cases}$$

These are evaluated in Sec. 6.3 for selected models.

COLD AND TEPID BIG-BANG MODELS

Although it is widely believed that the microwave background radiation points to a hot big-bang universe, some other ideas have been put forward. In 1977 Bernard Carr [C6] revived the theory which had been prevalent before the work of Gamov, that the universe originated in a cold big bang, with no radiation present initially. In the same year Carr and Martin Rees proposed the alternative that the universe may have initially contained far fewer photons than the 10^9 per baryon we see today (the "tepid" big bang) [C7]. The remaining photons of the observed microwave background would then have to be generated from astrophysical processes, for example, accretion of gas onto massive black holes, and be thermalized either by free–free absorption or by dust with unusual properties.

CHAOTIC COSMOLOGY

In the conventional hot big-bang model, explaining the density fluctuations from which galaxies grow has been extremely difficult, except as due to initial conditions. Charles Misner suggested in 1967 [M4] that the universe may have started off with an arbitrary degree of anisotropy and inhomogeneity and that most of this was smoothed out during the early Compton era (see Sec. 5.1). However, it has proved difficult to find a sufficiently efficient mechanism. A further difficulty with this type of model is that the observed entropy per baryon ($S = 4aT^3/3nk = 74.0\ T^3/n \sim 10^9$) does not seem large enough if arbitrarily large amounts of inhomogeneity and anisotropy have been smoothed out [B3]. John Barrow, elaborating an idea by Rees, has suggested a partially chaotic model in which inhomogeneities are generated by "shear" fluctuations, that is, those associated with differential rotation between different parts of the universe [B2],[R1]. The high degree of isotropy of the microwave background shows that any rotation or anisotropy in the expansion of the universe must be very small, and even stronger limits follow from attempts to model the observed primordial abundances of helium and deuterium in anisotropically expanding models [B1],[O1].

VARIABLE G AND OTHER MODELS

The idea that the gravitational constant might vary over cosmological time scales goes back to Milne and Dirac in the 1930s. In 1961 C. Brans and R.H. Dicke proposed a detailed cosmology in which the gravitational constant G changes with time [B5]. For a time this was the main serious rival to general relativity, but solar-system experiments have shown that any deviations from general relativity are so small as to make the Brans–Dicke models uninteresting. Variable G models proposed by Paul Dirac [D2],[D3] to explain coincidences in large dimensionless numbers (see Sec. 6.4) also appear to be in conflict with observations, though V. Canuto and co-workers have developed models of this type which they claim to be consistent with available observations, including the microwave background radiation [C1]–[C5].

The steady-state model of Bondi, Gold, and Hoyle [B4],[H3] has long ago been refuted by evidence for evolution in quasars and radio galaxies and by the discovery of the microwave background radiation, but remains of historical interest. The cosmological principle was strengthened to include the postulate that the universe retains the same appearance at all times. The expansion of the universe then requires that matter be created continuously to maintain the mean density of the universe constant.

I.E. Segal [S4] has put forward a *chronometric cosmology* which, because of its abstractness and its prediction of a quadratic dependence of redshift on distance, has found few supporters.

5.4 Absolute motion of the earth

In order to estimate the size of the universe using the velocity–distance law, it is necessary to correct for the effects of any local motions of which we are part, for example, the earth's motion around the sun, the sun's motion around the Galaxy, and the Galaxy's motion with respect to other nearby galaxies. We essentially have to try to determine the absolute motion of the earth or, more precisely, its motion with respect to the local cosmological frame of reference (the one in which the expansion of the universe looks perfectly isotropic).

The earth's motion around the sun can be detected both through the phenomenon of aberration (see Sec. 1.3) and through the parallax of nearby stars (see Sec. 2.1). The motion of the solar system relative to the other nearby stars was discussed in Sec. 2.3, where the Local Standard of Rest was defined. The sun is believed to be rotating around the Galaxy at a speed of between 200 and 300 km s^{-1}. Estimates of the sun's rotation velocity are summarized in Table 5.2.

The Local Group of galaxies consists more or less of our Galaxy, M31, and their dwarf satellites. The radial velocity of M31 is -300 km s^{-1}, but its transverse velocity cannot be measured. The net motion of the Local Group can therefore only be found by observing more distant sources of radiation.

Numerous efforts have been made to determine the velocity of our Galaxy with respect to other nearby galaxies, after correction for the sun's rotation around the Galaxy. These measurements are summarized in Table 5.3, where it can be seen that agreement between them is poor. Vera Rubin and co-workers attempted to measure the velocity of the Galaxy with respect to a distant sample of Sc I galaxies, and obtained an extremely high velocity of ~ 600 km s^{-1} [R5]. Sandage and Tammann, however, argued that the Galaxy could not have such a high velocity with respect to the galaxies within 20 Mpc [S1].

Since the Local Group is part of the Local Supercluster, with its strong concentration of mass at the Virgo cluster, we may expect the motion of the Local Group to be affected by the collective gravitational attraction of the Supercluster. Basically we may expect to find that our velocity away from Virgo is lower than would be expected from the general cosmological expansion. This effect was first reported by de Vaucouleurs in 1958 [V1]. The pattern of velocities around Virgo has been calculated on the basis of simple spherical models introduced by Joe Silk and James Peebles [S5],[S6],[P3], in which the density profile in the cluster is assumed to fall off

TABLE 5.2

*Velocity of the sun around the Galaxy, Θ_0.**

Method	*Authors*	V_0 *(km s⁻¹)*	l *(deg)*	b *(deg)*	Θ_0 *(km s⁻¹)*
1 *Velocity of sun relative to the halo*					
Globular clusters	Mayall (1946) [M1]	200 ± 25	87	0	
Globular clusters $>$ 9 kpc,	Kinman (1959) [K1]	182 ± 50	87	2	
Globular clusters	Woltjer (1975) [W2]				$200-225$ ($R_0 = 9$)
Globular clusters	Frenk and White (1980) [F2]				$200-224$
RR Lyrae stars	Wooley and Savage (1971) [W3]	225 ± 25	90	6	
Highest velocity stars, on verge of escape	Isobe (1974) [I1]				275 ± 20
2 *Solar motion relative to velocity centroid of Local Group*					
	Mayall (1946) [M1]	300 ± 25	93 ± 6	-14 ± 6	
	Humason and Wahlquist (1955) [H4]	292 ± 32	106 ± 6	-7 ± 4	
	Byrnes (1966) [B7]	280 ± 23	107 ± 5	-7 ± 4	
	de Vaucouleurs and Peters (1968) [V4]	315 ± 15	95 ± 6	-8 ± 3	
	Yahil et al. (1977) [Y2]	308 ± 23	105 ± 5	-7 ± 4	
	de Vaucouleurs and Peters (1981) [V5]	336 ± 17	107 ± 6	-16 ± 6	
	Average	305 ± 25	105 ± 5	-10 ± 4	
3 *21-cm line profiles*	Knapp et al. (1978) [K3]				220
4 *Relative motion of M31 and the Galaxy*					
	Lynden-Bell and Lin (1977) [L1]				$230-340$ ($R_0 = 10$) $180-270$ ($R_0 = 8$)
5 *Oort's constants*		*A*	*B*	R_0 *(kpc)*	
	IAU values	15	-10	10	250
	Mihalas and Binney (1981) [M3]	16	-11	9	245 ± 40

TABLE 5.3

*The Galaxy's absolute motion in space.**

Method	Authors	V ($km\ s^{-1}$)	l (deg)	b (deg)
1 Galaxies	Rubin et al. (1976) [R5]	454 ± 125	163	-11
	de Vaucouleurs (1978) [V3]	310 ± 60	180	50
	Sandage et al. (1979) [S2]	245 ± 60	205	31
	de Vaucouleurs and Peters (1981) [V5]	230 ± 60	194	37
	de Vaucouleurs et al. (1981) [V6]	205 ± 40	207	47
2 Microwave background	Smoot and Lubin (1979) [S10]	520 ± 75	264 ± 10	33 ± 10
	Cheng et al. (1979) [C8]	540 ± 36	280 ± 6	30 ± 6
	Broughn et al. (1981) [B6]†	653 ± 33	273 ± 6	27 ± 5
	Gorenstein and Smoot (1981) [G1]†	567 ± 60	256 ± 9	41 ± 9
3 HI fluxes, Sbc galaxies ($1000 < V_0 < 5500$ km s^{-1}, $i > 30°$, excluding high Virgocentric velocities)	Hart and Davis (1982) [H1]†	436 ± 55	264 ± 18	45 ± 12
HI fluxes, Sc galaxies ($1000 < V_0 < 5000$ km s^{-1})	Hart and Davis (1982) [H1]†	550 ± 62	245 ± 8	35 ± 9
4 Infrared Tully–Fisher	Aaronson et al. (1982) [A2]	366 ± 41	28	70
	Average of values by authors labeled †	**546 ± 70**	**261 ± 9**	**39 ± 7**

* Corrected for an assumed solar rotation around the Galaxy of $V_0 = 300$ km s^{-1} toward $l = 90°$ and $b = 0°$.

with the distance r from the center of the Virgo cluster as $r^{-\gamma}$, the mean density of the cluster interior to the position of the Galaxy being ρ_1, and the whole cluster being superposed on a uniform background density ρ_u. The free parameters of the model, γ and ρ_1/ρ_u, are generally taken as 2 and 2–4, respectively. Different estimates of the total velocity of our Galaxy toward the Virgo cluster are given in Table 5.4, and it is indicated there whether an infall model has been used.

Recently a very important advance has been made in our efforts to determine the "absolute" motion of the earth. Groups at Berkeley, Princeton, and Florence have measured our velocity with respect to the microwave background radiation through the dipole anisotropy in the intensity, or in the blackbody temperature of the radiation. Due to the Doppler shift the intensity and the temperature are raised in the direction toward which the earth is traveling and reduced in the opposite direction. If θ is the angle between the direction of the earth's velocity V and that of observation, the special theory of relativity predicts that the temperature will depend on direction as

$$T(\theta) = T_0 \frac{1 + V \cos \theta/c}{(1 - V^2/c^2)^{1/2}}. \tag{5.48}$$

Fig. 5.10 shows some observations of $T(\theta)$, and part 2 of Table 5.3 summarizes the velocities obtained by the different groups, which are in good agreement with each other.

The problem immediately arises that these measurements do not agree very well with those found in most attempts to measure our velocity with respect to nearby galaxies. This could be because the measurements with respect to nearby galaxies are in error, or because the dipole anisotropy in the background radiation is not caused solely by our motion. The existence of galaxies and clusters shows that density inhomogeneities are present, with an amplitude that declines as the mass scale increases. Density perturbations on very large scales ($\gtrsim 100$ Mpc) are capable of causing intrinsic dipole anisotropy in the background unconnected with our local peculiar motion: these have been discussed by Peebles [P6] and by Silk and Wilson [S8]. A further consequence of density perturbations on very large scales is that a quadrapole anisotropy should also be present in the background, but this has not yet been observed, and this suggests that the dipole anisotropy is due to our Galaxy's motion rather than to inhomogeneity.

Recently L. Hart and Rod Davis [H1] have completed a very interesting radio study of the motion of our Galaxy with respect to nearby galaxies. Their method of estimating the distances of the galaxies is similar to that of Tully and Fisher (see Sec. 4.4) in that the 21-cm line width is used as a luminosity indicator, but in place of the optical luminosity they use the luminosity of the galaxy in the 21-cm line. Their Tully–Fisher relation shows considerably more scatter ($\sigma \sim 1$ mag) than those based on optical or infrared luminosities. In spite of this, Hart and Davis's value for the magnitude and the direction of our Galaxy's motion (Table 5.3) agrees well with that deduced from the microwave background. They note that their estimate of our velocity toward Virgo (Table 5.4) agrees well with other estimates which are

TABLE 5.4

*Component of our Galaxy's peculiar velocity in the direction of Virgo.**

	Authors	V_{Virgo}
1 *Methods using galaxies*		
Luminosity class and HII region diameters of spirals	Sandage and Tammann (1975) [S1]	-39 ± 162
Infall model	Peebles (1976) [P3]	250
Surface brightness and diameter of ellipticals	Kormendy (1976) [K4]	-39 ± 114
Redshifts of Sc galaxies	Rubin et al. (1976) [R5]	-138
	De Vaucouleurs (1978) [V3]	219 ± 50
Luminosity class of spirals	Tammann et al. (1979) [T1]	60 ± 132
	Sandage et al. (1979) [S2]	136 ± 60
Infall model	Hoffman et al. (1980) [H2]	240 ± 70
Absolute magnitude versus velocity dispersion in E	Schecter (1980) [S3]	190 ± 130
	Yahil et al. (1980) [Y3]	125 ± 40
Infrared Tully–Fisher	Aaronson et al. (1980) [A1]	$480 \pm 75^{\dagger}$
Absolute magnitude versus velocity dispersion		
In E	Tonry and Davis (1981) [T2]	$470 \pm 75^{\dagger}$
In SO	Tonry and Davis (1981) [T2]	416 ± 90
Luminosity class and diameter of spirals	De Vaucouleurs and Peters (1981) [V5]	130 ± 60
Tully–Fisher relation in blue	De Vaucouleurs et al. (1981) [V6]	153 ± 40
HI measurements of Sbc and Sc galaxies ($1000 < V_0 < 5000$)	Hart and Davis (1982) [H1]	$410 \pm 55^{\dagger}$
Infall model	Yahil et al. (1980) [Y3]	230 ± 75
Infrared Tully–Fisher	Aaronson et al. (1982) [A2]	
Pure infall		$257 \pm 32^{\dagger}$
Infall plus peculiar velocity		$331 \pm 41^{\dagger}$
2 *Methods using microwave background*†		
	Smoot and Lubin (1979) [S10]	373 ± 25
	Cheng et al. (1979) [C8]	385 ± 28
	Broughn et al. (1981) [B6]	439 ± 23
	Gorenstein and Smoot (1981) [G1]	411 ± 36

* After correction for solar motion around Galaxy.
† Methods unaffected by interstellar extinction.

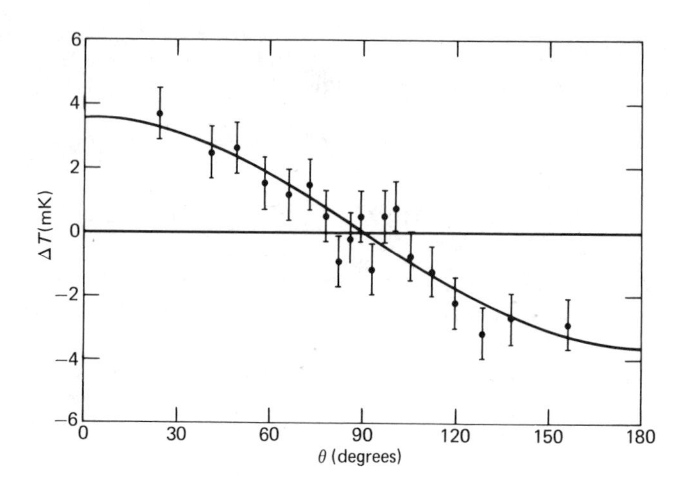

FIGURE 5.10
Plot of measured anisotropy in microwave background temperature ΔT *against angle* θ *between the direction of the earth's motion and that of observation. Solid line is the best-fitting curve of the form Eq. (5.48). (From [G1].)*

relatively free from the effects of interstellar and internal extinction, namely, those from the microwave background, that of M. Aaronson and co-workers using the infrared Tully–Fisher method [A1], and that of J.L. Tonry and M. Davis using the absolute magnitude–velocity dispersion relation in ellipticals (see Sec. 4.4). This suggests that the failure of other methods to give the correct absolute velocity of our Galaxy may be partly due to incorrect allowance for the effects of dust. A second factor, which may have masked this high infall velocity toward Virgo, is that methods which do not model the infall, and in which nearby galaxies ($V_0 \lesssim$ 1000 km s^{-1}) contribute a significant weight, will yield spuriously low values for our velocity toward Virgo because the nearby galaxies have an infall velocity that is similar to our own. Table 5.3 gives the combined best estimate of the velocity of our Galaxy in space.

5.5 Density inhomogeneities

The work of Hart and Davis, together with the small amplitude of the quadrupole anisotropy of the microwave background, suggest that the intrinsic component of the dipole anisotropy, that is, that associated with large-scale density inhomogeneities, is small (<20%, say). Most of the dipole anisotropy seems to be due to our motion with respect to the local cosmological frame. However, only part of this motion may be due to the attraction of the Local Supercluster. Studies of the local flow using infall models have consistently yielded infall velocities to Virgo of ~250 km s^{-1}, far lower than the velocity derived from studies of the microwave background anisotropy. A second contribution to our local motion could be a rotation of the Local Supercluster, an effect first claimed by de Vaucouleurs in 1958 and confirmed since by several investigators [V1],[V2],[S11],[A2]. This may be needed to explain the fact that the best estimate of our absolute velocity given in Table 5.3 has a substantial component of velocity in the supergalactic plane perpendicular to Virgo. A third contribution could be caused by density inhomogeneities on a scale of 100 Mpc or more.

Density inhomogeneities which were present when the microwave background radiation last interacted with matter ($z \sim 1000$) should also lead to anisotropies in the background on small angular scales. It is usually assumed that at the epoch of recombination, the amplitude of density perturbations on scale l has an amplitude given by

$$\left(\frac{\delta\rho}{\rho}\right)_l^2 \propto l^n \tag{5.49}$$

so that the corresponding dependence on the mass scale M is, say,

$$\frac{\delta\rho}{\rho} \propto M^{-1/2-n/6} = M^{-\alpha}. \tag{5.50}$$

For example $\alpha = \frac{1}{2}$ ($n = 0$) corresponds to Poissonian fluctuations (amplitude the same on all length scales), whereas $\alpha = \frac{2}{3}$ ($n = 1$) corresponds to "constant-curvature" fluctuations. (Fluctuations always have the same amplitude when they first come within the particle horizon, that is, the first moment when all parts of the fluctuation can be in communication with each other, so the fluctuation can be thought of as a coherent entity. See Sec. 6.3 for precise definition of particle horizon.)

Some support for a simple power-law spectrum of density fluctuations comes from studies of galaxy clustering, which can be described by the *covariance function* $\zeta(r)$, defined as the excess probability of finding a galaxy at a distance r from a random galaxy (compared with an unclustered distribution of the galaxies),

$$dp = n[1 + \zeta(r)]\, dV \tag{5.51}$$

where n is the average number density of galaxies and dV is a small element of volume. Peebles and co-workers have shown from studies of a variety of galaxy catalogs that $\zeta(r)$ has a simple power-law form,

$$\zeta(r) = \left(\frac{r}{r_0}\right)^{-1.77}, \qquad \text{for } 0.1 \leq r \leq 10 \text{ Mpc} \tag{5.52}$$

where $r_0 = 5(100/H_0)$ Mpc, corresponding to a mass scale of $\sim 10^{15} M_\odot$ [P4].

J.R. Gott and E.L. Turner have shown that this relation extends down to scales of 10 kpc [G3]. The galaxy redshift survey of Marc Davis et al. [D1] shows that the galaxy distribution has considerable structure on scales larger than 10 Mpc, which may also satisfy Eq. (5.52). However, very large voids in the galaxy distribution have also been found [K2], so that $\zeta(r)$ may go negative on very large scales (> 20 Mpc).

Computer simulations of the gravitational clustering of galaxies, the so-called N-body calculations, show that to generate a covariance function of the form given by Eq. (5.52), we require n in Eq. (5.50) to satisfy $-2 \leq n \leq 0$ ($\frac{1}{6} \leq \alpha \leq \frac{1}{2}$) for isothermal fluctuations and $2 \leq n \leq 4$ ($\frac{5}{6} \leq \alpha \leq \frac{7}{6}$) for adiabatic fluctuations [G2],[P5],[S7]. Fig. 5.11 summarizes attempts to measure small-scale fluctuations in the microwave background radiation on different angular scales. The combination of observations of the dipole and quadrupole anisotropies together with the detection of small-scale anisotropies would allow conclusions to be drawn about n, Ω_0, and the type of perturbation. The present position has been reviewed by Bruce Partridge [P2] and by Joe Silk [S7]:

1 The observed dipole anisotropy and the limit on the quadrupole anisotropy are consistent with isothermal fluctuations if $n \sim 0$, but not with adiabatic perturbations unless the density of the universe is dominated by massive neutrinos (or other nonbaryonic matter), in which case $0 \lesssim n \lesssim \frac{1}{2}$ would be acceptable.
2 The possible detection of small-scale anisotropies, $\Delta T/T \sim (3.7 \pm 0.7) \times 10^{-5}$, with a 3° telescope beam at millimeter wavelengths by the Italian group led by Francesco Melchiorri [F1],[M2] requires $n \gtrsim 1$ for adiabatic and $n \gtrsim 0$ for isothermal fluctuations if $\Omega_0 = 1$, and $n \gtrsim 3$ for adiabatic and $n \gtrsim 1$ for isothermal fluctuations if $\Omega_0 = 0.1$.
3 The best centimetric wavelength limits require $\Omega_0 = 1$.

Thus all the available information on density fluctuations in the early universe is consistent with there having been isothermal density fluctuations with $n = 0$ (that is, Poissonian) in an $\Omega_0 = 1$ (Einstein–de Sitter) universe, and there are some difficulties in constructing a consistent adiabatic fluctuation picture.

An important modification to all these arguments applies if between the epochs of decoupling ($z \sim 1000$) and galaxy formation, the matter in the universe was reionized, for example, as a result of heating by Population III objects. In this case information about the initial fluctuation spectrum would be lost, except on the very largest scales, but there would also be the interesting possibility of small-scale anisotropies in the background intensity whose amplitude depended on the wavelength observed, for example because of the wavelength-dependent opacity of dust ejected from Population III stars.

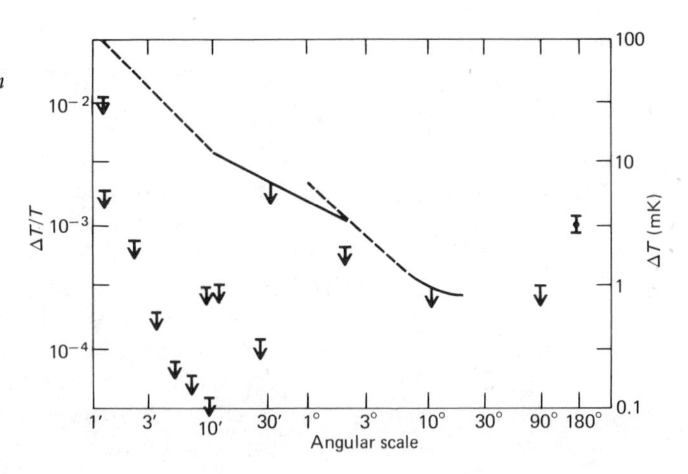

FIGURE 5.11
Amplitude of anisotropy in the cosmic microwave background ΔT/T on different angular scales. (Adapted from [S9].)

BIBLIOGRAPHY

[A1] Aaronson, M., Mould, J., Huchra, J., Sullivan, W.T., III, Schomer, R.A., and Bohun, G.S., "A Distance Scale from the Infrared Magnitude/HI Velocity Width Relation. III—The Expansion Rate outside the Local Supercluster." *Astrophys. J.*, 1980, **239,** pp. 12–37.

[A2] Aaronson, M., Huchra, J., Mould, J., Schechter, P.L., and Tully, R.B., "The Velocity Field in the Local Supercluster," *Astrophys. J.*, 1982, **258,** pp. 64–76 .

[A3] Arp, H., and Hazard, C., "Peculiar Configurations of Quasars in Two Adjacent Areas of the Sky," *Astrophys. J.*, 1980, **240,** pp. 726–736.

[A4] Arp, H., "Quasars near Companion Galaxies," *Astrophys. J.*, 1981, **250,** pp. 31–42.

[A5] Arp, H., "Further Observations and Analysis of Quasars near Companion Galaxies," *Astrophys. J.*, 1983, **271,** pp. 479–506.

[B1] Barrow, J.D., "Light Elements and the Isotropy of the Universe," *Month. Not. R. Astron. Soc.*, 1976, **175,** pp. 359–370.

[B2] Barrow, J.D., "A Chaotic Cosmology," *Nature,* 1977, **267,** pp. 117–120.

[B3] Barrow, J.D., and Matzner, R.A., "The Homogeneity and Isotropy of the Universe," *Month. Not. R. Astron. Soc.*, 1977, **181,** pp. 719–727.

[B4] Bondi, H., and Gold, T., "The Steady-State Theory of the Expanding Universe," *Month. Not. R. Astron. Soc.*, 1948, **108,** pp. 252–270.

[B5] Brans, C., and Dicke, R.H., "Mach's Principle and a Relativistic Theory of Gravitation," *Phys. Rev.*, 1961, **124,** pp. 929–935.

[B6] Broughn, S.P., Cheng, E.S., and Wilkingson, D.T., "Dipole and Quadrupole Anisotropy of the 2.7K Radiation," *Astrophys. J.*, 1981, **243,** pp. L113–L117.

[B7] Byrnes, D.V., "New Values of the Solar Motion with Respect to the Local Group and External Galaxies," *Publ. Astron. Soc. Pacific,* 1966, **78,** pp. 46–48.

[C1] Canuto, V., Hsieh, S.H., and Adams, P., "Scale-Covariant Theory of Gravitation and Astrophysical Applications," *Phys. Rev. Lett.*, 1977, **39,** pp. 429–432.

[C2] Canuto, V., and Hsieh, S.H., "Scale Covariance and *G*-Varying Cosmology. II—Thermodynamics, Radiation, and the 3K Background," *Astrophys. J.*, suppl. ser., 1979, **41,** pp. 243–262.

[C3] Canuto, V., Hsieh, S.H., and Owen, J.R., "Scale Covariance and *G*-Varying Cosmology. III—The (m,z), (θ_m,z), (θ_j,z), and $(N(m),m)$ Tests," *Astrophys. J.*, suppl. ser., 1979, **41,** pp. 263–300.

[C4] Canuto, V., and Owen, J.R., "Scale Covariance and *G*-Varying Cosmology. IV—The log N–log S Relation," *Astrophys. J.*, suppl. ser., 1979, **41,** pp. 301–326.

[C5] Canuto, V., and Hsieh, S.H., "Cosmological Variation of *G* and the Solar Luminosity," *Astrophys. J.*, 1980, **237,** pp. 613–615; "Primordial Nucleosynthesis and Dirac's Large Number Hypothesis," *Astrophys. J.*, 1980, **239,** p. L91.

[C6] Carr, B.J., "Primordial Star Formation in a Cold Early Universe," *Astr. Astrophys.*, 1977, **60,** pp. 13–26.

[C7] Carr, B.J., and Rees, M.J., "A Tepid Model for the Early Universe," *Astr. Astrophys.*, 1977, **61,** pp. 705–709.

[C8] Cheng, E.S., Saulson, P.R., Wilkinson, D.T., and Corey, B.E., "Large Scale Anisotropy in the 2.7K Radiation," *Astrophys. J.*, 1979, **232,** pp. L139–L143.

[C9] Coleman, G.D., Wu, C.-C., and Weedman, D.W., "Colors and Magnitudes Predicted for High Redshift Galaxies," *Astrophys. J.*, suppl. ser., 1980, **43,** pp. 393–416.

[D1] Davis, M., Geller, M.J., and Huchra, J., "The Local Mean Mass Density of the Universe: New Methods for Studying Galaxy Clustering," *Astrophys. J.*, 1978, **221,** pp. 1–18.

[D2] Dirac, P.A.M., "The Cosmological Constants," *Nature,* 1937, **139,** p. 323.

[D3] Dirac, P.A.M., "Long Range Forces and Broken Symmetries," *Proc. R. Soc. London A,* 1973, **333,** pp. 403–418.

[E1] Ellis, R.S., Fong, R., and Phillips, S., "On the Optical Observations of Galaxies with Large Redshifts," *Nature,* 1977, **265,** pp. 313–314.

[F1] Fabbri, R., Guidi, L., Melchiorri, F., and Natale, V., "Measurements of the Cosmic-Background Large-Scale Anisotropy in the Millimeter Range," *Phys. Rev. Lett.,* 1980, **44,** pp. 1563–1566.

[F2] Field, G., Arp, H., and Bahcall, J.N., *The Redshift Controversy,* Benjamin, 1973.

[F3] Frenk, C.S., and White, S.D.M., "The Kinematics and Dynamics of the Galactic Globular Cluster System," *Month. Not. R. Astron. Soc.,* 1980, **193,** pp. 295–311.

[G1] Gorenstein, M.V., and Smoot, G.F., "Large-Angular-Scale Anisotropy in the Cosmic Background Radiation," *Astrophys. J.,* 1981, **244,** pp. 361–381.

[G2] Gott, J.R. III, Turner, E.L., and Aarseth, S.J., "*N*-Body Simulations of Galaxy Clustering. III—The Covariance Function," *Astrophys. J.,* 1979, **234,** pp. 13–26.

[G3] Gott, J.R., III, and Turner, E.L., "An Extension of the Galaxy Covariance Function To Small Scales," *Astrophys. J.,* 1979, **232,** pp. L79–L81.

[H1] Hart, L., and Davis, R.D., "Motion of the Local Group of Galaxies and Isotropy of the Universe," *Nature,* 1982, **297,** pp. 191–196.

[H2] Hoffman, G.L., Olson, D.W., and Salpeter, E.E., "Dynamical Models and the Mass of the Virgo Cluster," *Astrophys. J.,* 1980, **242,** pp. 861–878.

[H3] Hoyle, F., "A New Model for the Expanding Universe," *Month. Not. R. Astron. Soc.,* 1948, **108,** 372–382; "On the Cosmological Problem," *Month. Not. R. Astron. Soc.,* 1949, **109,** pp. 365–371.

[H4] Humason, H.L., and Wahlquist, H.D., "Solar Motion with Respect to the Local Group of Nebulae," *Astr. J.,* 1955, **60,** pp. 254–259.

[H5] Hutchings, J.B., Crampton, D., Campbell, B., and Pritchet, C., "Optical Morphology of 13 QSOs," *Astrophys. J.,* 1981, **247,** 743–749.

[I1] Isobe, S., "The Local Velocity of Rotation in the Galaxy," *Astr. Astrophys.,* 1974, **36,** pp. 327–332.

[K1] Kinman, T.D., "Globular Clusters. III—An Analysis of the Cluster Radial Velocities," *Month. Not. R. Astron. Soc.,* 1959, **119,** pp. 559–575.

[K2] Kirschner, R.P., Oemler, A., Jr., Schecter, P.L., and Schectman, S.A., "A Million Cubic Megaparsecs Void in Bootes?" *Astrophys. J.,* 1981, **248,** pp. L57–L60.

[K3] Knapp, G.R., Tremaine, S.D., and Gunn, J.E., "The Global Properties of the Galaxy. I—The HI Distribution Outside the Solar Circle," *Astr. J.,* 1978, **83,** pp. 1585–1593.

[K4] Kormendy, J., "Brightness Distribution in Compact and Normal Galaxies. II—Structure Parameters of the Spheroidal Component," *Astrophys. J.,* 1976, **218,** pp. 33–346.

[L1] Lynden-Bell, D., and Lin, D.N.C., "On the Motions of the Sun, the Galaxy, and the Andromeda Nebula," *Month. Not. R. Astron. Soc.,* 1977, **181,** pp. 37–57.

[M1] Mayall, N.U., "The Radial Velocity of Fifty Globular Clusters," *Astrophys. J.,* 1946, **104,** pp. 290–323.

[M2] Melchiorri, F., Melchiorri, B.O., Ceccarelli, C., and Pietronera, L., "Fluctuations in the Microwave Background at Intermediate Angular Scales," *Astrophys. J.,* 1981, **250,** pp. L1–L4.

[M3] Mihalas, D., and Binney, J., *Galactic Astronomy,* Freeman, 1981.

[M4] Misner, C.W., "Transport Processes in the Primordial Fireball," *Nature,* 1967, **214,** pp. 40–41.

[M5] Misner, C.W., Thorne, K.S., and Wheeler, J.A., *Gravitation,* Freeman, 1973.

[N1] Narlikar, J.V., *Introduction to Cosmology,* Jones and Bartlett Publishers, 1983.

[O1] Olsen, D.W., "Helium Production and Limits on the Anisotropy of the Universe," *Astrophys. J.,* 1978, **219,** pp. 777–780.

[P1] Pagel, B.E.J., "Abundances of Elements of Cosmological Interest," *Phil.Trans. R. Soc. London A,* 1982, **307,** pp. 19–35.

[P2] Partridge, R.B., "The Evolution of Structure in the Universe: Observational Considerations," in *Cosmology and Particles,* Audouze, J., Crane, P., Gaisser, T., Hegyi, D., and Tran Thanh Van, J., Eds., Editions Frontières, Dreux, France, 1982, pp. 273–295.

[P3] Peebles, P.J.E., "The Peculiar Velocity Field in the Local Supercluster," *Astrophys. J.,* 1975, **205,** pp. 318–328.

[P4] Peebles, P.J.E., *The Large Scale Structure of the Universe,* Princeton University Press, 1980.

[P5] Peebles, P.J.E., "Primeval Adiabatic Perturbations: Constraints from the Mass Distribution," *Astrophys. J.,* 1981, **248,** pp. 885–897.

[P6] Peebles, P.J.E., "Large-Scale Fluctuations in the Microwave Background and the Small-Scale Clustering of Galaxies," *Astrophys. J.,* **243,** pp. L119–L122.

[P7] Pence, W., "*K*-Corrections for Galaxies of Different Morphological Types," *Astrophys. J.,* 1976, **203,** pp. 39–51.

[P8] Penzias, A.A., and Wilson, R.W., "A Measurement of Excess Antenna Temperature at 4080 Mc/s," *Astrophys. J.,* 1965, **142,** pp. 419–421.

[R1] Rees, M.J., "Origin of the Cosmic Microwave Background Radiation in a Chaotic Universe," *Phys. Rev. Lett.,* 1972, **28,** pp. 1669–1671.

[R2] Rowan-Robinson, M., "Quasars and the Cosmological Distance Scale," *Nature,* 1976, **262,** p. 97.

[R3] Rowan-Robinson, M., *Cosmology,* 2nd ed., Oxford University Press, 1981.

[R4] Rowan-Robinson, M., Negroponte, J., and Silk, J., "Distortions of the Cosmic Microwave Background Spectrum by Dust," *Nature,* 1979, **281,** pp. 635–638.

[R5] Rubin, V.C., Thonnard, N., Ford, W.K., and Roberts, M.S., "Motion of the Galaxy and the Local Group Determined from the Velocity Anisotropy of Distant Sc I Galaxies. II—The Analysis for the Motion," *Astr. J.,* 1976, **81,** pp. 719–737.

[S1] Sandage, A., and Tammann, G.A., "Steps Toward the Hubble Constant. V—The Hubble Constant from nearby Galaxies and the Regularity of the Local Velocity Field," *Astrophys. J.,* 1975, **196,** pp. 313–328.

[S2] Sandage, A., Tammann, G.A., and Yahil, A., "The Velocity Field of Bright nearby Galaxies. I—The Variation of Mean Absolute Magnitude with Redshift for Galaxies in a Magnitude-Limited Sample," *Astrophys. J.,* 1979, **232,** pp. 352–364.

[S3] Schecter, P.L., "Mass-to-Light Ratios for Elliptical Galaxies," *Astr. J.,* 1980, **85,** pp. 801–811.

[S4] Segal, I.E., *Mathematical Cosmology and Extragalactic Astronomy,* Academic Press, 1976.

[S5] Silk, J., "Large-Scale Inhomogeneity of the Universe: Implications for the Deceleration Parameter," *Astrophys. J.,* 1974, **193,** pp. 525–527.

[S6] Silk, J. "Large-Scale Inhomogeneity of the Universe: Spherically Symmetric Models," *Astr. Astrophys.,* 1977, **59,** pp. 53–58.

[S7] Silk, J., "Anisotropy of the Cosmic Microwave Background Radiation," in *Cosmology and Particles,* Audouze, J., Crane, P., Gaisser, T., Hegyi, D., and Tran Thanh Van, J., Eds., Editions Frontières, Dreux, France, 1982, pp. 253–271.

[S8] Silk, J., and Wilson, M.L., "Large-Scale Anisotropy of the Cosmic Microwave Background Radiation," *Astrophys. J.,* 1981, **244,** pp. L37–L41.

[S9] Smoot, G.F., "Fluctuations in the Microwave Background at Large Angular Scales," *Physica Scripta,* 1980, **21,** pp. 619–623.

[S10] Smoot, G.F., and Lubin, P.M., "Southern Hemisphere Measurements of the Anisotropy in the Cosmic Microwave Background Radiation," *Astrophys. J.,* 1979, **234,** pp. L83–L86.

[S11] Stewart, J.M., and Sciama, D.W., "Peculiar Velocity of the Sun and Its Relation to the Cosmic Microwave Background," *Nature,* 1967, **216,** pp. 748–753.

[S12] Stockton, A., "The Nature of QSO Redshifts," *Astrophys. J.,* 1978, **223,** 747–757.

[T1] Tammann, G.A., Sandage, A., and Yahil, A., "The Determination of Cosmological Parameters," in *Physical Cosmology,* Balain, R., Audouze, J., and Schramm, D.N., Eds., North-Holland, 1979, pp. 53–125.

[T2] Tonry, J.L., and Davis, M., "Velocity Dispersions of Elliptical and SO Galaxies. II—Infall of the Local Group to Virgo," *Astrophys. J.,* 1981, **246,** pp. 680–695.

[V1] Vaucouleurs, G. de, "Further Evidence for a Local Super-cluster of Galaxies: Rotation and Expansion," *Astr. J.,* 1958, **63,** pp. 253–265.

[V2] Vaucouleurs, G. de, "The Velocity-Distance Relation and the Hubble Constant for Nearby Groups of Galaxies," in *External Galaxies and Quasi-stellar Objects* (IAU Symp. 44), Reidel, 1972, pp. 353–366.

[V3] Vaucouleurs, G. de, "The Local Supercluster," in *Large-Scale Structure of the Universe* (IAU Symp. 79), Longair, M.S., and Einasto, J., Eds., Reidel, 1978, pp. 205–213.

[V4] Vaucouleurs, G. de, and Peters, W.L., "Motion of the Sun with Respect to the Galaxies and the Kinematics of the Local Supercluster," *Nature,* 1968, **220,** pp. 868–874.

[V5] Vaucouleurs, G. de, and Peters, W.L., "Hubble Ratio and Solar Motion from 200 Spiral Galaxies Having Distances Derived from the Luminosity Index," *Astrophys. J.,* 1981, **248,** pp. 395–407.

[V6] Vaucouleurs, G. de, Peters, W.L., Bottinelli, L., Gouguenheim, L., and Paturel, G., "Hubble Ratio and Solar Motion from 300 Spirals Having Distances Derived from HI Line Widths," *Astrophys. J.,* 1981. **248,** pp. 408–422.

[W1] Weinberg, S., *Gravitation and Cosmology,* 1972.

[W2] Woltjer, L., "The Galactic Halo: Globular Clusters," *Astr. Astrophys.*, 1975, **42,** pp. 109–118.

[W3] Wooley, R., and Sacage, A., "Masses, Radii and Luminosities of RR Lyrae Variable Stars," Royal Observatory Bull. 170, 1971.

[W4] Wyckoff, S., Wehinger, P.A., and Gehren, T., "Resolution of Quasar Images," *Astrophys. J.*, 1981, **247,** pp. 750–761.

[Y1] Yahil, A., Sandage, A., and Tammann, G.A., "The Deceleration of Nearby Galaxies," in *Physical Cosmology,* Balian, R., Audouze, J., and Schramm, D.N., Eds., North-Holland, 1972, pp. 127–159.

[Y2] Yahil, A., Tammann, G.A., and Sandage, A., "The Local Group: The Solar Motion Relative to Its Centroid," *Astrophys. J.*, 1977, **217,** pp. 903–915.

[Y3] Yahil, A., Sandage, A., and Tammann, G.A., "The Velocity Field of Bright Nearby Galaxies. III—The Distribution in Space of Galaxies within 80 Megaparsecs: The North Galactic Density Anomaly," *Astrophys. J.*, 1980, **242,** pp. 448–468.

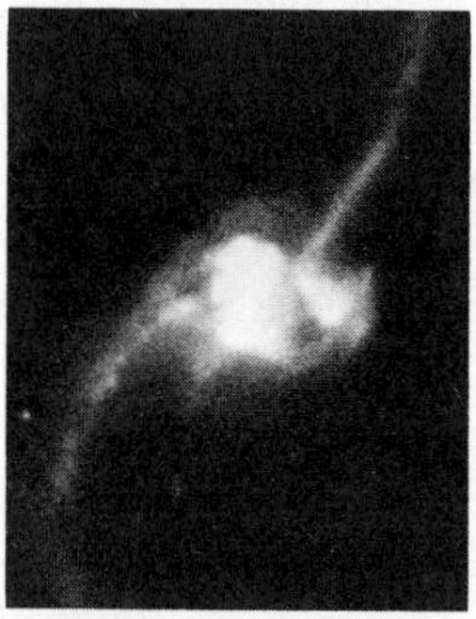

CHAPTER SIX

THE SIZE AND AGE OF THE UNIVERSE

The size of the universe, as measured by the Hubble constant, has been the subject of the fiercest controversy during the past decade. The Hubble constant, the coefficient of proportionality in Hubble's velocity–distance law, is usually measured in units of kilometers per second per megaparsec, but the reciprocal of it has the dimensions of a time and can be interpreted as the size of the universe (in light-years, say). The battle lines in this controversy were drawn up at the International Astronomical Union Colloquium on "Redshift and the Expansion of the Universe", held in Paris in 1976. There Tammann argued on behalf of himself and Sandage for a value of 50 km s^{-1} Mpc^{-1} for H_0, while de Vaucouleurs argued for a value of 100. This disagreement by a factor of 2 about the size of the universe was surprising since, for the most part, the same galaxies and often the same observational data were used by both protagonists. When during the subsequent discussion at this Paris meeting I asked them to define the range outside which they could not imagine the Hubble constant to lie, the two ranges given did not even overlap. Since that time both Sandage and Tammann as well as de Vaucouleurs have brought new arguments, new methods, and new observations to bear on the problem, but still arrive at the same discrepant conclusions.

In Sec. 6.1 I discuss the values of the Hubble constant found by many different workers in the field and try to explain the origin of the controversy. The main cause, in my view, has been overestimation of the accuracy of the results. Different treatments of interstellar and internal extinction, of primary calibrators, and of our local motion also play a part. Circular arguments have unfortunately crept into the arguments of both main camps, though these can easily be eliminated.

I also give an estimate of the Hubble constant based on the distance scale I adopted in Chaps. 3 and 4, using only groups and clusters of galaxies with well-determined distances and which are more than 40 Mpc from the Virgo cluster. The large value of our local velocity found from observations of the microwave background anisotropy, and from other studies, shows that we cannot hope to see a true cosmological flow until we look out to distances well beyond the Local Supercluster. My estimate at least has the merit that most reliable distance estimates contribute to it, though the calibrations, corrections, and weightings I have used are, of course, open to debate.

To determine distances to galaxies with significant redshifts, we have to try to determine the density and deceleration parameters Ω_0 and q_0. These are discussed in Sec. 6.2. In Sec. 6.3 we explore aspects of distance and time on the large scale, for example, the age of the universe, the concept of a horizon, and how we can define the size of the universe. Finally in Sec. 6.4 we investigate whether there are any characteristic length scales in the universe and discuss the novel and controversial "anthropic principle."

6.1 The redshift–distance law and the Hubble constant

In Chaps. 3 and 4 we reviewed the methods by which the extragalactic distance scale has been built up. Many of these methods have been used to obtain estimates of the Hubble constant. Table 6.1, part 1, taken from the 1981 review by Paul Hodge [H5], gives estimates that are based on the distance to the Virgo cluster or on samples of galaxies which are predominantly within 20 Mpc or so.

Two effects tend to make these estimates unreliable. The first (as we saw in Sec. 5.4) is that our Galaxy has a substantial peculiar velocity relative to the cosmological frame. In fact all galaxies that, like our own, lie within 20 Mpc or so of the Virgo cluster are likely to have their cosmological expansion velocities modified by the Local Supercluster. Efforts to model the local deviations from a pure Hubble expansion in terms of a spherically symmetric infall toward Virgo (see Sec. 5.4) have not been totally convincing. Such models, which incorporate the effects of infall, account for only part of the velocity of our Galaxy which we infer from the microwave background anisotropy. This seems to imply that the whole Local Supercluster has a substantial peculiar velocity that will distort measurements of the cosmological recession velocity of galaxies closer than 20 Mpc or so. We therefore have to look on a larger scale to see a true cosmological flow and to find the true cosmological expansion rate. Part 2 of Table 6.1 gives estimates of the Hubble constant where at least a substantial proportion of the galaxies are at distances greater than 20 Mpc. It is encouraging, and perhaps surprising, that the estimates in parts 1 and 2 of Table 6.1 both fall in the range of 50–100.

A second effect that tends to make some of the estimates for H_0 given in parts 1 and 2 of Table 6.1 unreliable (whether based on galaxies up to or beyond 20 Mpc) is that the galaxies selected for study tend to be drawn from magnitude-limited samples like the *Shapley–Ames Catalog* and the *Reference Catalogue of Bright Galaxies*, and are therefore subject to the Malmquist effect. This is the effect that in a sample of sources selected to be brighter than some apparent magnitude limit, the more distant sources are forced to be amongst the most luminous members of the source population (App. A.17). Tammann tends to attribute most high estimates of the Hubble constant H_0 to Malmquist bias [T1],[T2], but this is hotly disputed by de Vaucouleurs [V1].

Part of the difference between the "long" ($H_0 = 50$) distance scale of Sandage

TABLE 6.1

The Hubble Constant H_0.

Method	*Authors*	H_0 *(km* s^{-1} *Mpc*$^{-1}$*)*
1 *Estimates based on Virgo cluster or dominated by galaxies within 20 Mpc**		
M101 group velocity and distance	Sandage and Tammann (1974) [S3]	55.5 ± 8.7
Virgo cluster	Sandage and Tammann (1974) [S4]	57 ± 6
Nearby Sc galaxies	Sandage and Tammann (1975) [S5]	57 ± 3
Various	van den Bergh (1975) [B6]	95 (+ 15, − 12)
21-cm width	Tully and Fisher (1977) [T6]	80
Virgo cluster, Tully – Fisher relation	Heidmann (1977) [H3]	83 ± 19
Virgo cluster	Peebles (1977) [P1]	42 - 77
Virgo cluster, E and SO	Visvanathan and Sandage (1977) [V5]	49.3 ± 4
Tully-Fisher relation	Shostak (1978) [S14]	80 ± 3
Hubble diagram, E and SO	Visvanathan (1979) [V4]	50.8
Globular clusters	Hanes (1979) [H1]	80 ± 11
HII region fluxes and galaxy diameters	Kennicutt (1979) [K2]	60 (+ 15, − 10)
HII region isophotal diameters	Kennicutt (1979) [K1]	65
Infrared Tully – Fisher relation	Mould et al. (1980) [M3]	65 ± 4
Luminosity classification of Sb galaxies	Stenning and Hartwick (1980) [S15]	75 ± 15
Virgo Sc HII luminosities	Kennicutt (1981) [K3]	55
2 *Other estimates using galaxies well beyond Virgo*†*		
Remote ScI galaxies	Sandage and Tammann (1975) [S6]	55 ± 6
Fisher – Tully relation	Sandage and Tammann (1976) [S7]	50.3 ± 4.3
$D(\Lambda_c)$, $M_{pg}(\Lambda_c)$	de Vaucouleurs and Bollinger (1979) [V2]	100 ± 10
Type I supernovae	Branch (1979) [B8]	56 ± 15
Infrared Tully-Fisher relation	Aaronson et al. (1980) [A1]	95 ± 4
$\langle M_{pg}(\Lambda_c), D(\Lambda_c)\rangle$ and Tully – Fisher relation	de Vaucouleurs et al. (1981) [V3]	100 ± 10
Type I supernovae	Sandage and Tammann (1982) [S8]	50 ± 7
Infrared Tully – Fisher relation	Aaronson and Mould (1983) [A2]	82 ± 10

* Compiled from [H6].

† Estimates not included are those of Birkinshaw [B7], based on a marginal detection of the Zeldovich – Sunyaev effect in clusters, and of Lynden-Bell [L3], based on a particular model for superluminal expansion in quasars.

TABLE 6.1 *(Continued)*

	Number of groups	$H_0^‡$	*External error*	$\sum W/\Sigma^2$
3 *Groups and clusters with reliable distances*				
All groups, no velocity correction	49	72.9	3.4	430.9
All groups, velocity corrected for Galaxy's motion in space [546 km s^{-1} to $(l, b) = (261, 39)$]	49	74.6	5.2	430.9
Groups with $d > 30$ Mpc, velocity corrected	16	67.7	5.3	73.7
Groups with $d_V > 25$ Mpc, velocity corrected	24	65.9	4.5	146.5
Groups with $d_V > 30$ Mpc, velocity corrected	20	70.1	3.9	113.5
Groups with $d_V > 40$ Mpc, velocity corrected	14	65.0	4.9	69.4
Effect of $\Delta\mu_0 = -1.0$ for type I supernova method		79.8		
Effect of $\Delta\mu_0 = +1.0$ for 21-cm infrared method		55.7		
4 *Brightest cluster galaxies with* $V_0 \lesssim 10{,}000$ *km s*$^{-1}$				
All clusters, no velocity correction	49	67.7	1.7	72.5
All clusters, velocity corrected	49	68.3	1.9	72.5
Clusters with $V_c > 2500$ km s^{-1}, velocity corrected	42	69.4	2.0	62.2
Effect of $\Delta\mu_0 = -1.0$ for type I supernova method		75.9		
Effect of $\Delta\mu_0 = +1.0$ for 21-cm infrared method		56.1		

‡ H_0—weighted mean V/d.
d_V—distance from Virgo.

and Tammann and the "short" ($H_0 = 100$) distance scale of de Vaucouleurs has already been accounted for in Chaps. 3 and 4. Table 6.2 summarizes some of the differences in the two approaches. In the distance scale that I adopted in Chaps. 3 and 4 I have corrected as far as possible circular arguments and known systematic errors, as well as giving due weight to the more reliable distance methods investigated by other workers.

The best way to control the Malmquist effect is to use only the most reliable distance estimates. Table 4.13 gives the weighted mean distances to groups and clusters for which the total weight $\sum_i W_i/\Sigma_i^2$ is greater than 2.5. Here Σ_i is the uncertainty and W_i the weighting factor (see Sec. 4.9) for the ith distance estimate. Fig. 6.1 shows a plot of the mean recession velocity of these groups V_0, corrected for the sun's rotation around the Galaxy, against their weighted mean distance d. Fig. 6.2 shows the effect of correcting these data for the Galaxy's absolute velocity in space. Comparison of Figs. 6.1 and 6.2 shows an interesting difference. In Fig. 6.1 galaxies within 20 Mpc show a tight velocity–distance relation, whereas in Fig. 6.2 they show much more scatter. Possible explanations are: (1) the absolute motion of our Galaxy is much smaller than assumed; (2) the nearby galaxies are falling together toward Virgo; (3) there is some observational selection or bias present. Sandage and Tammann adopted explanation (1) in their 1975 paper [S5], but the evidence discussed in Sec. 5.4 favors (2).

TABLE 6.2

Some of the causes of the difference between Sandage and Tammann's "long" ($H_0 \sim 50$) and de Vaucouleurs's "short" ($H_0 \sim 100$) distance scales.

	Sandage and Tammann		*de Vaucouleurs*
Method or factor	*1970–1974*	*1982*	*1976–1979*
Interstellar extinction law	Low values		High values
Internal extinction	Neglected in some applications (bright stars, N2403 cepheids)		Included
Cepheids	Use period–luminosity–color relation		Uses period–luminosity relation
RR Lyrae, novae	Used to justify $\mu_{\mathrm{Hya}} = 3.03$		Given equal weight to Cepheids
Local Group distances	Based entirely on Cepheids		Based on several methods, including secondary methods (i.e., circular)
Brightest red stars	Not used	Sole secondary indicator	Used
Brightest blue stars	Application to M101 involves extrapolation	Not used	Used
HII region diameters	Systematic error present; application to M101 involves extrapolation	Not used	Sandage–Tammann error corrected
HII rings	Not used	Not used	Almost all galaxies in which indicator has been studied are involved in calibration
HII velocity dispersion	Not used	Not used	Extrapolation in application to M101
Brightest globular clusters	Used to confirm scale		Used
Tully–Fisher method	Used to confirm scale	Not used	Used as linearity test
Supernovae	Used to confirm scale	Type I supernovae sole tertiary indicator	Used as linearity test
Luminosity class or index	Sole tertiary indicator	Not used	Main tertiary indicator (with isophotal diameters)

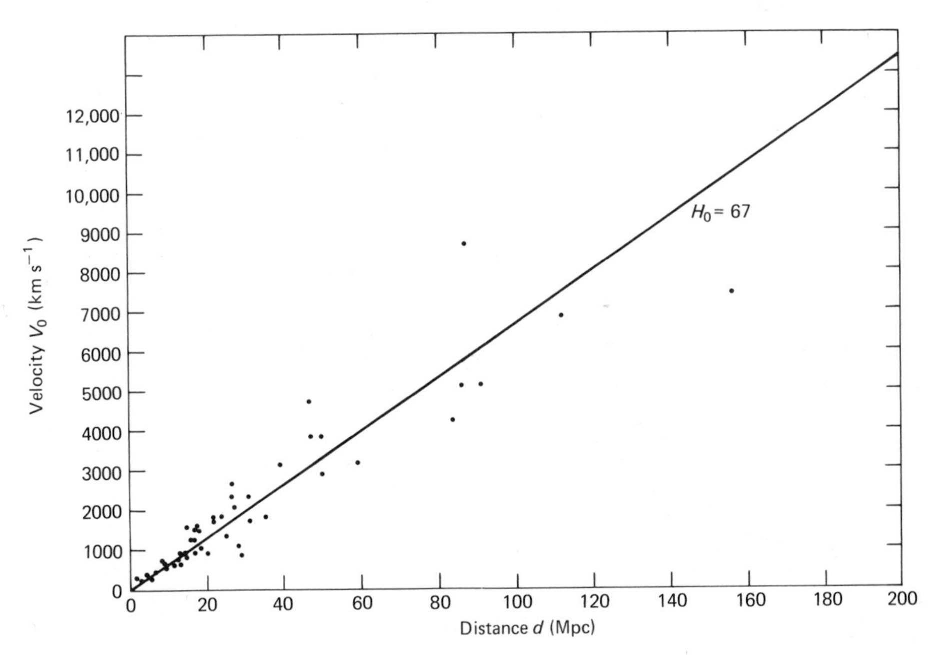

FIGURE 6.1

Velocity–distance relation for groups and clusters of galaxies. Velocities corrected for galactic rotation. Data are taken from Table 4.15. Straight line corresponds to $H_0 = 67$ *[Eq. (6.3)].*

Sandage has explored the use of the magnitude of the brightest galaxies in clusters as a distance indicator on the large scale (see Sec. 4.8). Fig. 6.3 shows the recession velocity for these cluster galaxies, corrected for the Galaxy's absolute motion in space, against their distance derived from the calibration of Table 4.11. Only the brightest galaxies in clusters with mean recession velocities less than 10,000 km s^{-1} are shown to avoid the cosmological deviations from a linear Hubble law predicted when the redshift is no longer small [Eq. (5.33)].

Parts 3 and 4 of Table 6.1 summarize my estimates of the Hubble constant under different assumptions. Using all groups and clusters of galaxies with reliably determined distances, the Hubble constant is found to be 74.6 or 72.9, depending on whether or not velocities are corrected for the Galaxy's absolute motion in space. However, if we restrict attention to groups and clusters well outside the Local Supercluster, then lower values of the Hubble constant are found. To be really safe from the effects of the Local Supercluster, we consider those groups and clusters of galaxies which have distances from Virgo, d_V, greater than 40 Mpc. Correcting their mean recession velocities for the sun's rotation around the Galaxy and for the Galaxy's absolute motion in space, I find a weighted mean Hubble constant

$$H_0 = 65.0 \text{ km s}^{-1} \text{ Mpc}^{-1} \tag{6.1}$$

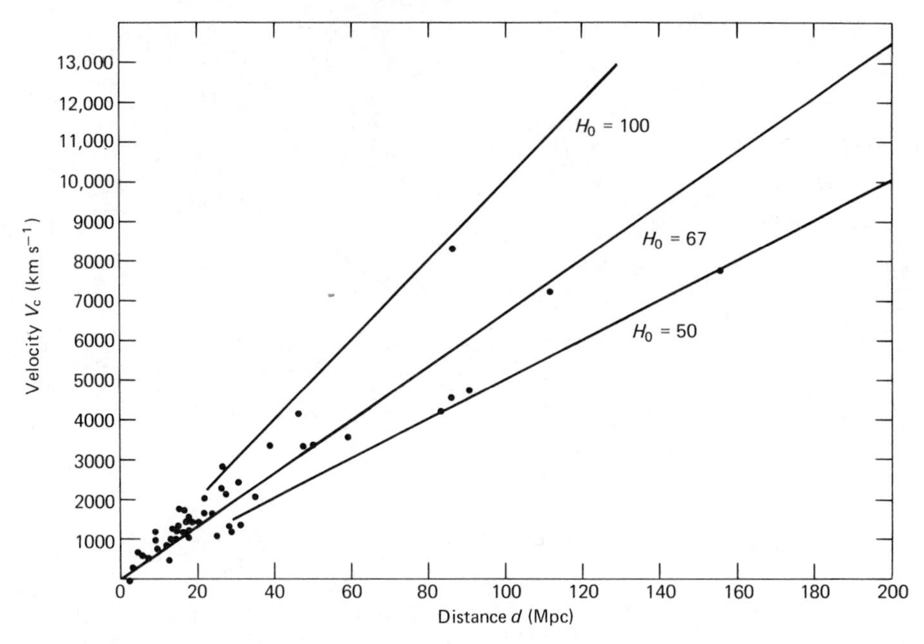

FIGURE 6.2

As Fig. 6.1, but velocities corrected also for our Galaxy's absolute motion in space (taken to be 546 km s^{-1} toward l = *261°,* b = *39°; Table 5.3). Straight lines correspond to* $H_0 = 67$ *[Eq. (6.3)], 50, and 100.*

Since the brightest cluster galaxies (see Fig. 6.3) are already restricted to recession velocities $V_c \leq 10{,}000$ km s^{-1}, we restrict attention to those with V_c greater than 2500 km s^{-1} in order to eliminate the effects of the Local Supercluster. (A distance limit would introduce a bias in H_0.) The mean value of the Hubble constant obtained for the brightest cluster galaxies which have $V_c > 2500$ km s^{-1} is

$$H_0 = 69.4 \text{ km s}^{-1} \text{ Mpc}^{-1}. \tag{6.2}$$

The external errors* for these estimates are ± 4.9 and ± 2.0, respectively. The internal errors associated with the distances, defined by $\sigma = (\sum_i \sigma_{\text{int},i}^{-2})^{-1/2}$, where $\sigma_{\text{int},i}$ is the internal error for a single group or cluster, are ± 0.12 and ± 0.13 mag, or ± 3.6 and ± 4.1 in H_0, respectively. (The error associated with the velocities is likely to be smaller than this.) However, the true uncertainty in H_0 is considerably larger because of the possibility of systematic errors, errors that always bias distance estimates from a particular method in the same direction. By their very nature, the magnitude of unknown systematic errors cannot be estimated, but if we assume that the systematic error in each rung of the cosmological distance ladder is on the order

* See App. A.17 for a definition of external and internal errors.

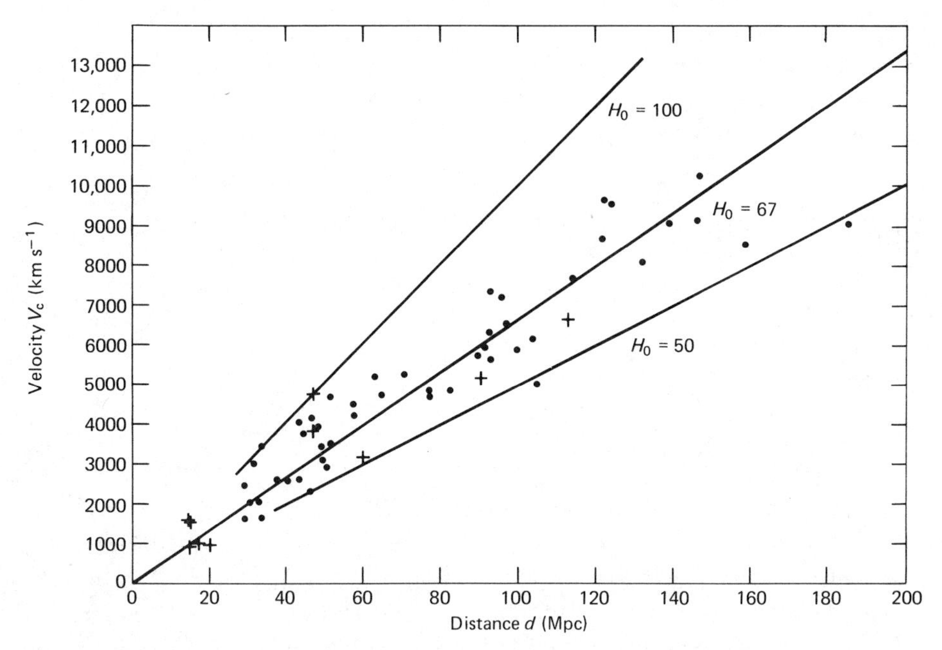

FIGURE 6.3

Velocity–distance relation for brightest cluster galaxies. Velocities corrected for galactic rotation and for the Galaxy's absolute motion in space. ●—from App. A.15; +—calibrating cluster galaxies, from Table 4.11. Straight lines correspond to $H_0 = 67$ *[Eq. (6.3)], 50, and 100.*

of the zero-point error for that rung, then the total uncertainty is on the order of ± 0.5 in μ_0, or $\pm 23\%$ in H_0 (Table 6.3). My adopted value for the Hubble constant is therefore

$$\boldsymbol{H_0 = 67 \pm 15 \text{ km s}^{-1} \text{ Mpc}^{-1}} \qquad (6.3)$$

It can be seen from part 2 of Table 6.1 that my adopted value for H_0 differs by about 2–3.5 standard deviations from the values quoted by Sandage and Tammann and by de Vaucouleurs (which differ from each other by 5–10 standard deviations) according to their quoted errors. However, their values and the others quoted in parts 1 and 2 of Table 6.1 lie within 1 or 2 standard deviations of my value, given in Eq. (6.3).

The most discrepant estimates are those of de Vaucouleurs and his co-workers, $H_0 = 100 \pm 10$, based on the absolute magnitude and the diameter of galaxies as a function of his luminosity index Λ_c and on the Tully–Fisher method. But note that the sample used by de Vaucouleurs includes a rather high proportion of galaxies with $V_0 < 1500$ km s^{-1} (55%) and very few galaxies with $V_0 > 3000$ km s^{-1} (9%). This high proportion of galaxies likely to be affected by the Local Supercluster may account for part of the difference between my value of H_0 and that of de Vaucouleurs. Another factor is the lower Local Group distances used by de Vaucouleurs

TABLE 6.3

Factors contributing to uncertainty in H_0.

	Possible systematic errors in primary indicators (mag)	*Impact on Local Group distance moduli (mag)*
Distance to Hyades	0.1	0.08
Main-sequence correction due to Hyades abundance	0.15	0.12
Interstellar extinction	0.12	0.12
Internal obscuration	$\gtrsim 0.12$	$\gtrsim 0.12$
Abundance effects in Cepheids	$\sim 0.08\ \Delta Y/Y$	$\lesssim 0.01$
	$\sim -0.06\ \Delta Z/Z$	± 0.06
(+similar effects for RR Lyrae stars)		

Net uncertainty in mean Local Group distances	0.2 mag.
Net uncertainty in M81 group distance modulus	0.3 mag.
Net uncertainty in Virgo group distance modulus	0.4 mag.
Net uncertainty in distant groups and clusters	0.5 mag.
Uncertainty in H_0 due to uncertainty in absolute motion of our Galaxy	$\lesssim 5\%$
Net uncertainty in H_0	**$\pm 23\%$**

than those I have adopted in this book. If de Vaucouleurs's distances are adjusted by the $\Delta\mu_0 = +0.29$ average difference between my adopted Local Group distance moduli and his, then his Hubble constant becomes 87. This is not in fact very different from my estimate of 75 using *all* groups and clusters with reliable distances, corrected for the effect of the Galaxy's absolute motion in space.

The most important potential source of error in my estimate of H_0 is the possibility that the type I supernova method or the infrared Tully–Fisher method, which contribute substantially to the distance scale at large distances, may contain systematic errors on the order of 0.5–1 mags (see Sec. 4.9). I have therefore estimated the effects of (1) decreasing all distance moduli from type I supernovae by 1 mag and (2) increasing all distance moduli beyond the M81 and M101 groups from the infrared Tully–Fisher method by 1 mag. The results are shown in parts 3 and 4 of Table 6.1. If the systematic error discussed in Sec. 4.9 lies entirely in the supernova method, my Hubble constant is increased from 67 to 78. If the error lies entirely in the infrared Tully–Fisher method, H_0 is reduced to 56. This gives an indication of the range over which my estimate of H_0 could shift in the immediate future.

Other factors that might result in significant shifts in my estimate of H_0 are (1) revision of interstellar and internal extinction corrections, which could lead to a shift of 0.1–0.2 mag (5–10%) in H_0 if either the low extinction correction of Sandage or the high extinction correction of de Vaucouleurs (see Box 2.5) proves to be correct, and (2) revision of the estimate of our absolute motion in space, which is still the subject of intensive research. However, parts 3 and 4 of Table 6.1 show that my estimate for H_0 does not depend too strongly on the adopted velocity.

6.2 *The deceleration and density parameters*

By studying the behavior of the brightness and angular diameter of different classes of astronomical objects as a function of redshift, we might hope to estimate the cosmological parameters q_0 and Ω_0, since different models make different predictions about how these quantities vary. The results of several attempts to use the magnitude–redshift relation for brightest cluster galaxies to measure q_0 (assuming $\Lambda = 0$, so $\Omega_0 = 2q_0$) are summarized in Table 6.4. The data

TABLE 6.4

Estimates of Ω_0 and q_0, for $\Lambda = 0$.

Method	q_0 *(no evolution correction, unless stated)*	*Authors*
Brightest cluster galaxies	0.96 ± 0.4	Sandage (1972) [S1]
First three ranked cluster galaxies	1 ± 1	Sandage and Hardy (1973) [S2]
Brightest cluster galaxies (with correction for stellar evolution	0.31 ± 0.68 -0.43 ± 0.54)	Gunn and Oke (1975) [G5]
Brightest cluster galaxies	1.6 ± 0.4	Kristian et al. (1978) [K4]
Brightest cluster galaxies	-0.55 ± 0.45	Hoessel et al. (1980) [H7]
Angular size of cluster galaxies	0.15 ± 0.3	Baum (1972) [B5]
Angular size of clusters	-0.9 ± 0.3	Hickson (1977) [H4]
Angular size of clusters	0.25 ± 0.5	Bruzual and Spinrad (1978) [B9]
Mass-to-light ratios in groups and clusters	0.05 ± 0.01	Gott et al. (1974) [G1]
Mass-to-light ratios in groups	0.08	Gott and Turner (1976) [G2]
Mass-to-light ratios in groups	0.06	Gott and Turner (1977) [G3]
Binary galaxies	0.08	Turner and Ostriker (1977) [T7]
Cross-correlation of Abell and Shane counts	0.7 ± 0.1	Seldner and Peebles (1977) [S11]
Cosmic virial theorem	$0.2 - 0.7$	Davis et al. (1978) [D1]
Clusters of galaxies	0.4 ± 0.2	Peebles (1979) [P2]
Virgo infall model	0.4 ± 0.1	Davis et al. (1980) [D2]
Virgo infall model	0.5 ± 0	Davis and Huchra (1982) [D3]
Groups and small clusters	0.15	Press and Davis (1982) [P4]
Superclusters	$0.06 - 0.16$	Ford et al. (1981) [F4]
Cosmological nucleosynthesis (in baryons)	$0.015 - 0.1$	Schramm (1982) [S10]

of J. Kristian, Sandage, and J.A. Westphal [K4] for brightest cluster galaxies are illustrated in Fig. 6.4, together with theoretical curves for $\Lambda = 0$ models with different values of q_0. To first order, the relationship between magnitude and redshift depends only on q_0 in fact (and not on Ω_0). Although a formal solution of $q_0 = 1.6 \pm 0.4$ was obtained by Kristian et al., the true uncertainty is much larger because we are looking back in time a significant fraction of the galaxies' lifetimes and the galaxies may have changed in the intervening time. Most importantly, we must allow for the change in luminosity of a galaxy due to the evolution of its stars and for galaxy "cannibalism," in which the most massive galaxy in a cluster swallows several of its smaller neighbors because of dynamical friction.

Beatrice Tinsley showed that with simple assumptions about the star-formation rate and the initial mass function for the stars, the evolutionary correction for the elliptical galaxies which are the brightest galaxies in rich clusters increases q_0 by the amount

$$\Delta q_0 \sim \frac{1 - 0.2x}{H_0 t} \tag{6.4}$$

where t is the age of the galaxy ($\simeq t_0$), H_0 is the Hubble constant, and the stellar birth rate is assumed to have the dependence on mass m,

$$\frac{dN}{dm} = N_0 m^{-1-x}. \tag{6.5}$$

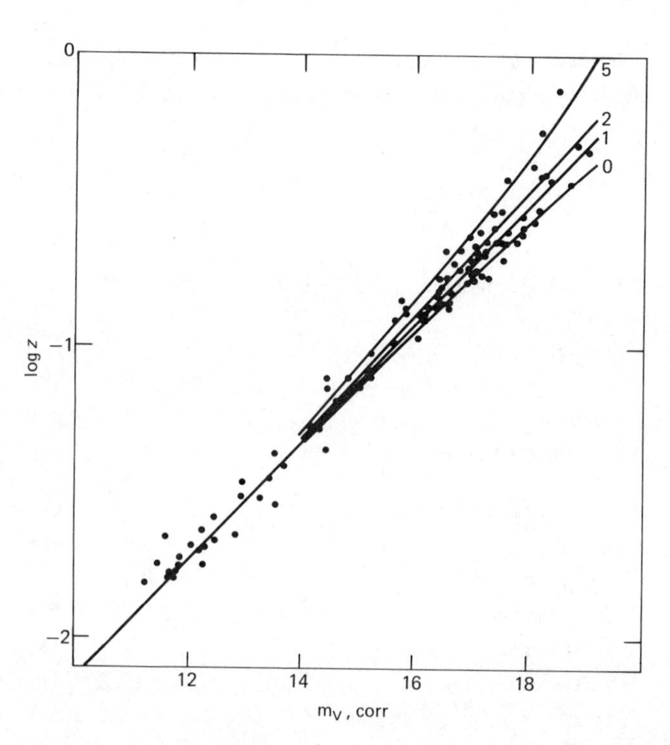

FIGURE 6.4

Redshift–magnitude curves log z versus $m_{V,corr}$ for cosmological models with $\Lambda = 0$ and different values of q_0, compared with observational data of Kristian, Sandage, and Westphal [K4]. The K-*correction has been subtracted from the observed magnitudes.*

She argued that $0 \lesssim x \lesssim 2$, so $0.6 \lesssim \Delta q_0 \lesssim 1.5$ for $\Omega_0 \leq 1$, $\Lambda = 0$ [T4].

The effects of galaxy cannibalism have been studied by J.P. Ostriker and S.D. Tremaine [O2] and by J.E. Gunn and Tinsley [G6]. The correction for this effect is difficult to estimate, but it is of the same order of magnitude as Eq. (6.4), in the opposite sense. The net result of the uncertainty in the corrections for galaxy evolution and for cannibalism is that the true q_0 can have any value between -1 and $+2$, from this method, and is essentially undetermined.

Several authors have looked for direct proof of evolution of the stellar populations in galaxies by studying how the colors of galaxies change with redshift [T4],[K5],[B10],[L1],[L2]. Fig. 6.5 shows observations of B-V and V-R against redshift for distant galaxies compared with what would be expected purely due to the redshifting of the light from the galaxy. Evidence for color changes due to galaxy evolution is not very strong. However, H. Butcher and A. Oemler have claimed that the galaxies in the clusters 0024+1654 and 3C295 are much bluer than galaxies in comparable nearby clusters. This suggests that clusters such as these two had a much higher proportion of spiral galaxies at earlier times [B11]. Presumably interactions between galaxies, or the motion of galaxies through intracluster gas, gradually strips away gas from the blue spiral galaxies so that they end up as red lenticular galaxies. However, not all lenticulars may have formed in this way. Moreover, the evidence of Butcher and Oemler has been challenged by R.D. Mathieu and H. Spinrad, who believe that most of the blue galaxies around 3C295 are foreground galaxies unconnected with the cluster [M1]. A clear demonstration of evolutionary effects in galaxies would be of the greatest interest to cosmology.

The variation of angular diameters with redshift has been studied in the brightest galaxies in clusters by W.A. Baum [B5] (Fig. 6.6), and in whole clusters of galaxies by several authors (Table 6.4 and Fig. 6.7). Baum's galaxy diameter method has not been applied by other workers and its usefulness has been questioned by Hoessel et al. [H7]. Possible evolutionary effects on galaxy diameters have been studied by D.O. Richstone and M.D. Potter [R4] and by S. Djorgovski and Spinrad [D7]. The use of a characteristic cluster size involves the assumption that the density profile of different clusters is similar, and seems contrary to the findings of covariance function studies (see Sec. 5.5), which show no structure in galaxy clustering on any length scale from 0.1 to 10 Mpc. This shows that strong selection effects are present in the cluster samples studied. Indeed, they are taken from Abell's catalog of rich clusters [A3], and we cannot be sure how this selection depends on redshift.

Various efforts have been made to determine q_0 using quasars, assuming that their redshifts are cosmological (see Box 5.1). For example, J. Bahcall and R.E. Hills suggest that there is a redshift-independent maximum luminosity that quasars achieve [B1], while J.A. Baldwin uses an empirical correlation between optical luminosity and the intensity in the CIV line at 1549 Å for a restricted class of quasar with flat radio spectra [B2]. These efforts to use quasars to measure q_0 are seriously undermined by the finding that the average optical luminosity of quasars was probably considerably greater at earlier epochs than it is today [C3]. H.H. Fliche and J.M. Souriau [F3] claim, however, that no evolution is required in certain cosmological models with a cosmological repulsion ($\Lambda > 0$).

To summarize, no convincing evidence on the value of q_0 has come from

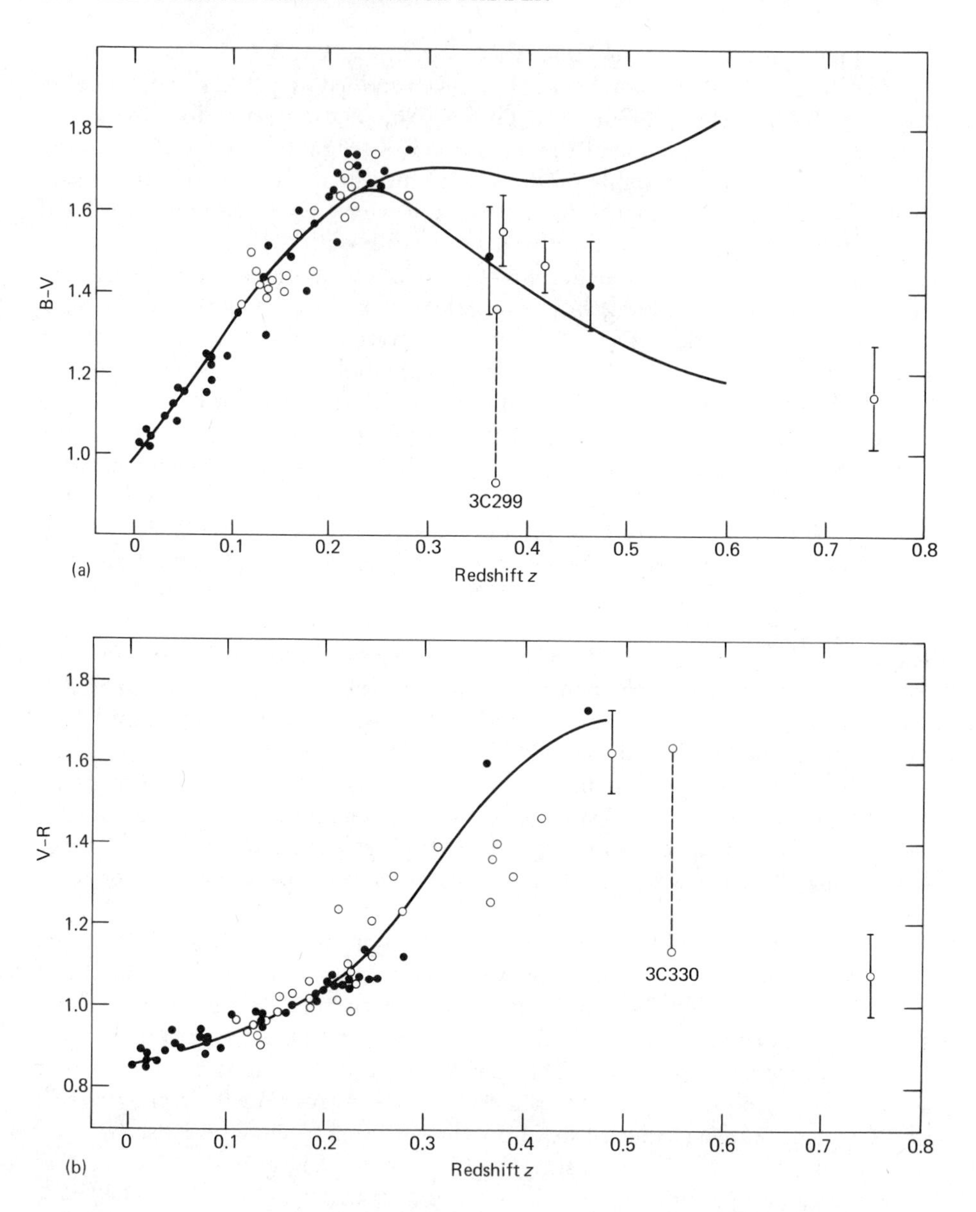

FIGURE 6.5

Color–redshift diagrams for brightest cluster galaxies. (a) B-V versus z. The two curves are predictions based on the visible and ultraviolet continua of nearby galaxies. The radio galaxy 3C299, which is also the brightest galaxy in its cluster, is 0.5 mag bluer than the other bright galaxies in the cluster, but the second brightest galaxy in the cluster (shown connected to 3C299 by a broken line) is normal. (From [K4].) (b) V-R versus z. The solid line is a prediction based on the near infrared and visible continua of nearby galaxies. The radio galaxy 3C330 is bluer than the other bright galaxies in the cluster, but the second brightest galaxy is normal. (From [K4])

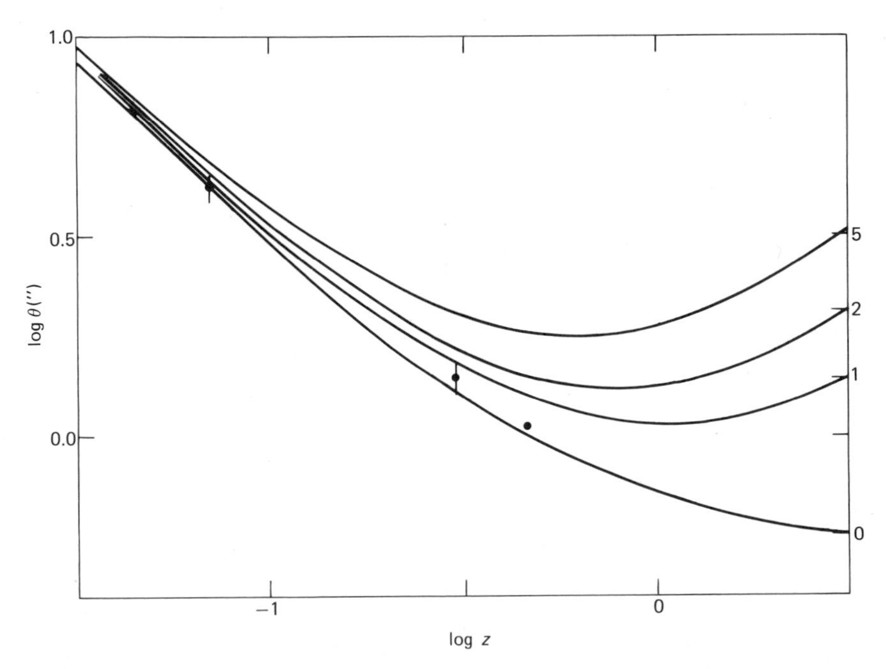

FIGURE 6.6

Angular-diameter–redshift diagram, log θ versus log z, for galaxies in clusters. θ is in seconds of arc. Theoretical curves are labeled with the deceleration parameter, q_0, assuming $\Lambda = 0$. (From [B5].)

studying the dependence of the luminosity and the diameter distances on redshift, and values in the range of -1 to $+2$ are acceptable. This eliminates only cosmological models in which a "bounce" has occurred recently or oscillating models in which the expansion is soon to be reversed. Note that for the case of zero cosmological constant, the deceleration parameter q_0 is always positive, so that in Table 6.4 solutions giving negative values for q_0 should either have been quoted as giving upper limits on q_0 or have allowed $\Lambda \neq 0$.

Attempts to estimate the density parameter Ω_0 have had more success. Rather few galaxies have had their masses estimated directly (from binary galaxy systems, analogously to the method of Sec. 2.2), so we must depend to a great extent on estimating the mass-to-light ratio M/L for different types of galaxy. Table 6.5 summarizes some of these estimates. Within the visible outline of a galaxy, values of M/L in the range of 1–10 (in solar units) are found. However, the flat rotation curves of giant spirals (Fig. 2.44), combined with evidence from pairs of galaxies and small groups, suggests that the outer parts of galaxies contain much dark matter and that the net value of M/L for the entire galaxy is 50–100. Finally when rich clusters of galaxies are studied, even larger values of M/L are found, in the range of 300–1000, suggesting additional dark matter spread through these clusters. Numerical simulations of the dynamics of galaxies in rich clusters show that this additional dark matter must be spread more smoothly than the galaxies, so it cannot

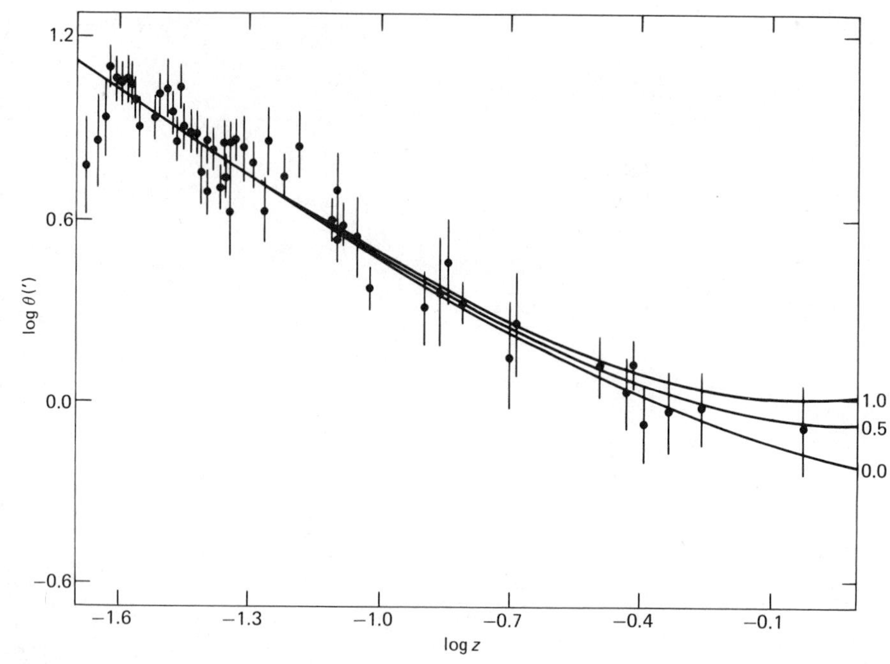

FIGURE 6.7

Angular-diameter–redshift diagram, log θ versus log z, for clusters of galaxies. θ is in minutes of arc. Theoretical curves are labeled with the deceleration parameter, q_0, assuming $\Lambda = 0$. (From [B9].)

be due to a few massive black holes, for example. This tendency for M/L to increase with scale size is illustrated in Fig. 6.8.

If we now measure the average luminosity emitted per unit volume by galaxies in the general field and multiply by the value for M/L found from groups, we can

TABLE 6.5

Mass-to-light ratios M/L in galaxies.

Method	M/L*
Solar neighborhood	2 ± 1
Inner luminous parts of spirals and ellipticals	$8-20\ h$
Binary galaxies and small groups	$60-180\ h$
Clusters of galaxies	$300-1000\ h$

* $h = H_0/100$.

SOURCE: Compiled from [O1],[F1].

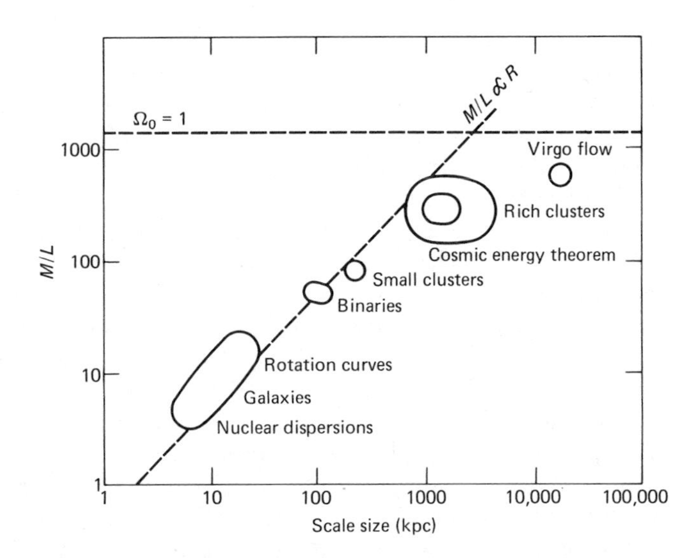

FIGURE 6.8

Typical measured mass-to-light ratios M/L *as a function of measuring scale, assuming* $H_0 = 100$*. Apart from values measured in their nuclei, the* M/L *values for galaxies are dominated by the unseen heavy halo. The sloping broken line corresponds to* M/L *increasing with* R*. The horizontal broken line is the* M/L *ratio required for an Einstein–de Sitter universe* ($\Lambda = 0$, $\Omega_0 = 1$). *(From [D2].)*

obtain an estimate for the average matter density in the universe due to galaxies

$$\rho_{m,0} \sim 2 \times 10^{-30} h^2 \quad [\text{g cm}^{-3}] \tag{6.6}$$

where $h = H_0/100$, and H_0 is the Hubble constant in km s^{-1} Mpc^{-1}). This corresponds to a value for the density parameter of $\Omega_0 \sim 0.1$. However, we can only be sure that 10% of this is due to ordinary baryonic matter (that is, protons, neutrons, and electrons). The dark matter in the halos of galaxies could consist of neutrinos with a nonzero rest mass, which J.E. Gunn estimated would have to be about 10^{-4} of the mass of an electron in order to be able to bind together gravitationally on the scale of galaxy masses [G4]. Other possible forms for the dark matter in galaxy halos might be other types of nonbaryonic particles, dim dwarf stars or Jupiter-sized objects ($0.001 - 0.1 M_\odot$), or black holes of up to $10^6 M_\odot$. The latter might be the remnants of a pregalactic generation of objects, population III (see Sec. 5.4 and [W2],[R5]). Maartin Schmidt showed that not more than 10% of the dark halo of our Galaxy could be made up of normal main-sequence stars with $M \geq 0.1 M_\odot$, and that the mass function for stars would have to be extrapolated to $0.01 M_\odot$ to make up the required mass [S9]. In a more recent study D.O. Richstone and F.G. Graham [R3] found that normal stars failed by a factor of 3 to account for the inferred mass of the halo. Black holes more massive than $10^6 M_\odot$ are ruled out because of the tidal interaction they would have with the disk of the Galaxy [C1].

Table 6.4 summarizes recent estimates of Ω_0 by different workers. Although some estimates derived from large-scale dynamical studies give Ω_0 values of 0.5 or more, most studies suggest that Ω_0 lies in the range of 0.05 – 0.2. For $\Lambda = 0$, this implies an open monotonically expanding universe.

Many theoretical studies have been made of the possibility that there exists a relatively uniformly distributed intergalactic gas (see, for example, the detailed study by R.D. Sherman [S13]). Rich clusters of galaxies are known from their X-ray emission to contain a mass of very hot ($\sim 10^8$ K) gas, comparable to the total mass associated with the visible light from the galaxies. This gas is believed to have originated in the galaxies in the cluster, because X-ray lines due to highly ionized iron has been observed in the gas [M2],[S12], suggesting that the gas has been processed in stars. The isotropic X-ray background radiation has been interpreted as arising from hot gas ($T \sim 10^7$ K) with a density corresponding to $\Omega_0 \sim 0.5 - 1$, heated by ultraviolet radiation from quasars and active galaxies [F2],[S13]. If the temperature of the intergalactic gas is less than about 10^6 K, then the contribution of this gas to Ω_0 is negligible because very strong limits can be set on the intergalactic density of neutral hydrogen. A substantial contribution to Ω_0 is possible only if the gas is very highly ionized.

Perhaps the strongest evidence on Ω_0 comes from the big-bang nucleosynthesis calculations. We saw in Chap. 5 that the primordial abundances of both helium and, especially, deuterium can be understood in a standard hot big-bang model with a low value of Ω_0, and a value as high as 1 due to baryonic matter is completely ruled out. Most modifications of the standard hot big-bang model (for example, variable G, anisotropic expansion, or tepid models) lead to the complete elimination of primordial deuterium and/or helium, and although Population III scenarios can be constructed which can produce the helium in very massive stars [R5],[T3], the deuterium would be exceedingly hard to account for astrophysically.

The successful prediction of the helium and deuterium abundances remains strong evidence for a hot big-bang model with $\Omega_0 \ll 1$. If $\Omega_0 = 1$, then the bulk of the matter in the universe must be in nonbaryonic form, for example, neutrinos with nonzero rest mass. In the absence of such nonbaryonic matter we must take $\Omega_0 \sim 0.03 - 0.1$ in the standard hot big-bang models. The latest primordial helium estimates, 24 or 25% by mass (Table 2.11), favor values at the lower end of this range of Ω_0.

6.3 The age and size of the universe

We saw that there are two methods for estimating the age of our Galaxy, from nucleocosmochronology and from stellar-evolution arguments (see Sec. 2.11). The study of radioactive isotopes, as noted, leads to age estimates for our Galaxy in the range of $9 - 16 \times 10^9$ years. Estimates of the ages of the oldest star clusters are slightly larger, $14 - 20 \times 10^9$ years. The age of our Galaxy is therefore about $14 - 16 \times 10^9$ years. Other nearby galaxies appear to be about the same age as our Galaxy, and since the formation of galaxies probably took place less than 10^9 years after the big bang, the age of the universe is probably not much greater than the age of our Galaxy.

If we take $\Omega_0 = 0.05$ and $H_0 = 67$ (the best values found in Secs. 6.1 and 6.2) and assume $\Lambda = 0$, we arrive at an age for the universe of $t_0 = 14 \times 10^9$ years, in good agreement with the estimate given above for the age of the Galaxy. Alternatively if we take $\Omega_0 = 1$ and $\Lambda = 0$, then to get an age for the universe in the observed range, we need $H_0 \sim 40-45$. From Eq. (6.3) this is an improbable but not inconceivable value. If de Vaucouleurs's value of $H_0 = 100$ is correct, on the other hand, we are forced toward models with a cosmological repulsion, $\Lambda > 0$, in which the age of the universe can be arbitrarily longer than the Hubble time. Models with $\Omega_0 = 0.05$ and q_0 in the range of -0.9 to -1.0 give values of t_0 in the range of $14-16 \times 10^9$ years if $H_0 = 100$, consistent with the age of the Galaxy (Table 6.6).

For many cosmologists the attractiveness of the big-bang models would be greatly reduced if it were necessary to introduce a new arbitrary parameter Λ. However, since we can select a model with the curvature constant $k = 0$ (namely, $\Omega_0 = 0.05$ and $q_0 = -0.925$, for which $t_0 = 14.6 \times 10^9$ years if $H_0 = 100$), which has similar theoretical attractiveness in eliminating an arbitrary parameter, the objections to $H_0 = 100$ on the grounds that $\Lambda > 0$ is required do not seem very strong. On the other hand if $\Omega_0 = 1$ and $H_0 = 100$, no models can be found which give a long enough age without running into the problem that the "antipole" (see below) would fall into the redshift range spanned by known quasars, resulting in a redshift distribution for quasars completely inconsistent with that observed. Thus the most satisfactory parameter set is $(\Omega_0, q_0, H_0) = (0.05, 0.025, 67)$ with $(0.05, -0.925, 100)$ and $(1, 0.5, 40)$ as more extreme possibilities.

As we look out to larger distances, and hence to larger redshifts, we look back in time. Table 6.7 gives the look-back time for different redshifts in the three cosmological models whose parameters are given above. We can now ask: at what redshift did galaxies form? The direct evidence from studies of the correlation of galaxy colors with redshift (see Sec. 6.2) shows that the redshift at which galaxies formed, z_f, must be much greater than 1. O.J. Eggen, D. Lynden-Bell, and Sandage have estimated that our Galaxy collapsed from its most extended state in about 10^8 years [E1]. Allowing a similar time for the evolution of the protogalaxy from the epoch of decoupling until the epoch of maximum extent, we may expect that galaxies formed about 2×10^8 years after the epoch $z \sim 1000$. From Table 6.7 this corresponds to a redshift $z_f \sim 20-30$ for all three models giving the right age for the universe.

One objection that might be raised to such a large value for z_f is that searches for quasars have shown that it is extremely hard to find quasars with redshift $z > 4$, though they ought to have been detectable if they were similar to those found with lower redshifts. It has therefore been speculated that $z \sim 4$ could correspond to the epoch when galaxies (and quasars) formed. However, the current view of the nature of quasars is that they are events in the nuclei of galaxies, probably due to the feeding of a massive black hole there with gas and stars, and so we do not particularly expect quasars to predate galaxies. In fact, it may take some time for the black hole in the galactic nucleus to build up to a significant mass. Quasars may therefore reach their peak power some time after galaxies form, and may then gradually decline in power as the gas available in the galactic nucleus dwindles. The failure to find quasars with $z > 4$ may indicate that $z \sim 4$ corresponds to the epoch of peak quasar power. From Table 6.7 this corresponds to an epoch about 2×10^9 years after galaxies form.

TABLE 6.6

Age of the universe (in units of 10^9 years) for different cosmological models, as a function of H_0.

	Model		Hubble constant H_0 ($km\ s^{-1}\ Mpc^{-1}$)							
	Ω_0	q_0	30	40	50	60	70	80	90	100
$\Lambda = 0$	0	0	32.59	24.45	19.56	16.30	13.97	12.22	10.86	9.78
	0.01	0.005	31.93	23.95	19.16	15.97	13.69	11.97	10.64	9.58
	0.03	0.015	31.11	23.33	18.66	15.55	13.33	11.66	10.37	9.33
	0.1	0.05	29.27	21.96	17.56	14.64	12.55	10.98	9.76	8.78
	0.5	0.25	24.56	18.42	14.74	12.28	10.53	9.21	8.19	7.37
	1.0	0.5	21.73	16.30	13.04	10.86	9.31	8.15	7.24	6.52
	2	1	18.60	13.95	11.16	9.30	7.97	6.98	6.20	5.58
	5	2.5	14.41	10.80	8.64	7.20	6.17	5.40	4.80	4.32
	10	5	11.46	8.59	6.87	5.73	4.91	4.30	3.82	3.44
$k = 0$	0.01	−0.985*	65.37	49.03	39.22	32.68	28.01	24.51	21.79	19.61
	0.03	−0.955	53.81	40.36	32.28	26.90	23.06	20.18	17.94	16.14
	0.1	−0.85	41.65	31.24	24.99	20.83	17.85	15.62	13.88	12.50
	0.5	−0.25	27.08	20.31	16.25	13.54	11.61	10.16	9.03	8.13
	0.8	0.2	23.38	17.54	14.03	11.69	10.02	8.77	7.79	7.01
$\Omega_0 = 0.01$	0.01	−1.05	86.26	64.70	51.76	43.13	36.97	32.35	28.75	25.88
	0.01	−1.0	68.23	51.17	40.94	34.12	29.24	25.59	22.74	20.47
	0.01	−0.5	39.24	29.43	23.55	19.62	16.82	14.72	13.08	11.77
	0.01	1.0	25.34	19.00	15.20	12.67	10.86	9.50	8.45	7.60
	0.01	2.0	21.86	16.40	13.12	10.93	9.37	8.20	7.29	6.56

$\Omega_0 = 0.1$	0.1	−1.2	62.00	46.50	37.20	31.00	26.57	23.25	20.67	18.60
	0.1	−1.1	52.22	39.17	31.33	26.11	22.38	19.58	17.41	15.67
	0.1	−1.0	46.80	35.10	28.08	23.40	20.06	17.55	15.60	14.04
	0.1	−0.5	34.86	26.14	20.92	17.43	14.94	13.07	11.62	10.46
	0.1	1.0	24.21	18.16	14.53	12.10	10.38	9.08	8.07	7.26
	0.1	2.0	21.14	15.86	12.69	10.57	9.06	7.93	7.05	6.34
$\Omega_0 = 1.0$	1	−2*	53.16	39.87	31.80	26.58	22.78	19.94	17.72	15.95
	1	−1.5	34.38	25.79	20.63	17.19	14.74	12.89	11.46	10.32
	1	−1.	28.78	21.59	17.27	14.39	12.34	10.79	9.59	8.64
	1	−0.5	25.57	19.18	15.34	12.79	10.96	9.59	8.52	7.67
	1	0	23.37	17.53	14.02	11.69	10.02	8.77	7.79	7.01
	1	1	20.43	15.32	12.26	10.21	8.75	7.66	6.81	6.13
	1	2	18.47	13.85	11.08	9.23	7.92	6.93	6.16	5.54

* Antipole visible at $z < 4$ (see text).

TABLE 6.7

Look-back times to different redshifts (in units of 10^9 years).

Model			*Redshift z*										
Ω_0	q_0	H_0	*0.1*	*0.5*	*1*	*2*	*3*	*4*	*5*	*10*	*30*	*100*	*1000*
0.05	0.025	67	1.332	4.863	7.264	9.626	10.780	11.457	11.899	12.856	13.496	13.673	13.713
1	0.5	40	2.171	7.426	10.535	13.161	14.260	14.839	15.188	15.850	16.202	16.281	16.296
0.05	−0.925	100	0.928	3.875	6.467	9.467	11.085	12.023	12.613	13.771	14.400	14.540	14.568

We have seen that in certain models, those with $\Lambda < 0$, for example, or with $\Lambda = 0$ and $k > 0$, the universe will then eventually recollapse to a second singularity, at time t_s, say. To calculate t_s, we note that from Eq. (5.14)

$$\frac{t_s}{\tau_0} = 2 \int_0^{\Omega_0/(\Omega_0-1)} \frac{dx}{(\Omega_0/x + 1 - \Omega_0)^{1/2}} \tag{6.7}$$

if $\Lambda = 0$ (since $\dot{R} = 0$ when $x = R/R_0 = \Omega_0/(\Omega_0 - 1)$). The substitution $x = [\Omega_0/(\Omega_0 - 1)] \sin^2 \theta$ gives

$$\frac{t_s}{\tau_0} = \frac{2\Omega_0^{1/2}}{\Omega_0 - 1}[\theta - \sin\theta \cos\theta]_0^{\pi/2} = \frac{\pi\Omega_0^{1/2}}{\Omega_0 - 1} \tag{6.8}$$

Thus for $\Omega_0 = 2$, 1.5, 1.1, 1.01, and $1 + \epsilon$, $\epsilon \ll 1$, we have
$t_s/\tau_0 = 4.5$, 7.8, 33, 316, and π/ϵ, respectively. As ϵ becomes arbitrarily small, t_s/τ_0 becomes arbitrarily large.

Roger Penrose has pointed out that t_s should have some physical meaning and that there is no reason why it should be just a few multiples of τ_0, the Hubble time at the present epoch [P3]. Therefore it seems plausible to suppose that if we are in an oscillating universe (that is, $\Omega_0 > 1$ if $\Lambda = 0$ and $k = 1$), then in fact Ω_0 is extremely close to 1, and the universe is indistinguishable from an Einstein–de Sitter model at the present epoch. Thus the much-vaunted question of whether the universe will keep on expanding forever or will recollapse to a second singularity may turn out to be rather academic.

What is the size of the universe? Is it finite or infinite in extent? The answers to these questions turn out to be complex, but the differences between different cosmological models are not nearly so clear-cut as they are often made out to be. In models with zero or negative curvature ($k \leq 0$), the universe is clearly infinite in extent (though, as we shall see below, only a finite part of it can be observed). The case of positive curvature ($k = 1$) is more complicated. In these models the radial proper distance to a galaxy is

$$d_{pr} = R_0 \int_0^{r_0} \frac{dr}{(1 - r^2)^{1/2}} = R_0 \sin^{-1} r_0 = R_0\chi_0 \tag{6.9}$$

Now d_{pr} can be arbitrarily large (though only a finite range of χ_0 may be observable at the present epoch—see below). However, r_0 traverses only the range $-1 \leq r_0 \leq 1$, and since r_0 is a comoving label attached to galaxies, this could be interpreted as implying that there are only a finite number of galaxies in the universe. Whether this is correct depends on the topology of the universe, how it connects up on the large scale. The two simplest possibilities are as follows:

1 *The spherical topology*, where the point $\chi_0 = \pi/2$, $r_0 = 1$, in the direction (θ, ϕ) is identical to the point $\chi_0 = \pi/2$, $r_0 = 1$, in the direction $(-\theta, -\phi)$. Then as χ_0 goes from $\pi/2$ to π and r_0 goes from 1 to 0, in direction (θ, ϕ), we traverse the same galaxies as we would in going from $\chi_0 = \pi/2$ to $\chi_0 = 0$ in direction $(-\theta, -\phi)$.

Although we would be seeing the same galaxy in two opposite directions if sufficient time were available since the big bang, it would not appear the same since the look-back time would be different for the two images (Fig. 6.9).

2 *The elliptical topology,* where the point $\chi_0 = \pi$ in the direction (θ, ϕ) is identical to the point $\chi_0 = \pi$ in the direction $(-\theta, -\phi)$. Since $r_0 = 0$ for this point, it is called the *antipole.* As χ_0 goes from π to 2π in direction (θ, ϕ), the same galaxies are traversed as if χ_0 varied from π to 0 in direction $(-\theta, -\phi)$.

Now general relativity is essentially a local theory and makes no predictions about the large-scale topology of the universe. In particular there is no reason that points in the direction (θ, ϕ) should be identified with those in the opposite direction. Each time χ_0 passes a multiple of π we may in effect be looking into a new universe. The fact that general relativity does not specify the large-scale topology of the universe seems to me to be a serious incompleteness of the theory in its application to cosmology.

We now need to consider the far more interesting question: how much of the universe is actually observable? This involves the concept of horizon. Two types of horizon can be defined for an observer O:

1 THE EVENT HORIZON This divides all events in the universe into two classes, those that can at some time be observable by O and those that can never be observable by O. Since as time goes by O observes more and more events, this horizon is essentially defined (if it exists) by the last events he or she observes as the universe ends, either at t_s in an oscillating model or as $t \to \infty$ in a model where the universe keeps on expanding monotonically. Let us call this final epoch $t = t_f$. We can show that if $\Lambda = 0$, there is an event horizon if $\Omega_0 > 1$, but not if $\Omega_0 \leq 1$. If $\Lambda \neq 0$, there is an event horizon in all models for which $\Omega_0 > 0$. (The case $\Omega_0 = 0$, with no matter in the universe, is obviously unphysical.)

[The most distant galaxy from which O has received a signal at $t = t_f$ is one

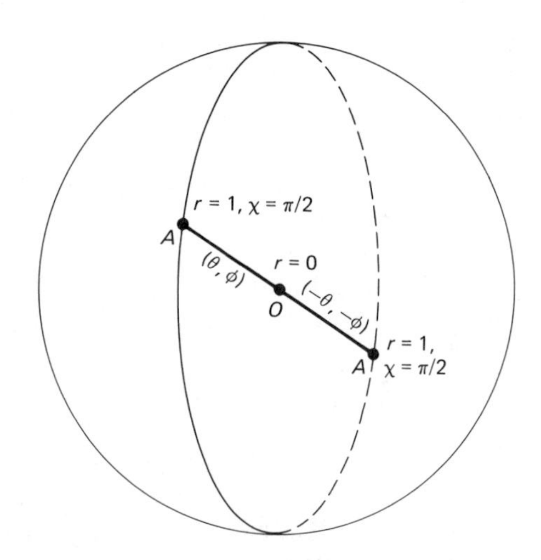

FIGURE 6.9

k = +1 *model with spherical topology. The observer at* O *sees the same point* A *in two opposite directions.*

whose light sets off at the initial instant $t = t_i$ ($t_i = 0$ for big-bang models, $t_i = -\infty$ for bounce models) and arrives at O at $t = t_f$, that is, using Eq. (5.22),

$$\int_{t_i}^{t_f} \frac{c\,dt}{R(t)} = \int_0^{r_{eh}} \frac{dr}{(1-kr^2)^{1/2}} = \chi_{eh} \tag{6.10}$$

where r_{eh} is the radial coordinate of a point on O's event horizon.

The existence of an event horizon is determined by whether the integral on the left-hand side of Eq. (6.10) converges as $t \rightarrow t_f$. From Eq. (5.28) this is true provided $\Omega_0 > 1$ for $\Lambda = 0$. If $\Lambda \neq 0$, there is an event horizon for all $\Omega_0 > 0$. Values for χ_{eh}, r_{eh}, and the proper distance to the event horizon in units of the Hubble distance, $d_{pr}(r_{eh})/c\tau_0$, are given in Table 6.8 for a number of models.]

2 THE PARTICLE HORIZON AT TIME t_0 This divides the particles in the universe into those that have been observed by O up to t_0 and those that have not. We can show that there is a particle horizon for all models with $\Omega_0 > 0$ (that is, all physically realistic models).

[Using Eq. (5.22), the particle horizon is defined by

$$\int_{t_i}^{t_0} \frac{c\,dt}{R(t)} = \int_0^{r_{ph}} \frac{dr}{(1-kr^2)^{1/2}} = \chi_{ph} \tag{6.11}$$

where r_{ph} is the radial coordinate of a point on O's particle horizon.

The existence of a particle horizon is determined by whether the integral on the left-hand side of Eq. (6.11) converges as $t \rightarrow t_i$. From Eq. (5.28) this is true provided $\Omega_0 > 0$ (and this is so regardless of the value of Λ). Values for χ_{ph}, r_{ph}, and $d_{pr}(r_{ph})/c\tau_0$ in a number of models are given in Table 6.8, and the size of the particle horizon at different epochs is illustrated in Fig. 6.10 for several models.]

Thus the answer to the question "How big is the universe?" is that today *we can only observe a finite part of it, whatever the cosmological model.* In an oscillating model we will forever observe only a finite part of the universe, while in a monotonically expanding model we will eventually see everything if we wait long enough and if $\Lambda = 0$ (but not if $\Lambda > 0$). However, in the light of what we said above about the likelihood of t_s, the time when recollapse occurs in an oscillating model, being extremely large, the difference between the two cases may again be academic.

Another interesting possibility arises in the Lemaître models. These are models with positive curvature ($k = 1$) in which the cosmological constant is close to but just larger than the critical value for a static universe ($\Lambda = \Lambda_c(1 + \epsilon)$, $\epsilon \ll 1$). In these models the long look-back time available means that we have the possibility of seeing the radial distance $\chi_0 = \pi, 2\pi, \ldots$. Since $r_0 = 0$ at these points [Eq. (6.9)], sources near them will appear very bright and large. In any magnitude-limited sample of objects (quasars, galaxies) we would therefore expect to see concentrations of sources with redshifts near the values at which $\chi_0 = \pi, 2\pi, \ldots$.

At one time it was thought that quasars exhibited such concentrations of redshift, but this has not been supported by subsequent observations. Thus we can probably eliminate models in which $\chi_0 = \pi$ falls at redshift <4 (see Table 6.8).

TABLE 6.8

The horizon distance.

	Model		*Particle horizon*			*Event horizon*	
	Ω_0	q_0	χ_{ph}	$\frac{d_{pr}}{c\tau_0}$	$\chi(z=4)$	χ_{eh}	$\frac{d_{pr}}{c\tau_0}$
$\Lambda=0$	0	0	13.36	13.36	1.609	—	—
	0.01	0.005	5.98	6.01	1.590	—	—
	0.03	0.015	4.88	4.95	1.552	—	—
	0.1	0.05	3.64	3.83	1.430	—	—
	0.5	0.25	1.76	2.49	0.896	—	—
	1	0.5	2	2	1.106	—	—
	2	1	1.57	1.57	0.927	6.28	6.28
	5	2.5	2.21	1.11	1.391	6.28	3.14
	10	5	2.50	0.83	1.622	6.28	2.09
$k=0$	0.01	−0.985	12.03	12.03	3.483*	13.04	13.04
	0.03	−0.955	8.06	8.06	2.982	9.07	9.07
	0.05	−0.925	6.66	6.66	2.699	7.68	7.68
	0.1	−0.85	5.11	5.11	2.294	6.15	6.15
	0.5	−0.25	2.68	2.68	1.415	3.97	3.97
	0.8	0.2	2.20	2.20	1.200	3.97	3.97
$\Omega_0=0.01$	0.01	−1.05	4.04	15.87	1.169	4.30	16.85
	0.01	−1.0	1.54	12.56	0.446	1.66	13.56
	0.01	−0.5	5.09	7.31	1.394	5.96	8.56
	0.01	1.0	6.83	4.85	1.751	15.43	10.95
	0.01	2.0	7.32	4.24	1.832	15.96	9.24
$\Omega_0=0.1$	0.1	−1.2	4.20	7.10	2.256	4.75	8.03
	0.1	−1.1	3.08	6.15	1.525	3.56	7.11
	0.1	−1.0	2.18	5.62	1.028	2.56	6.61
	0.1	−0.5	2.61	4.42	1.094	3.33	5.63
	0.1	1.0	4.47	3.29	1.651	10.67	7.84
	0.1	2.0	4.97	2.95	1.761	11.24	6.66
$\Omega_0=1$	1	−2	6.07	3.84	4.509*	7.25	4.58
	1	−1.5	3.93	2.78	2.567	5.08	3.59
	1	−1	3.00	2.45	1.838	4.10	3.35
	1	−0.5	2.25	2.25	1.321	3.29	3.29
	1	0	1.49	2.11	0.847	2.43	3.44
	1	1	1.35	1.91	0.731	1.91	2.70
	1	2	2.18	1.78	1.133	2.68	2.19

* Models for which $\chi=\pi$ falls at $z<4$.

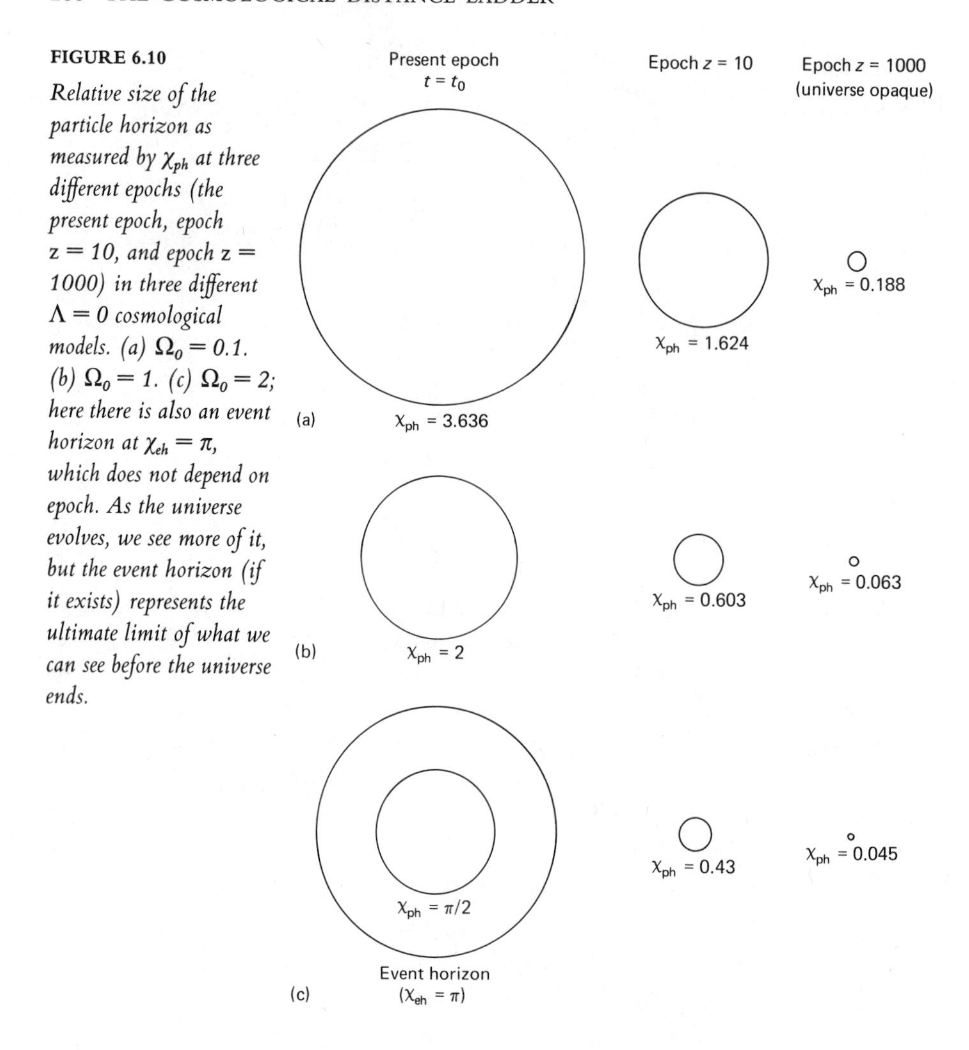

FIGURE 6.10

Relative size of the particle horizon as measured by χ_{ph} at three different epochs (the present epoch, epoch z = 10, and epoch z = 1000) in three different $\Lambda = 0$ cosmological models. (a) $\Omega_0 = 0.1$. (b) $\Omega_0 = 1$. (c) $\Omega_0 = 2$; here there is also an event horizon at $\chi_{eh} = \pi$, which does not depend on epoch. As the universe evolves, we see more of it, but the event horizon (if it exists) represents the ultimate limit of what we can see before the universe ends.

The importance of the particle horizon is that it encloses all the matter that can be interacting with the observer gravitationally (or indeed by any field propagating at the speed of light). The fact that the horizon was much smaller in the past (Fig. 6.10), and hence that no causal explanation can be given for the smoothness and isotropy of the universe on the large scale, has been called the horizon problem. A.H. Guth has proposed a novel solution to this problem, the *inflationary universe* [G7], in which our whole observable universe would have grown from a single tiny perturbation present during the phase transition predicted by Grand Unified Theories to take place at $t \sim 10^{-35}$ (see Sec. 1.6). The inflation depends on a very high value for the cosmological repulsion associated with the energy density of a "false" vacuum during the phase transition. Progress with this model, and some of the problems associated with it, have been reviewed by John Barrow and M.S. Turner [B4].

6.4 Are there natural length scales in the universe?

Are there length or mass scales present in the universe today which have a cosmological significance? The range of sizes and masses on which stars and planets are formed are clearly determined by astrophysical processes at work today—the interaction between gravitation and other physical forces. The same is true for gas clouds and open clusters in irregular galaxies and in the disks of spiral galaxies. When we come to the very oldest objects, globular clusters, galaxies themselves, and clusters of galaxies, the position is less clear. In Sec. 4.3 we saw that it has been proposed that globular clusters form a homogeneous population in all galaxies, with a characteristic mean total luminosity and spread in luminosity, which presumably translates into a characteristic mean total mass and spread in mass. These characteristic mass scales for globular clusters could have been determined by astrophysical processes at the epoch when they formed. In particular it is interesting that the typical mass of a globular cluster (10^4–$10^6 M_\odot$) is comparable to the minimum mass at the epoch of decoupling which is capable of collapsing under its own gravitation (the Jeans mass), $\sim 10^6(\Omega_0 h^2)^{-1/2} M_\odot$, where $h = H_0/100$. The latter is one of the characteristic mass scales of the hot big-bang universe, and we might expect to find structures on this scale, dating from the epoch of decoupling. If the globular clusters do not represent this structure, then there may be dark remnants of Population III objects on this mass scale (presumably black holes).

Galaxies show a wide range of masses, from 10^8 to 10^{12} or $10^{13} M_\odot$. The range of sizes, 1–300 kpc, is not so great and may tell us more about the physics of galaxy formation than about initial conditions in the universe. Groups and clusters of galaxies range in mass from a few multiples of $10^{11} M_\odot$ for aggregations like the Local Group to 10^{14}–$10^{15} M_\odot$ for rich clusters like Virgo or Coma. The characteristic size of these clusters ranges from 1 Mpc for the Local Group to 100 Mpc for the largest superclusters. However, the covariance function studies of galaxy clustering described in Sec. 5.5 show that there is no characteristic length scale for clustering.

A scale that might have been expected to be present if the density fluctuations from which clusters of galaxies formed were adiabatic is the mass scale below which adiabatic fluctuations would have been dissipated during the fireball era, namely, about $2 \times 10^{12}(\Omega_0 h^2)^{-5/4} M_\odot$. Since a small group of galaxies like the Local Group has about this mass, we can imagine that groups and clusters form from primordial adiabatic fluctuations, and that galaxies condense out later. This is the scenario proposed by the Soviet cosmologists A.G. Doroshkevich, R.A. Sunyaev, and Ya. B. Zeldovich [D8].

Turning to the larger scale, are there characteristic sizes associated with the universe itself? One obvious example is the size of the universe when it reaches its maximum extent in an oscillating model. To study this and other characteristic length scales in the universe, we need to compare the magnitude of the different terms in the equation for the scale factor,

$$\dot{R}^2 = \Omega_0 R_0^2 H_0^2 \frac{R_0}{R} - kc^2 + \Lambda \frac{R^2}{3}. \qquad (6.12)$$

There are two characteristic sizes arising from this. The first, denoted by $R = R_c$, say, is associated with the epoch when the first two terms on the right-hand side are of the same order and is given by

$$\frac{R_c}{R_0} = \frac{\Omega_0 R_0^2 H_0^2}{c^2} = \frac{\Omega_0}{|1 + q_0 - 3\Omega_0/2|}, \qquad k \neq 0$$

using Eq. (5.12). Thus

$$R_c = c\tau_0 \frac{\Omega_0}{|1 + q_0 - 3\Omega_0/2|^{3/2}}. \tag{6.13}$$

If $k > 0$ and $\Lambda = 0$, this is just the epoch when the universe reaches its maximum size and starts to recontract. We argued that there is no a priori reason why this should happen to be of the same order of magnitude as $c\tau_0$, so it is likely that $1 + q_0 - 3\Omega_0/2$ is very close to 0 and that $R_c \gg c\tau_0$. In general, R_c is simply the epoch when the Einstein–de Sitter model ceases to be a good description of the evolution of the universe. If $\Lambda = 0$ and the density parameter lies in the range of $0.01 \lesssim \Omega_0 \ll 1$ indicated by observations (see Sec. 6.2), then $R_c \sim c\tau_0\Omega_0$. This characteristic size is not very much smaller than $c\tau_0$. A priori we might find this surprising and prefer to have $\Lambda > 0$, $1 + q_0 - 3\Omega_0/2 \ll 1$, so that $R_c \gg c\tau_0$, and this epoch lies far in the future. If $\Omega_0 \ll 1$, then q_0 would be close to -1, a value not excluded by observations.

The second characteristic scale is associated with the magnitude of the third term on the right-hand side of Eq. (6.12) and defines the size of the universe at which the cosmological term takes on a major role. At the present epoch the magnitude of this term compared with the left-hand side is $\Lambda/3H_0^2 = \Omega_0/2 - q_0$, by Eq. (5.12). Thus we will be close to the epoch when the cosmological term takes on a major role, unless $\Omega_0 \sim 2q_0$.

The fact that Ω_0 is not a very tiny number ($\gtrsim 0.01$ rather than 10^{-40}, say) tells us something quite significant. Either we are close to one of the special epochs when the terms in Eq. (6.12) are of comparable size, or we must have $\Omega_0 \simeq 2q_0 \simeq 1$. We have seen that if the estimate of the age of our Galaxy as $14–16 \times 10^9$ years is correct (see Sec. 2.11), then $\Omega_0 \simeq 2q_0 \simeq 1$ implies that H_0 would have to be in the range of 40–45, not an inconceivable value. This "fine-tuning" of the universe is illustrated in Fig. 6.11, which shows the narrow range of cosmological models in which we could have existed.

Are we at a special epoch? A completely different argument suggests that we may be, though there does not seem to be any connection with the specialness already discussed. Arthur Eddington noticed that the ratio of the electromagnetic and gravitational forces between an electron and a proton in a hydrogen atom gives a very large dimensionless number,

$$\frac{e^2}{Gm_e m_p} = 2.3 \times 10^{39} \tag{6.14}$$

where e is the charge on the electron, G is the gravitational constant, and m_e and m_p

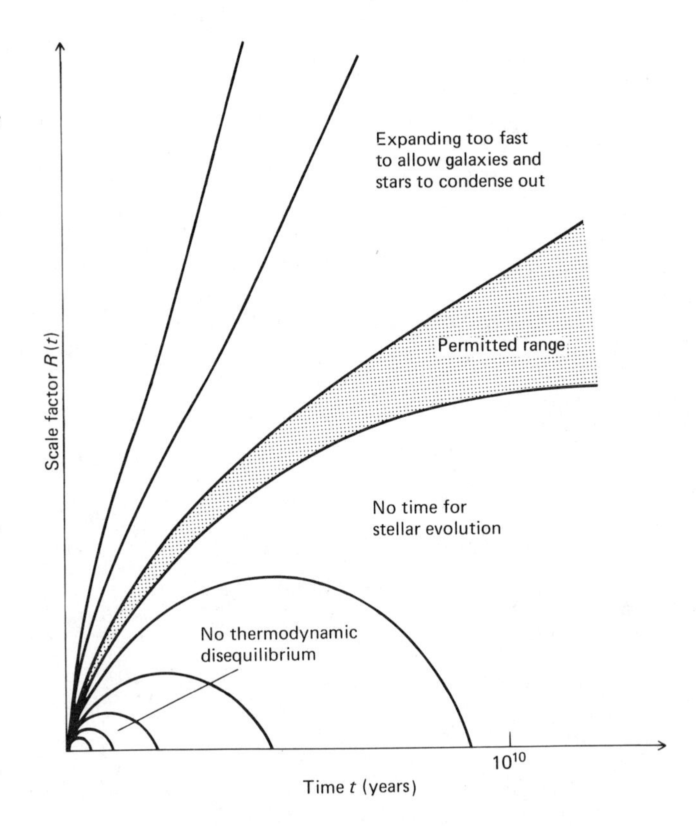

FIGURE 6.11

"Fine-tuning" of the universe. If the universe is oscillating and the time for one cycle is too short, there will be insufficient time for stellar evolution processes. If the time for one cycle is less than a million years, there will not even be time for matter and radiation to come out of thermal equilibrium. If the universe is monotonically expanding and the matter density at the epoch of recombination is too low, it will not be possible for stars and galaxies to condense out. (From [R2].)

are the masses of the electron and the proton. If the present "radius" of the universe $c\tau_0$ is compared with the classical electron radius, we find almost the same large dimensionless number

$$\frac{c\tau_0}{(e^2/m_e c^2)} \simeq 3.3 \times 10^{40}\, h^{-1} \tag{6.15}$$

where $h = H_0/100$. If $G\rho_0\tau_0^2 \sim 1$, then a third large dimensionless number can be deduced, $[\rho_0(c\tau_0)^3/m_p]^{1/2}$, which is roughly the square root of the number of particles in the observable universe. Since τ_0 changes with time in big-bang models, there are only two possibilities: either (1) the constants of nature change with time in such a way as to keep these large numbers the same, or (2) we are at, or close to, a special epoch when these numbers happen to be the same.

Explanation (1) is Dirac's large-number hypothesis. In his 1937 theory Paul Dirac suggested $G \propto t^{-1}$, $R \propto t^{1/3}$, $\tau \propto t$, and $G\rho t^2 =$ constant, so all three numbers are proportional to t and remain in proportion at all times [D5]. However, this leads to an age for the universe which is too short ($t_0 = \tau_0/3$). George Gamow suggested $e^2 \propto t^{-1}$, $G =$ constant, as an alternative, but this runs into difficulties with isotope abundances and the fact that the lines in quasar and galaxy spectra show that the fine-structure constant e^2/hc does not vary with the epoch. In Dirac's 1973 theory

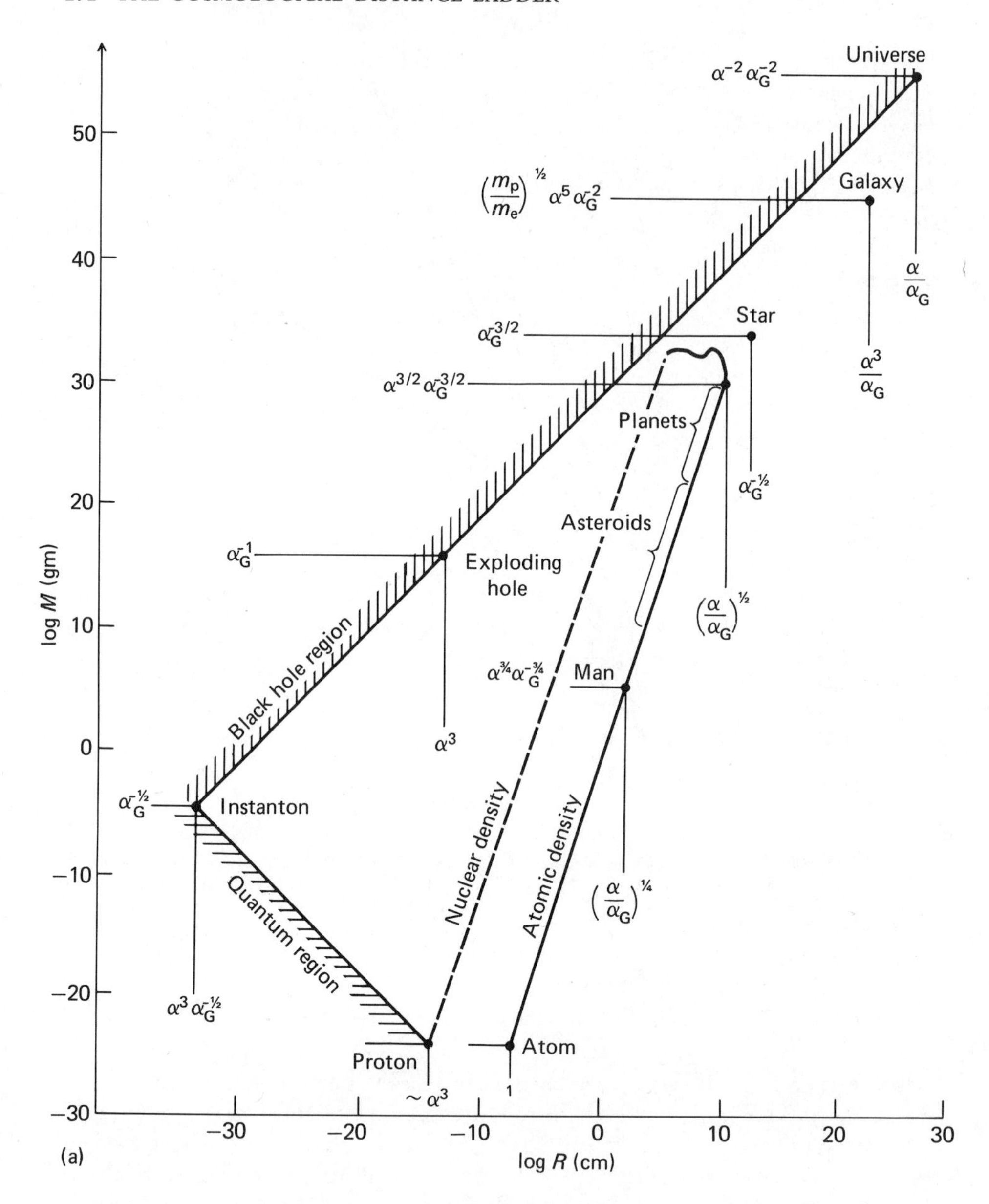

Universe
Galaxy
Star
Planets
Asteroids
Exploding hole
Man
Instanton
Proton
Atom
Black hole region
Quantum region
Nuclear density
Atomic density
log M (gm)
log R (cm)
(a)

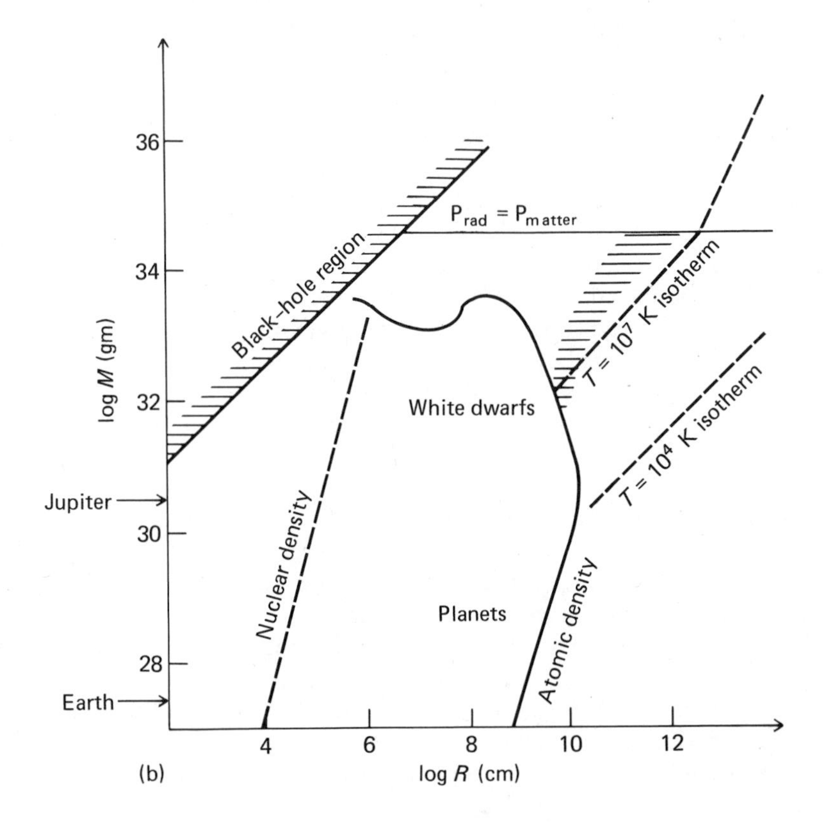

FIGURE 6.12

(a) Mass and length scales of various natural structures expressed in terms of the electromagnetic and gravitational fine-structure constants $\alpha = e^2/hc \simeq 1/137$ and $\alpha_G = Gm_p^2/hc \simeq 5 \times 10^{-39}$. Some scales also depend on the electron-to-proton mass ratio m_e/m_p, but this has been eliminated in most cases using $m_e/m_p \simeq 10\alpha^2$. The asteroid scale also depends on the molecular weight A of the rocky material. All scales can be deduced from known physics, except the mass and length scales of the universe itself, which depend on the age of the universe being (coincidentally?) α_G^{-1} times the electron time scale h/m_ec^2. Also shown are the atomic-density line, the nuclear-density line, the black-hole line, and the "quantum" line corresponding to the Compton wavelength. Most characteristic scales depend on simple powers of α_G; the wide span of so many orders of magnitude is a consequence of the huge numerical value of α_G^{-1}, which reflects the weakness of gravity on the microscopic scale. Formulae for masses are given in units of the proton mass, for radii in units of a_0, the radius of the hydrogen atom. (b) Enlargement of part of (a). All solid planets lie on the atomic-density line and all main-sequence stars lie in the narrow mass range of 0.1–100 times $\alpha_G^{-3/2}\, m_p$ (shaded). Regions with mass below this would never get hot enough to ignite their nuclear fuel; regions with mass above this range would be radiation-pressure dominated and consequently unstable. The precise form of the main-sequence band depends on the opacity and nuclear reaction rates. All stars smaller than the Chanraskhar mass ($\simeq 1.4 M_\odot$) must eventually end up on the electron-degeneracy-supported white dwarf line. Stars bigger than this either shed some of their mass in a supernova explosion and end up as a neutron star or collapse to a black hole. Also shown are the $T \simeq 10^7$ K nuclear ignition isotherm and the $T \simeq 10^4$ K ionization isotherm. (From [C2].)

$G \propto t^{-1}$, $R \propto t$, $\tau \propto t$, and $G\rho\tau^2 =$ constant, and there is the additional feature of continuous particle creation, either uniformly or in proportion to existing matter [D6]. This rate of variation of G would destroy the agreement between the abundances of helium and deuterium produced in the big bang and the observed primordial abundances of these elements [B3],[Y1]. It is also, probably, inconsistent with solar-system evidence on how fast G can be varying [R1].

The second explanation of the Eddington numbers involves the suggestion that we are indeed at a special epoch, namely, the one at which life has evolved to a state where these questions can be asked. Certain conditions are necessary to produce life (existence of galaxies, stars, planets), and these involve relationships between the fundamental constants. This has been called the anthropic principle and was first proposed by Robert Dicke [D4]. Fig. 6.12 illustrates how some of these arguments work.

The problem with this line of thought is, where does it stop? Anything in the universe has to be as it is because if it were not, we would not be here (shades of Aristotle!). Science becomes a totally anthropocentric pursuit, and there is no motive to look for deeper structure to try to explain difficult phenomena.

The argument can also be turned on its head. There is no reason why stars like the sun with planets like the earth should not have formed several billion years before the solar system formed. If the development of intelligent life is not a freak occurrence and it does not immediately destroy itself on becoming advanced, life elsewhere should have reached an unimaginably advanced state and should have colonized the Galaxy by now. Where are They then?*

However, I would like to put forward a different argument. The first of the Eddington large numbers [defined by Eq. (6.14)] would have to be explained by any "superunified" theory amalgamating gravity and other forces (strong and weak nuclear forces and electromagnetic force). Such a theory, which does not yet exist, might make other predictions, and in particular it might explain the number of photons per baryon in the universe. This in turn determines a special epoch, namely, the one where the universe changes from radiation dominated to matter dominated, and at that epoch the second of the two Eddington large numbers [Eq. (6.15)] would be $\sim 10^{35}$, within a few powers of π, e, and so on, of the first large number (and even closer to the ratio of the electromagnetic and gravitational forces between two protons, $\sim 10^{36}$). Thus the similarity of Eqs. (6.14) and (6.15) does not necessarily tell us anything about the present epoch, but may be due to deeper structure than is revealed by our present theories.

If the orthodox big-bang picture presented in Chap. 5 is correct, what does the future of the universe hold for us? Fig. 6.13 outlines the possible futures of the universe for the two cases of models which expand monotonically forever and models which recollapse to a second singularity. In either case it appears that life in

* De Vaucouleurs remarks that They have descovered La Fontaine's principle: pour vivre heureux, vivons cachés! See also the article by Frank Tipler [T5].

The far future of an ever-expanding universe

10^{14} yr	Ordinary stellar activity completed
10^{17} yr	Significant dynamic relaxation in galaxies
10^{20} yr	Gravitational radiation effects in galaxies
$10^{31}-10^{36}$ yr	Proton decay
$10^{64}(M/M_{\odot})^3$ yr	Quantum evaporation of black holes

(a)

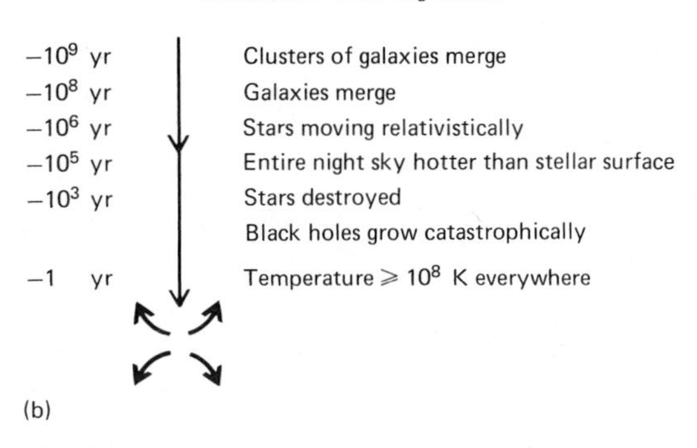

FIGURE 6.13

(a) Time scales for the far future of an everexpanding universe. (Adapted from [R2].) (b) Stages in the contraction of a recollapsing universe. Times are measured backward from the "crunch." (From [R2].)

the universe, as well as having a finite past, has a finite future. To some people this is a troubling thought, but the time scales are so long as to be virtually infinite compared with the human life span or even with the entire time that life has existed on earth.

Postscript

We have seen how the cosmological distance ladder is painstakingly built up, from the solar system up to the largest cosmological scale. Fig. 6.14 gives a summary of the cosmological distance ladder in graphic form. Many of these methods can be pushed to greater distances with existing ground-based techniques. The parallax and proper-motion satellite Hipparcos, due to be launched in the late 1980s, should extend the parallax method by a factor of 3 – 10 in distance. While it will improve the calibration of the main-sequence-fitting method for open clusters and of the spectroscopic and photometric parallax methods enormously, it could also lead to a more fundamental calibration of the Cepheid, RR Lyrae, and nova methods. The Hubble Space Telescope, also due to be launched during the 1980s, has the potential to extend almost all of the distance methods of Fig. 6.14.

Two methods which have the potential, in theory, to measure distances to very remote galaxies and clusters and hence to derive not only the Hubble constant H_0,

FIGURE 6.14

The cosmological distance ladder, showing the range of distances over which different distance indicators have been applied. The lower half illustrates the indisputably primary methods; the upper half the secondary and tertiary methods. The supernova method has been placed here arbitrarily, since there is still some controversy about its validity as a primary method.

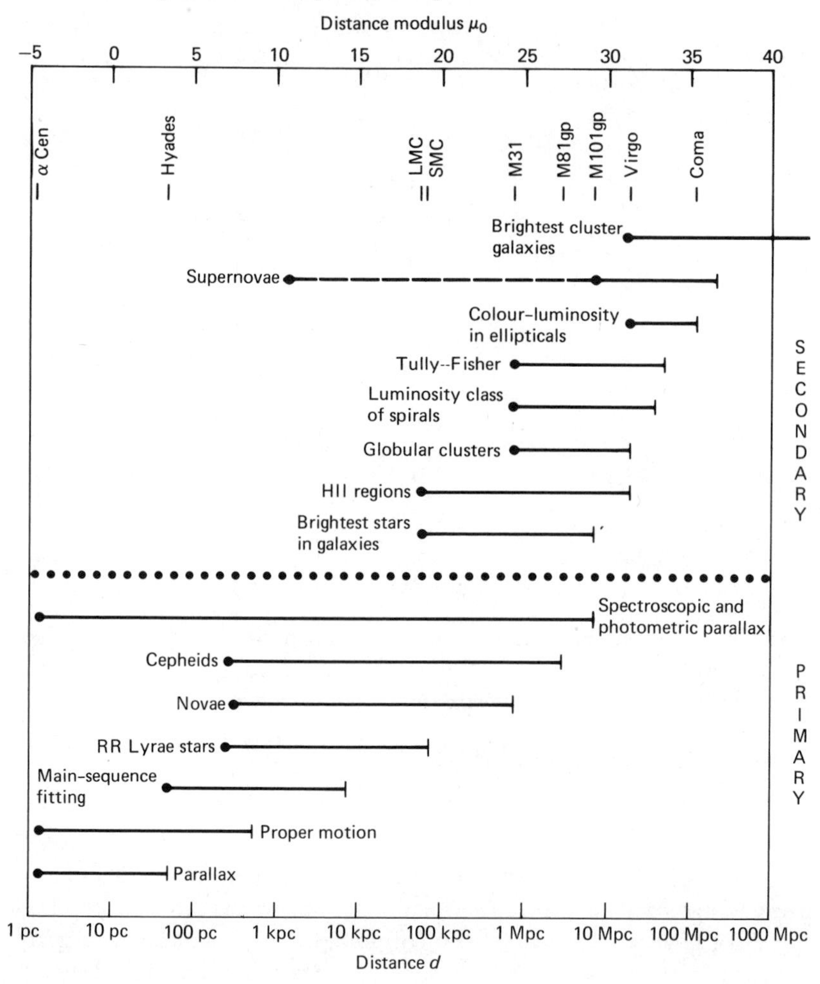

but also the deceleration parameter q_0, are the Baade–Wesselink method for type I supernovae (see Sec. 3.2 and [W2]) and the Zeldovich–Sunyaev effect in clusters [B7]. The latter arises when the very hot X-ray-emitting gas in rich clusters Compton scatters the photons of the microwave background, causing a dark patch in the background at millimeter and centimeter wavelengths and a bright patch at submillimeter wavelengths. My own prediction is that it will be many years before these methods give believable estimates of q_0.

BIBLIOGRAPHY

[A1] Aaronson, M., Mould, J., Huchra, J., Sullivan, W.T., III, Schomer, R.A., and Bothun, G.D., "A Distance Scale from the Infrared Magnitude/HI Velocity Width Relation. III—The Expansion Rate outside the Local Supercluster," *Astrophys. J.*, 1980, **239,** pp. 12–37.

[A2] Aaronson, M., and Mould, J., "A Distance Scale from the Infrared Magnitude/HI Velocity-Width Relation. IV—The Morphological Type Dependence and Scatter in the Relation: The Distances to nearby Groups," *Astrophys. J.*, 1983, **265,** pp. 1–17.

[A3] Abell, G.O., "The Distribution of Rich Clusters of Galaxies," *Astrophys. J.*, suppl. ser., 1958, **3,** pp. 211–288.

[B1] Bahcall, J.N., and Hills, R.E., "The Hubble Diagram for the Brightest Quasars," *Astrophys. J.*, 1973, **179,** pp. 699–703.

[B2] Baldwin, J.A., "Luminosity Indicators in the Spectra of Quasi-Stellar Objects," *Astrophys. J.*, 1977, **214,** pp. 679–684.

[B3] Barrow, J.D., "A Cosmological Limit on the Possible Variations of *G*" *Month. Not. R. Astron. Soc.*, 1978, **184,** pp. 677–682.

[B4] Barrow, J.D., and Turner, M.S., "The Inflationary Universe—Birth, Death and Transfiguration," *Nature*, 1982, **298,** pp. 801–805.

[B5] Baum, W.A., "The Diameter Redshift Relation," in *External Galaxies and Quasi-stellar Objects* (IAU Symp. 44), Evans, D.S., Ed., Reidel, 1972, pp. 393–396.

[B6] Bergh, S. van den, "The Extragalactic Distance Scale," in *Stars and Stellar Systems*, vol. 9: "Galaxies and the Universe," Sandage, A., Sandage, M., and Kristian, J., Eds., University of Chicago Press, 1975, pp. 505–530.

[B7] Birkenshaw, M., "Limits to the Value of the Hubble Constant Deduced from Observations of Clusters of Galaxies," *Month. Not. R. Astron. Soc.*, 1979, **187,** pp. 847–862.

[B8] Branch, D., "On the Use of Type I Supernovae to Determine the Hubble Constant," *Month. Not. R. Astron. Soc.*, 1979, **186,** pp. 609–616.

[B9] Bruzual, G., and Spinrad, H., "The Characteristic Size of Clusters of Galaxies: A Metric Rod Used for a Determination of q_0," *Astrophys. J.*, 1978, **220,** pp. 1–7.

[B10] Bruzual, G., and Kron, R.G., "On the Interpretation of Colors of Faint Galaxies," *Astrophys. J.*, 1981, **241,** pp. 25–40.

[B11] Butcher, H., and Oemler, A. Jr., "The Evolution of Galaxies in Clusters. I—SIT Photometry of C10024+1654 and 3C295," *Astrophys. J.*, 1978, **219,** pp. 18–30.

[C1] Carr, B.J., "On the Cosmological Density of Black Holes," *Comments on Astrophys.*, 1978, **7,** pp. 161–173.

[C2] Carr, B.J., and Rees, M.J., "The Anthropic Principle and the Structure of the Physical World," *Nature*, 1978, **278,** pp. 605–612.

[C3] Cheney, J.E. and Rowan-Robinson, M., "The Interpretation of Optical Counts of Quasars," *Month. Not. R. Astron. Soc.*, 1981, **195,** pp. 497–504.

[D1] Davis, M., Gellar, M.J., and Huchra, J., "The Local Mean Mass Density of the Universe: New Methods for Studying Galaxy Clustering," *Astrophys. J.*, 1978, **221,** pp. 1–18.

[D2] Davis, M., Tonry, J.L., Huchra, J., and Latham, D.W., "On the Virgo Supercluster and The Mean Mass Density of the Universe," *Astrophys. J.*, 1980, **238,** pp. L113–L116.

[D3] Davis, M., and Huchra, J., "A Survey of Galaxy Redshifts. III—The Density Field and the Induced Gravity Field," *Astrophys. J.*, 1982, **254,** pp. 437–450.

[D4] Dicke, R.H., "Dirac's Cosmology and Mach's Principle," *Nature,* 1961, **192,** pp. 440–441.

[D5] Dirac, P.A.M., "The Cosmological Constants," *Nature,* 1937, **139,** p. 323.

[D6] Dirac, P.A.M., "Long Range Forces and Broken Symmetries," *Proc. Roy. Soc. London A,* 1973, **333,** pp. 403–418.

[D7] Djorgovski, S., and Spinrad, H., "Toward the Application of a Metric Size Function in Galactic Evolution and Cosmology," *Astrophys. J.*, 1981, **251,** pp. 417–423.

[D8] Doroshkevich, A.G., Sunyaev, R.A., and Zeldovich, Ya.B., "The Formation of Galaxies in a Friedmannian Universe," in *Confrontation of Cosmological Models with Observational Data* (IAU Symp.), Longair, M.S., Ed., Reidel, 1974, pp. 213–225.

[E1] Eggen, O.J., Lynen-Bell, D., and Sandage, A.R., "Evidence from the Motions of Old Stars that the Galaxy Collapsed," *Astrophys. J.*, 1962, **136,** pp. 748–766.

[F1] Faber, S.M., and Gallacher, J.S., "Masses and Mass-to-Light Ratios of Galaxies," *Ann. Rev. Astron. Astrophys.*, 1979, **17,** pp. 135–187.

[F2] Field, G.B., and Perrenod, S.C., "Constraints on a Dense Hot Intergalactic Medium," *Astrophys. J.*, 1977, **215,** pp. 717–722.

[F3] Fliche, H.H., and Souriau, J.M., "Quasars et Cosmologie," *Astr. Astrophys.*, 1979, **78,** pp. 87–99.

[F4] Ford, H.C., Harms, R.J., Ciardullo, R., and Bartko, F., "The Dynamics of Superclusters: Initial Determinations of the Mass Density of the Universe at Large Scales," *Astrophys. J.*, 1981, **245,** pp. L53–L57.

[G1] Gott, J.R., III, Gunn, J.E., Schramm, D.N., and Tinsley, B.M., "An Unbound Universe?" *Astrophys. J.*, 1974, **194,** pp. 543–553.

[G2] Gott, J.R., III, and Turner, E.L., "The Mean Luminosities and Mass Densities in the Universe," *Astrophys. J.*, 1976, **209,** pp. 1–5.

[G3] Gott, J.R., III, and Turner, E.L., "Groups of Galaxies. III—Mass-to-Light Ratios and Crossing Times," *Astrophys. J.*, 1977, **213,** pp. 309–322.

[G4] Gunn, J.E., "Massive Galactic Halos. I—Formation and Evolution," *Astrophys. J.*, 1977, **218,** pp. 529–598.

[G5] Gunn, J.E., and Oke, J.B., "Spectrophotometry of Faint Cluster Galaxies and the Hubble Diagram: An Approach to Cosmology," *Astrophys. J.*, 1975, **195,** pp. 255–268.

[G6] Gunn, J.E., and Tinsley, B.M., "Dynamical Friction: The Hubble Diagram as a Cosmological Test," *Astrophys. J.*, 1976, **210,** pp. 1–6.

[G7] Guth, A.H., "Inflationary Universe: A Possible Solution to the Horizon and Flatness Problems," *Phys. Rev. D,* 1981, **23,** pp. 347–356.

[H1] Hanes, D.A., "A New Determination of the Hubble Constant," *Month. Not. R. Astron. Soc.*, 1979, **188,** pp. 901–909.

[H2] Hart, L., and Davis, R.D., "Motion of the Local Group of Galaxies and Isotropy of the Universe," *Nature,* 1982, **297,** pp. 191–196.

[H3] Heidmann, J., in *Décalages vers le rouge et expansion de l'univers—L'évolution des galaxies et ses implications cosmologiques* (IAU Colloq. 37), Balkowski, C., and Westerlund, B.E., Eds., Editions CNRS, Paris, 1977, pp. 487–495.

[H4] Hickson, P., "The Angular-Size-Redshift Relation. II—A Test for the Deceleration Parameter," *Astrophys. J.*, 1977, **217,** pp. 964–975.

[H5] Hodge, P.W., "The Extragalactic Distance Scale," *Ann. Rev. Astron. Astrophys.*, 1981, **19,** pp. 357–372.

[H6] Hoessel, J.G., "The Photometric Properties of Brightest Cluster Galaxies. II—SIT and CCD Surface Photometry," *Astrophys. J.*, 1980, **241,** pp. 493–506.

[H7] Hoessel, J.G., Gunn, J.E., and Thuan, T.X., "The Photometric Properties of Brightest Cluster Galaxies. I—Absolute Magnitudes in 116 nearby Abell Clusters," *Astrophys. J.*, 1980, **241,** pp. 486–492.

[K1] Kennicutt, R.C., Jr., "HII Regions as Extragalactic Distance Indicators. II—Application of Isophotal Diameters," *Astrophys. J.*, 1979, **228,** pp. 696–703.

[K2] Kennicutt, R.C., Jr., "HII Regions as Extragalactic Distance Indicators. III—Application of HII Region Fluxes and Galaxy Diameters," *Astrophys. J.*, 1979, **288,** pp. 704–711.

[K3] Kennicutt, R.C., Jr., "HII Regions as Extragalactic Distance Indicators. IV—The Virgo Cluster," *Astrophys. J.*, 1981, **247,** pp. 9–16.

[K4] Kristian, J., Sandage, A., and Westphal, J.A., "An Extension of the Hubble Diagram. II—New Redshifts and Photometry of Very Distant Clusters: First Indication of a Deviation of the Hubble Diagram from a Straight Line," *Astrophys. J.*, 1978, **221,** pp. 383–394.

[L1] Lebofsky, M.J., "Evolution of High-Redshift Galaxies," *Astrophys. J.*, 1981, **245,** pp. L59–L62.

[L2] Lilley, S.J., and Longair, M.S., "Infrared studies of a sample of 3C radio galaxies," 1982, *Month. Not. R. Astron. Soc.*, **199,** pp. 1053–1068.

[L3] Lynden-Bell, D., "Hubble's Constant Determined from Super-luminal Radio Sources," *Nature,* 1977, **270,** pp. 396–399.

[M1] Mathieu, R.D., and Spinrad, H., "Luminosity Function and Colours of the 3C295 Cluster of Galaxies," *Astrophys. J.*, 1981, **251,** pp. 485–496.

[M2] Mitchell, R.J., Culhane, J.L., Davison, P.J.N., and Ives, J.C., "Ariel 5 Observations of the X-Ray Spectrum of the Perseus Cluster," *Month. Not. R. Astron. Soc.*, 1976, **175,** pp. 29–34.

[M3] Mould, J., Aaronson, M., and Huchra, J., "A Distance Scale from the Infrared Magnitude/HI Velocity-Width Relation. II—The Virgo Cluster," *Astrophys. J.*, 1980, **238,** pp. 458–470.

[O1] Olive, K.A., Schramm, D.N., Steigman, G., Turner, M.S., and Yang, J., "Big-Bang Nucleosynthesis as a Probe of Cosmology and Particle Physics," *Astrophys. J.*, 1981, **246,** pp. 557–568.

[O2] Ostriker, J.P., and Tremaine, S.D., "Another Evolutionary Correction to the Luminosity of Giant Galaxies," *Astrophys. J.*, 1975, **202,** pp. L113–L117.

[P1] Peebles, P.J.E., *Comments on Astrophys. Space Phys.*, 1977, **7,** pp. 197–204.

[P2] Peebles, P.J.E., "The Mean Mass Density Estimated from the Kirshner, Oemler, Schechter Galaxy Redshift Sample," *Astr. J.*, 1979, **84,** pp. 730–734.

[P3] Penrose, R., Time-Asymmetry, Cosmological Uniformity and Space-Time Singularities," in *Progress in Cosmology,* Wolfendale, A., Ed., Reidel, 1982, pp. 87–88.

[P4] Press, W.H., and Davis, M., "How to Identify and Weigh Virialized Clusters of Galaxies in a Complete Redshift Catalogue," *Astrophys. J.*, 1982, **259,** pp. 449-473.

[R1] Reasenberg, R.D., and Shapiro, I.I., *Experimental Gravitation,* Academia Nazionale dei Lincei, Roma, 1977.

[R2] Rees, M.J., "Our Universe and Others," *Quart. J. R. Astron. Soc.,* 1981, **22,** pp. 109-124.

[R3] Richstone, D.O., and Graham, F.G., "A New Determination of the Halo Luminosity Density of the Galaxy," *Astrophys. J.,* 1981, **248,** pp. 516-523.

[R4] Richstone, D.O., and Potter, M.D., "Galactic Mass Loss: A Mild Evolutionary Correction to the Angular Size Test," *Astrophys. J.,* 1982, **254,** pp. 451-455.

[R5] Rowan-Robinson, M., and Tarbet, P., "Population III and the Microwave Background," in *Progress in Cosmology,* Wolfendale, A., Ed., Reidel, 1982, pp. 101-117.

[S1] Sandage, A., "The Redshift-Distance Relation. II—The Hubble Diagram and Its Scatter for First-Ranked Cluster Galaxies: A Formal Value for q_0," *Astrophys. J.,* 1972, **178,** pp. 1-24.

[S2] Sandage, A., and Hardy, E., "The Redshift-Distance Relation. VII—Absolute Magnitudes of the First Three Ranked Cluster Galaxies as a Function of Cluster Richness and Bautz-Morgan Cluster Type: The Effect on q_0," *Astrophys. J.,* 1973, **183,** pp. 743-757.

[S3] Sandage, A., and Tammann, G.A., "Steps toward the Hubble Constant. III—The Distance and Stellar Content of the M101 Group of Galaxies," *Astrophys. J.,* 1974, **194,** pp. 223-243.

[S4] Sandage, A., and Tammann, G.A., "Steps toward the Hubble Constant. IV—Distances to 39 Galaxies in the General Field Leading to a Calibration of the Galaxy Luminosity Classes and a First Hint of the Value of q_0," *Astrophys. J.,* 1974, **194,** pp. 559-568.

[S5] Sandage, A., and Tammann, G.A., "Steps toward the Hubble Constant. V—The Hubble Constant from nearby Galaxies and the Regularity of the Local Velocity Field," *Astrophys. J.,* 1975, **196,** pp. 313-328.

[S6] Sandage, A., and Tammann, G.A., "Steps toward the Hubble constant. VI—The Hubble Constant Determined from Redshifts and Magnitudes of Remote Sc I Galaxies: The Value of q_0," *Astrophys. J.,* 1975, **197,** pp. 265-280.

[S7] Sandage, A., and Tammann, G.A., "Steps toward the Hubble Constant. VII—Distances to NGC 2403, M101, and the Virgo Cluster Using 21 Centimeter Line Widths Compared with Optical Methods: The Global Value of H_0," *Astrophys. J.,* 1976, **210,** pp. 7-24.

[S8] Sandage, A., and Tammann, G.A., "Steps toward the Hubble Constant. VIII—The Global Value," *Astrophys. J.,* 1982, **256,** pp. 339-345.

[S9] Schmidt, M., "The Mass of the Galactic Halo Derived from the Luminosity Function of High-Velocity Stars," *Astrophys. J.,* 1975, **202,** pp. 22-29.

[S10] Schramm, D., "Constraints on the Density of Baryons in the Universe," *Phil. Trans. R. Soc. London A,* 1982, **307,** pp. 43-54.

[S11] Seldner, M., and Peebles, P.J.E., "A New Way to Estimate the Mean Mass Density Associated with Galaxies," *Astrophys. J.,* 1977, **214,** pp. L1-L4.

[S12] Serlemitsos, P.J., Smith, B.W., Boldt, E.A., Holt, S.S., and Swank, J.H., "X-Radiation from Clusters of Galaxies: Spectral Evidence for a Hot Evolved Gas," *Astrophys. J.,* 1977, **211,** pp. L63-L66.

[S13] Sherman, R.D., "Theory of the Intercluster Medium," *Astrophys. J.*, 1979, **232,** pp. 1–17.

[S14] Shostak, G.S., "Integral Properties of Late-Type Galaxies Derived from HI Observations," *Astr. Astrophys.*, 1978, **68,** pp. 321–341.

[S15] Stenning, M., and Hartwick, F.D.A., "The Local Value of the Hubble Constant from Luminosity Classification of Sb Galaxies," *Astr. J.*, 1980, **85,** pp. 101–116.

[T1] Tammann, G.A., "The value of H_0," *Highlights in Astronomy*, vol. 6, West, R.M., Ed., Reidel, 1983, pp. 301–313.

[T2] Tammann, G.A., Sandage, A., and Yahil, A., "The Determination of Cosmological Parameters," in *Physical Cosmology*, Balian, R., Audouze, J., and Schramm, D.N., Eds., North-Holland, 1979, pp. 53–125.

[T3] Tarbet, P.W., and Rowan-Robinson, M., "A Pregalactic Population III: Primordial Helium and Heavy Elements and the Microwave Background," *Nature*, 1982, **298,** pp. 711–715.

[T4] Tinsley, B.M., "A First Approximation to the Effect of Evolution on q_0," *Astrophys. J.*, 1972, **173,** pp. L93–97.

[T5] Tipler, F.J., "Extraterrestrial Intelligent Beings Do Not Exist," *Quart. J. R. Astron. Soc.*, 1980, **21,** pp. 267–281.

[T6] Tully, R.B., and Fisher, J.R., "A New Method of Determining Distances to Galaxies," *Astr. Astrophys.*, 1977, **54,** pp. 661–673.

[T7] Turner, E.L., and Ostriker, J.P., "The Mass-to-Light Ratio of Late-Type Binary Galaxies: Luminosity versus Number-Weighted Averages," *Astrophys. J.*, 1977, **217,** pp. 24–26.

[T8] Tyson, J.A., and Jarvis, J.F., "Evolution of Galaxies: Automated Faint Object Counts to 24th Magnitude," *Astrophys. J.*, 1979, **230,** pp. L153–L156.

[V1] Vaucouleurs, G. de, "Extragalactic Distance Scale, Malmquist Bias and Hubble Constant," *Month. Not. R. Astron. Soc.*, 1983, **201,** pp. 367–378.

[V2] Vaucouleurs, G. de, and Bollinger, G., "The Extragalactic Distance Scale. VII—The Velocity-Distance Relation in Different Directions and the Hubble Ratio within and without the Local Supercluster," *Astrophys. J.*, 1979, **233,** pp. 433–452.

[V3] Vaucouleurs, G. de, Peters, W.L., Bottinelli, L., Gouguenheim, L., and Paturel, G., "Hubble Ratio and Solar Motion from 300 Spirals Having Distances Derived from HI Line Widths," *Astrophys. J.*, 1981, **248,** pp. 408–422.

[V4] Visvanathan, N., "The Hubble Diagram for E and SO Galaxies in the Local Region," *Astrophys. J.*, 1979, **228,** pp. 81–94.

[V5] Visvanathan, N., and Sandage, A., "The Color-Absolute Magnitude Relation for E and SO Galaxies. I—Calibration and Tests for Universality Using Virgo and Eight Other Nearby Clusters," *Astrophys. J.*, 1977, **216,** pp. 214–226.

[W1] Wagoner, R.V., "Determining Cosmological Distances from Supernovae," in *Physical Cosmology*, Balain, R., Audouze, J., and Schramm, D.N., Eds., North-Holland, 1979, pp. 179–194.

[W2] White, S.D.M., and Rees, M.J., "Core Condensation in Heavy Halos: A Two-Stage Theory for Galaxy Formation and Clustering," *Month. Not. R. Astron. Soc.*, 1978, **183,** pp. 341–358.

[Y1] Yang, J., Schramm, D.N., Steigman, G., and Rood, D.T., "Constraints on Cosmology and Neutrino Physics from Big-Bang Nucleosynthesis," *Astrophys. J.*, 1979, **227,** pp. 697–707.

APPENDIXES

APPENDIX A1

STARS NEARER THAN FIVE PARSECS

(from K1)

No.	*Name*	*RA* (1950)	*Decl* (1950)	*Proper motion*	*Parallax*	*Distance light years*
1	Sun					
2	α Centauri[a]	14^h36^m2	$-60°38'$	3″68	0″760	4.3
3	Barnard's star	16 55.4	+ 4 33	10.31	.552	5.9
4	Wolf 359	10 54.1	+ 7 19	4.71	.431	7.6
5	BD +36°2147	11 00.6	+36 18	4.78	.402	8.1
6	Sirius	6 42.9	−16 39	1.33	.377	8.6
7	Luyten 726-8	1 36.4	−18 13	3.36	.365	8.9
8	Ross 154	18 46.7	−23 53	0.72	.345	9.4
9	Ross 248	23 39.4	+43 55	1.58	.317	10.3
10	ϵ Eridani	3 30.6	− 9 38	0.98	.305	10.7
11	Luyten 789-6	22 35.7	−15 36	3.26	.302	10.8
12	Ross 128	11 45.1	+ 1 06	1.37	.301	10.8
13	61 Cygni	21 04.7	+38 30	5.22	.292	11.2
14	ϵ Indi	21 59.6	−57 00	4.69	.291	11.2
15	Procyon	7 36.7	+ 5 21	1.25	.287	11.4
16	Σ 2398	18 42.2	+59 33	2.28	.284	11.5
17	BD +43°44	0 15.5	+43 44	2.89	.282	11.6
18	CD −36°15693	23 02.6	−36 09	6.90	.279	11.7
19	τ Ceti	1 41.7	−16 12	1.92	.273	11.9
20	BD +5°1668	7 24.7	+ 5 23	3.73	.266	12.2
21	CD −39°14192	21 14.3	−39 04	3.46	.260	12.5
22	Kapteyn's star	5 09.7	−45 00	8.89	.256	12.7
23	Krüger 60	22 26.3	+57 27	0.86	.254	12.8
24	Ross 614	6 26.8	− 2 46	0.99	.249	13.1
25	BD −12°4523	16 27.5	−12 32	1.18	.249	13.1
26	van Maanen's star	0 46.5	+ 5 09	2.95	.234	13.9
27	Wolf 424	12 30.9	+ 9 18	1.75	.229	14.2
28	G158-27	0 4.2	− 7 48	2.06	.226	14.4
29	CD −37°15492	0 02.5	−37 36	6.08	.225	14.5
30	BD +50°1725	10 08.3	+49 42	1.45	.217	15.0
31	CD −46°11540	17 24.9	−46 51	1.13	.216	15.1

No.	Name	Visual apparent magnitude and spectral type: A	B	C	Visual luminosity: A	B	C
1	Sun	−26.8 G2			1.0		
2	α Centauri[a]	0.1 G2	1.5 K6	11 M5e	1.3	0.36	0.00006
3	Barnard's star	9.5 M5	[b]		.00044	[b]	
4	Wolf 359	13.5 M8e			.00002		
5	BD +36°2147	7.5 M2	[b]		.0052	[b]	
6	Sirius	−1.5 A1	8.3 DA		23.	.0028	
7	Luyten 726-8	12.5 M6e	13.0 M6e		.00006	.00004	
8	Ross 154	10.6 M5e			.0004		
9	Ross 248	12.2 M6e			.00011		
10	ϵ Eridani	3.7 K2			.30		
11	Luyten 789-6	12.2 M6			.00012		
12	Ross 128	11.1 M5			.00033		
13	61 Cygni	5.2 K5	6.0 K7	[b]	.083	.040	[b]
14	ϵ Indi	4.7 K5			.13		
15	Procyon	0.3 F5	10.8		7.6	.0005	
16	Σ 2398	8.9 M4	9.7 M5		.0028	.0013	
17	BD +43°44	8.1 M1	11.0 M6		.0058	.00040	
18	CD −36°15693	7.4 M2			.012		
19	τ Ceti	3.5 G8			.44		
20	BD +5°1668	9.8 M4	[b]		.0014	[b]	
21	CD −39°14192	6.7 M1			.025		
22	Kapteyn's star	8.8 M0			.0040		
23	Krüger 60	9.7 M4	11.2 M6		.0017	.00044	
24	Ross 614	11.3 M5e	14.8		.0004	.00002	
25	BD −12°4523	10.0 M5			.0013		
26	van Maanen's star	12.4 DG			.00017		
27	Wolf 424	12.6 M6e	12.6 M6e		.00014	.00014	
28	G158-27	13.8 m			.00005		
29	CD −37°15492	8.6 M3			.00058		
30	BD +50°1725	6.6 K7			.040		
31	CD −46°11540	9.4 M4			.0030		

APPENDIX A1 CONTINUED

No.	Name	RA (1950)	Decl (1950)	Proper motion	Parallax	Distance light years
32	CD −49°13515	21 30.2	−49 13	.81	.214	15.2
33	CD −44°11909	17 33.5	−44 17	1.16	.213	15.3
34	Luyten 1159-16	1 57.4	+12 51	2.08	.212	15.4
35	BD +15°2620	13 43.2	+15 10	2.30	.208	15.7
36	BD +68°946	17 36.7	+68 23	1.33	.207	15.7
37	L145-141	11 43.0	−64 33	2.68	.206	15.8
38	BD −15°6290	22 50.6	−14 31	1.16	.206	15.8
39	40 Eridani	4 13.0	− 7 44	4.08	.205	15.9
40	BD +20°2465	10 16.9	+20 07	0.49	.202	16.1
41	Altair	19 48.3	+ 8 44	0.66	.196	16.6
42	70 Ophiuchi	18 02.9	+ 2 31	1.13	.195	16.7
43	AC +79°3888	11 44.6	+78 58	0.89	.194	16.8
44	BD +43°4305	22 44.7	+44 05	0.83	.193	16.9
45	Stein 2051	4 26.8	+58 53	2.37	.192	17.0

		Visual apparent magnitude and spectral type			Visual luminosity		
No.	Name	A	B	C	A	B	C
32	CD −49°13515	8.7 M3			.0058		
33	CD −44°11909	11.2 M5			.00063		
34	Luyten 1159-16	12.3 M8			.00023		
35	BD +15°2620	8.5 M2			.0076		
36	BD +68°946	9.1 M3.5	[b]		.0044	[b]	
37	L145-141	11.4			.0008		
38	BD −15°6290	10.2 M5			.0016		
39	40 Eridani	4.4 K0	9.5 DA	11.2 M4e	.33	.0027	.00063
40	BD +20°2465	9.4 M4.5	[b]		.0036	[b]	
41	Altair	0.8 A7			10.		
42	70 Ophiuchi	4.2 K1	6.0 K6		.44	.083	
43	AC +79°3888	11.0 M4			.0009		
44	BD +43°4305	10.1 M5e	[b]		.0021	[b]	
45	Stein 2051	11.1 M5	12.4 DC		.0008	.0003	

[a] Position of α Centauri C is 14h 26.3m, −62°28′.
[b] Unseen components.

APPENDIX A2

PROPERTIES OF SELECTED BINARY SYSTEMS

Type	*Star*	*P(day)*	*Separation* R_0	*Spectral type*
Detached eclipsing binary	α Cr B (Alphecca)	17.36	41.9	A0 G6
Semi-detached eclipsing triple	β Per (Algol)	2.87	15.7	B8 gK0
Visual binary	α C Ma (Sirius)	49.9 yr	2.10^3	A1 V DA
Visual triple	α Cen A,B (Rigel Kent)	80.1 yr	2.10^3	G4 K1
Eclipsing spectroscopic binary	α Vir (Spica)	4.01		B2 V B3 V
Eclipsing spectroscopic binary	β Aur (Menkalinan)	3.96		A2 IV A2 IV

Type	*Star*	$M/M_\odot$	$R/R_\odot$	M_{bol}	m_V	$d(pc)$
Detached eclipsing	α Cr B	2.5	2.9	−0.1	2.3	22
binary	(Alphecca)	0.89	0.87	+5.4		
Semi-detached	β Per	5.2	3.57	−1.0	2.2	27
eclipsing triple	(Algol)	1.01	3.76	+2.7		
Visual binary	α C Ma	2.28		0.8	−1.5	2.64
	(Sirius)	0.98		11.2	8.6	
Visual triple	α Cen A,B	1.08		4.4	0.1	1.32
	(Rigel Kent)	0.88		5.65	1.4	
Eclipsing spectroscopic	α Vir	$7.51/\sin^3 i$			1.0	65
binary	(Spica)	$4.52/\sin^3 i$				
Eclipsing spectroscopic	β Aur	$2.20/\sin^3 i$			1.9	26
binary	(Menkalinan)	$2.12/\sin^3 i$				

APPENDIX A3(a)

DISTANCES TO OPEN CLUSTERS WITHIN 750 pc (from M1)

Cluster	*Galactic coordinates* *l* (*degrees*)	*b* (*degrees*)	*Reddening* $E(B\text{-}V)$	*Distance* *pc*
U Ma cl.	110.2	+44.8	0.00	22
Hyades	179.1	−23.9	0.00	43
Coma Ber	221.1	+84.1	0.00	89
Pleiades	166.6	−23.5	0.04	130
IC 2391	270.4	−6.9	0.00	150
IC 2602	289.6	−4.9	0.04	150
α Per	147.0	−6.0	0.08	170
Praesepe	205.5	+32.5	0.00	174
NGC 2451	252.4	−6.7	0.00	200
ζ Sco	14.9	−79.2	0.00	240
NGC 6475	355.9	−4.5	0.06	260
NGC 7092	92.5	−2.3	0.01	305
IC 348	160.4	−17.7	non-uniform	316
NGC 2232	214.2	−8.1	0.02	320
Cr 132b	243.3	−9.2	0.03	330
IC 4665	30.6	+17.1	0.18	350
Rup 98	297.2	−2.2	0.17	350
NGC 6633	36.1	+8.3	0.11	360
Cr 140	245.0	−8.0	0.05	378
NGC 752	137.2	−23.4	0.03	380
NGC 2547	264.5	−8.5	0.05	380
Orion cl.	206.7	−20.7	0.06	400
NGC 2516	274.0	−15.9	0.11	400
IC 4756	36.4	+5.3	0.19	400
NGC 3228	280.7	+4.6	0.00	415
NGC 1662	187.7	−21.1	0.29	420
Tr 10	262.8	+0.6	0.02	440
NGC 3532	289.6	+1.5	0.02	445
NGC 2422	231.0	+3.1	0.07	456
NGC 6405	355.6	−0.7	0.15	460
NGC 1039	143.6	−15.6	0.09	470
NGC 2281	175.0	+17.1	0.07	496

Cluster	Galactic coordinates *l* (degrees)	*b* (degrees)	Reddening $E(B\text{-}V)$	Distance pc
NGC 6281	347.8	+2.0	0.14	525
NGC 1647	180.4	−16.8	0.39	550
Pi 4	262.7	−2.4	0.03	550
Cr 132a	243.3	−9.2	0.02	560
Ma 6	134.7	+0.0	0.64	560
NGC 6451	359.5	−1.6	0.09	565
NGC 6124	340.8	+6.0	0.68	580
NGC 2527	246.1	+1.9	0.08	590
NGC 6167	335.3	−1.3	0.89	590
NGC 1342	155.0	−15.4	0.30	600
NGC 6716	15.4	+9.6	0.13	600
IC 4725	13.6	−4.5	non-uniform	600
Ba 19	307.4	+1.3	—	600
Tr 2	137.4	−3.9	0.33	615
NGC 225	122.0	−1.1	0.29	630
NGC 5367	316.5	+21.1	non-uniform	630
NGC 7063	83.1	−9.9	0.08	630
NGC 2548	227.9	+15.3	0.00	640
NGC 3114	283.3	−3.8	0.10	640
Cr 463	127.4	+9.6	non-uniform	650
NGC 2287	231.1	−10.2	0.01	680
NGC 6494	9.8	+2.9	0.38	690
NGC 5662	316.9	+3.5	0.32	700
NGC 6134	334.4	−0.2	0.45	700
NGC 5460	315.8	+12.7	0.13	720
NGC 5823	321.2	+2.6	0.27	720
Rup 108	308.2	+4.0	0.18	730
NGC 5822	321.7	+3.6	0.14	740
NGC 7160	104.0	+6.5	0.36	740
NGC 2925	276.0	−1.2	0.06	745
Rup 46	283.4	+5.9	0.07	750

APPENDIX A3(b)

DISTANCE MODULI TO SELECTED OPEN CLUSTERS AND ASSOCIATIONS

	Distance modulus		
Cluster	$H\beta$ (C4,C5)	$H\gamma$ (C3)	*Associated OB association*
Pleiades	5.54*	5.5 (adopted)	
IC 2602	5.62	5.92 ± 0.05	
IC 2391	6.00 ± 0.06		
α Per	6.1	6.21 ± 0.07	Per OB3
Sco-Cen		6.49 ± 0.07	
IC 4665	7.5		
Ori OB1 subgp a	8.03	8.12 ± 0.07 (subgp a and b)	Ori OB1
subgp b	8.21		
III Cep = Cep OB3	9.3		Cep OB3
NGC 6913	10.2 ± 0.1		Cyg OB1
NGC 6910	10.6 ± 0.1		Cyg OB9
NGC 2362	11.0		
h and χ Per	11.3	11.46 ± 0.07	Per OB1
NGC 6231	11.5		Sco OB1
NGC 6871	11.63 ± 0.06		Cyg OB3
NGC 4755	11.7	11.75 ± 0.08	Cen OB1
IC 2944		12.04 ± 0.11	
NGC 3293		12.32 ± 0.09	

* Calibration based on trigonometric parallaxes.

APPENDIX A4(a)

MEAN ABSOLUTE VISUAL MAGNITUDE, M_V, VERSUS MK SPECTRAL TYPE (from M3)

Spectral class	Luminosity class						
	V	*IV*	*III*	*II*	*Ib*	*Ia*	*Ia0*
O5	−5.6						
O9	−4.8	−5.3	−5.7	−6.0	−6.1	−6.2	
B0	−4.3	−4.8	−5.0	−5.4	−5.8	−6.2	−8.1
B5	−1.0	−1.8	−2.3	−4.4	−5.7	−7.0	−8.3
A0	0.7	0.2	−0.4	−3.0	−5.2	−7.1	−8.4
A5	1.9	1.4	0.3	−2.7	−4.8	−7.7	−8.5
F0	2.5	1.9	0.8	−2.4	−4.7	−8.5	−8.7
F5	3.3	2.1	1.2	−2.3	−4.6	−8.2	−8.8
G0	4.4	2.8	0.9	−2.1	−4.6	−8.0	−9.0
G5	5.2	3.0	0.5	−2.1	−4.6	−8.0	
K0	5.9	3.1	0.6	−2.2	−4.5	−8.0	
K5	7.3		−0.2	−2.3	−4.5	−8.0	
M0	8.8		−0.4	−2.3	−4.6	−7.5	
M2	10.0		−0.6	−2.4	−4.8	−7.0	
M5	12.8		−0.8				

APPENDIX A4(b)

M_V FOR O AND B STARS (from W1, W2)

Luminosity class	*V*	*IV*	*III*	*II*	*Ib*	*Iab*	*Ia*
O3	−5.5						−7.0
O4	−5.5		−6.4				−7.0
O5	−5.5		−6.4				−7.0
O6	−5.3		−5.6		−6.3:		−7.0
O6.5	−5.3		−5.6		−6.3:		−7.0
O7	−4.8		−5.6	−5.9	−6.3:		−7.0
O7.5	−4.8		−5.6	−5.9	−6.3:		−7.0
O8	−4.4		−5.6	−5.9	−6.2:	−6.5	−7.0
O8.5	−4.4		−5.6	−5.9	−6.2:	−6.5	−7.0
O9	−4.3	−5.0	−5.6	−5.9	−6.2:	−6.5	−7.0
O9.5	−4.1	−4.7	−5.3	−5.9	−6.2:	−6.5	−7.0
B0	−3.6	(−4.3)	(−4.9)	−5.8	−6.0	−6.5	−7.0
B0.5	−3.6	(−4.3)	(−4.9)	(−5.5:)	−6.0	−6.5	−7.0
B1	−2.7	−3.5	−4.2	(−5.0:)	−6.0	−6.5	−7.0
B1.5	−2.7	(−3.5)	−4.2	(−5.0:)	−5.4	−6.2	−7.0
B2	(−2.2)	−2.9	(−3.5:)	(−4.5:)	−5.4	−6.2	−7.0
B2.5	−1.6	(−2.5:)	(−3.5:)	(−4.5:)	−5.4	−6.2	−7.0

( ) interpolated.
: uncertain.

APPENDIX A4(c)

M_V FOR LUMINOSITY CLASS V FOR B8–G8 STARS (from B6)

	A. Blaauw (1963)	*J. Jung*[a] *(1970)*	*J. Jung*[b] *(1971)*
B8	−0.5	−0.7	
9	0.0	0.0	
A0	+0.5	+0.5	+0.2
1	+0.8	+0.6	
2	+1.2	+0.7	+0.6
3	+1.5	+0.8	+1.4
5	+1.8:	+1.5	+1.2
7	+2.0:	+1.6	
F0	+2.4:	+2.3	+2.5
2	+2.8:	+2.5	
5	+3.2	+3.0	+3.3
6	+3.5	+3.2	
8	+4.0	+3.5	+3.4
G0	+4.4	+4.1	+4.3
2	+4.7	+4.3	
5	+5.1	+4.9	+4.8
8	+5.3	+5.5	

APPENDIX A4(d)

M_V FOR LUMINOSITY CLASS III FOR G5–M4 STARS (from B6)

<table>
<thead>
<tr><th></th><th>A. Blaauw (1963)</th><th>J. Jung[a] (1970)</th><th>J. Jung[b] (1971)</th></tr>
</thead>
<tbody>
<tr><td>G5</td><td>+0.4:</td><td>+0.3</td><td rowspan="2">+0.2</td></tr>
<tr><td>8</td><td>+0.4</td><td>+0.2</td></tr>
<tr><td>K0</td><td>+0.8</td><td>+0.1</td><td>+0.4</td></tr>
<tr><td>1</td><td>+0.8</td><td>+0.5</td><td>+0.4</td></tr>
<tr><td>2</td><td>+0.8</td><td>+0.2</td><td>+0.6</td></tr>
<tr><td>3</td><td>+0.1</td><td>−0.3</td><td>−0.2</td></tr>
<tr><td>4</td><td>−0.1</td><td>−0.8</td><td rowspan="2">−0.3</td></tr>
<tr><td>5</td><td>−0.3</td><td>−1.0</td></tr>
<tr><td>M0</td><td>−0.4:</td><td>−1.2</td><td></td></tr>
<tr><td>2</td><td>−0.4:</td><td>−1.5</td><td></td></tr>
<tr><td>4</td><td>−0.5:</td><td>−1.7:</td><td></td></tr>
</tbody>
</table>

[a] By proper motions and radial velocities.
[b] By trigonometric parallaxes.

APPENDIX A5

DISTANCES TO OB ASSOCIATIONS, d, AND THEIR DISTANCES FROM THE GALACTIC CENTER, R (from H3)

Association	*d (kpc)*	*R (kpc)*	*Spiral Arm**
Sgr OB 1	1.58	8.44	S-C
Sgr OB 7	1.74	8.30	S-C
Sgr OB 4	2.40	7.67	S-C
Sgr OB 6	2.00	8.07	S-C
Ser OB 1	2.19	7.92	S-C
Sct OB 3	1.66	8.43	S-C
Ser OB 2	2.00	8.13	S-C
Sct OB 2	1.00	9.09	L
Vul OB 1	2.00	9.18	L
Vul OB 4	1.02	9.54	L
Cyg OB 3	2.29	9.56	L
Cyg OB 1	1.82	9.71	L
Cyg OB 8	2.29	9.78	L
Cyg OB 9	1.20	9.82	L
Cyg OB 2	1.82	9.85	L
Cyg OB 4	1.00	10.00	L
Cyg OB 7	0.83	10.03	L
Lac OB 1	0.60	10.09	L
Cep OB 2	0.83	10.21	L
Cep OB 1	3.47	11.30	P
Cep OB 5	2.09	10.85	P
Cep OB 3	0.87	10.35	L
Cas OB 2	2.63	11.25	P
Cas OB 5	2.51	11.34	P
Cep OB 4	0.84	10.42	L
Cas OB 4	2.88	11.72	P
Cas OB 14	1.10	10.60	L
Cas OB 7	2.51	11.57	P
Cas OB 1	2.51	11.59	P
Cas OB 8	2.88	12.04	P
Per OB 1	2.29	11.73	P
Cas OB 6	2.19	11.67	P

APPENDIX A5 CONTINUED

Association	*d (kpc)*	*R (kpc)*	*Spiral Arm**
Cam OB 1..................	1.00	10.81	L
Cam OB 3..................	3.31	12.90	P
Per OB 2..................	0.40	10.38	L
Aur OB 1	1.32	11.31	L
Aur OB 2	3.16	13.14	P
Gem OB 1..................	1.51	11.49	L
Mon OB 1..................	1.71	10.66	L
Ori OB 1..................	0.50	10.45	L
Mon OB 2..................	1.51	11.37	L
CMa OB 1	1.32	10.99	L
Pup OB 1	2.51	11.33	L
Vel OB 1..................	1.82	10.32	L
Car OB 1..................	2.51	9.62	S-C
Car OB 2..................	2.00	9.50	S-C
Cru OB 1	2.40	9.26	S-C
Cen OB 1	2.51	8.84	S-C
Nor OB 1	3.47	7.29	N
Ara OB 1a..................	1.38	8.74	S-C
Sco OB 1..................	1.91	8.19	S-C
Sco OB 2..................	0.16	9.84	L
Sgr OB 5..................	3.02	6.98	N

* Spiral arms: Norma—N, Sagittarius-Carina—S-C, Local—L, Perseus—P.

APPENDIX A6(a)

DISTANCES TO GLOBULAR CLUSTERS BY DETAILED FIT TO HR DIAGRAM

(from C1)

Cluster	*Distance (kpc)*
47 Tuc	4.7
M3	9.2
M5	7.0
M13	7.0
M92	6.2
NGC 6752	4.1
M15	8.0

APPENDIX A6(b)

DISTANCES FOR GLOBULAR CLUSTERS

(adapted from H2)

NGC	*Name*	*Galactic coordinates*		*Reddening*	*Distance*	*[Fe/H]*
		l	*b*	*E(B-V)*	*(kpc)*	
104	47 Tuc	305.9	−44.9	0.04	4.7	−0.44
288		149.7	−89.4	0.03	8.3	−1.41
362		301.5	−46.3	0.04	9.0	−1.20
1261		270.6	−52.1	0.02	10.7	−1.24
	Pal 1	130.0	+19.1	0.12	46:	−1.9:
1851		244.5	−35.0	0.07	10.9	−1.29
1904	M79	227.2	−29.3	0.01	13.3	−1.58
2298		245.6	−16.0	0.11	12.2	−1.41
2419		180.4	+25.3	0.03	93	−2.00
2808		282.2	−11.3	0.22	9.3	−1.09
	Pal 3	240.3	+41.9	0.03	96:	−2.2:
3201		277.2	+8.6	0.21	5.0	−1.26
	Pal 4	202.3	+71.8	0.00	93	−2.4:

APPENDIX A6(b) CONTINUED

NGC	*Name*	*Galactic coordinates* *l*	*b*	*Reddening E(B-V)*	*Distance (kpc)*	*[Fe/H]*
4147		252.9	+77.2	0.02	17.6	−1.77
4372		301.0	−9.9	0.45	5.1	−1.7:
4590	M68	299.6	+36.0	0.03	9.6	−2.04
4833		303.6	−8.0	0.38	5.6	−2.15
5024	M53	333.0	+79.8	0.05	17.3	−1.85
5053		335.6	+79.0	0.03	15.4	−2.09
5139	Cen	309.1	+15.0	0.11	5.2	−1.6:
5272	M3	42.2	+78.7	0.01	9.9	−1.57
5286		311.6	+10.6	0.27	9.1	−1.38
5466		42.1	+73.6	0.05	14.5	−1.91
5634		342.2	+49.3	0.07	21.9	−1.70
5694		331.1	+30.4	0.08	32	−1.91
IC4499		307.4	−20.5	0.24	19.1	−1.0:
5824		332.6	+22.1	0.14	24.0	−1.67
	Pal 5	0.9	+45.9	0.03	21.6	−1.24
5897		342.9	+30.3	0.06	12.1	−1.45
5904	M5	3.9	+46.8	0.03	9.6	−1.25
5927		326.6	+4.9	0.55	7.8	−0.67
5946		327.3	+4.2	0.56	10:	−1.5:
5986		337.0	+13.3	0.27	10.4	−1.26
	Pal 14	28.8	+42.2	0.03	66:	−1.6:
6093	M80	352.7	+19.5	0.21	8.5	−1.54
6101		317.7	−15.8	0.08	12.4	−1.8:
6121	M4	351.0	+16.0	0.35	2.2	−1.30
6139		342.4	+6.9	0.68	9:	−1.27:
6144		351.9	+15.7	0.36	8.2	−0.9:
6171	M107	3.4	+23.0	0.37	6.1	−0.79
6205	M13	59.0	+40.9	0.02	7.2	−1.42
6218	M12	15.7	+26.3	0.19	5.4	−1.64
6229		73.6	+40.3	0.01	31	−1.44
6235		358.9	+13.5	0.38	11:	−1.2:
6254	M10	15.1	+23.1	0.26	4.5	−1.43
6266	M62	353.6	+7.3	0.46	6.3	−1.14
6273	M19	356.9	+9.4	0.38	11.0	−1.61
6284		358.4	+9.9	0.27	10.4	−1.01:
6287		0.1	+11.0	0.36	9.2	−0.39:
6293		357.6	+7.8	0.34	7.6:	−1.86:
6304		355.8	+5.4	0.58	5.8	−0.37

NGC	Name	Galactic coordinates *l*	*b*	Reddening E(B-V)	Distance (kpc)	[Fe/H]
6316		357.2	+5.8	0.48	13:	−0.44:
6325		1.0	+8.0	0.80	7.3	−0.7:
6333	M9	5.5	+10.7	0.36	7.1:	−1.81:
6341	M92	68.4	+34.9	0.01	7.9	−2.12
6342		4.9	+9.7	0.49	16:	−0.41
6352		341.4	−7.2	0.25	5.6	−0.06
6355		359.6	+5.4	0.76	6.9:	−1.05:
6356		6.7	+10.2	0.60	17.6	−0.37
6362		325.5	−17.6	0.12	7.2	−0.9:
6366		18.4	+16.0	0.65	4.4:	−0.1:
6388		345.5	−6.7	0.32	15.0	−0.48
6397		338.2	−12.0	0.18	2.3	−1.83
6401		3.5	+4.0	0.79	7.3:	−0.7:
6402	M14	21.3	+14.8	0.58	10.8	−1.28
	Pal 6	2.1	+1.8	1.8:	3.5:	—
6426		28.1	+16.2	0.40	16.7	−1.35:
6440		7.7	+3.8	1.11	4.1	−0.28
6441		353.5	−5.0	0.45	10.7	−0.24
6453		355.7	−4.0	0.67:	7.6:	−1.1:
6496		348.1	−10.0	0.07	9.1:	−0.1:
6517		19.2	+6.8	1.14	8.7:	−1.33
6522		1.0	−3.9	0.50	6.7	−1.04
6528		1.1	−4.2	0.65	7.8	−0.43
6535		27.2	+10.4	0.36	11.4	−1.9:
6539		20.8	+6.8	1.22	2.5:	−1.2:
6541		349.3	−11.2	0.13	7.0	−1.59
6544		5.8	−2.2	0.63	4.9	−1.02
6553		5.3	−3.1	0.79	6.4	−0.4:
6558		0.2	−6.0	0.40	9.6:	−0.9:
IC1276	Pal 7	21.8	+5.7	0.92	14.0	−0.8:
6569		0.5	−6.7	0.63	8.2:	−0.54:
6584		342.1	−16.4	0.11	14.8:	−1.40
6624		2.8	−7.9	0.25	8.7	−0.34
6626	M26	7.8	−5.6	0.33	6.3:	−1.08
6637	M69	1.7	−10.3	0.17	10.4	−0.47
6638		7.9	−7.2	0.36	8.2:	−0.6:
6642		9.8	−6.4	0.36	6.4	−0.88:
6652		1.5	−11.4	0.11	15:	−0.5:

APPENDIX A6(b) CONTINUED

NGC	*Name*	*Galactic coordinates* *l*	*b*	*Reddening E(B-V)*	*Distance (kpc)*	*[Fe/H]*
6656	M22	9.9	−7.6	0.35	3.2	−1.69
	Pal 8	14.1	−6.8	0.30	32:	—
6681	M70	2.9	−12.5	0.07	10.9	−1.17
6712		25.3	−4.3	0.35	7.8	−0.43
6715	M54	5.6	−14.1	0.14	21.9	−1.55
6717	Pal 9	12.9	−10.9	0.18	16:	—
6723		0.1	−17.3	0.03	8.8	−0.85
6752		336.5	−25.6	0.03	4.2	−1.62
6760		36.1	−3.9	0.91	4.3	−1.06
6779		62.7	+8.3	0.22	9.7	−1.79
	Pal 10	52.4	+2.7	1.2:	10:	—
6809	M55	8.8	−23.3	0.07	5.2	−1.78
	Pal 11	31.8	−15.6	0.35:	12:	−0.63
6838	M71	56.7	−4.5	0.28	4.1	−0.28
6864	M75	20.3	−25.8	0.17	18.6	−1.30
6934		52.1	−18.9	0.12	14.9	−1.38
6981	M72	35.2	−32.7	0.03	17.4	−1.27
7006		63.8	−19.4	0.13	35	−1.66
7078	M15	65.0	−27.3	0.12	9.6	−2.01
7089	M2	53.4	−35.8	0.06	11.3	−1.53
7099	M30	27.2	−46.8	0.01	8.2	−2.03
	Pal 12	30.5	−47.6	0.02	19.1	−1.55
	Pal 13	87.1	−42.7	0.05	24.6	−2.03
7492		53.3	−63.5	0.00	22.0	−2.0:

: uncertain values.
$[Fe/H] = \log_{10}\{(Fe/H)_{cluster}/(Fe/H)_{sun}\}$, where (Fe/H) is the abundance of iron relative to hydrogen, by mass.

APPENDIX A7(a)

THE ZERO-AGE MAIN SEQUENCE (ZAMS)

(from M3)

$(B-V)_0$	M_V	$(B-V)_0$	M_V
−0.30	−3.50	0.30	2.80
−0.25	−2.30	0.40	3.35
−0.20	−1.30	0.50	4.05
−0.15	−0.50	0.60	4.60
−0.10	0.30	0.70	5.20
−0.05	0.90	0.80	5.70
0.00	1.30	0.90	6.10
0.05	1.55	1.00	6.60
0.10	1.80	1.10	7.00
0.20	2.25	1.20	7.45
		1.30	7.90

APPENDIX A7(b)

PHYSICAL PROPERTIES OF MAIN-SEQUENCE STARS (from M3)

$\log(M/M_{\odot})$	Spectral class	$\log(L/L_{\odot})$	M_{bol}	M_V	$\log(R/R_{\odot})$
−1.0	M6	−2.9	12.1	15.5	−0.9
−0.8	M5	−2.5	10.9	13.9	−0.7
−0.6	M4	−2.0	9.7	12.2	−0.5
−0.4	M2	−1.5	8.4	10.2	−0.3
−0.2	K5	−0.8	6.6	7.5	−0.14
0.0	G2	0.0	4.7	4.8	0.00
0.2	F0	0.8	2.7	2.7	0.10
0.4	A2	1.6	0.7	1.1	0.32
0.6	B8	2.3	−1.1	−0.2	0.49
0.8	B5	3.0	−2.9	−1.1	0.58
1.0	B3	3.7	−4.6	−2.2	0.72
1.2	B0	4.4	−6.3	−3.4	0.86
1.4	O8	4.9	−7.6	−4.6	1.00
1.6	O5	5.4	−8.9	−5.6	1.15
1.8	O4	6.0	−10.2	−6.3	1.3

APPENDIX A7(c)

THE EFFECTIVE-TEMPERATURE AND BOLOMETRIC-CORRECTION SCALES

(from M3)

	Luminosity class					
	V		*III*		*I*	
Spectral class	T_{eff}	*B.C.*	T_{eff}	*B.C.*	T_{eff}	*B.C.*
O5	47,000	−4.0				
O7	38,000	−3.5				
O9	34,000	−3.2			30,000	−2.9
B0	30,500	−3.00			25,500	−2.6
B2	23,000	−2.30				
B3	18,500	−1.85				
B5	15,000	−1.40			13,500	−0.9
B7	13,000	−0.90				
B8	12,000	−0.70				
A0	9,500	−0.20				
A5	8,300	−0.10				
F0	7,300	−0.08			6,400	−0.2
F5	6,600	−0.01				
G0	5,900	−0.05	5,400	−0.1	5,400	−0.3
G5	5,600	−0.10	4,800	−0.3	4,700	−0.6
K0	5,100	−0.2	4,400	−0.5	4,000	−1.0
K5	4,200	−0.6	3,600	−1.1	3,400	−1.6
M0	3,700	−1.2	3,300	−1.5	2,800	−2.5
M5	3,000	−2.5	2,700:	−3:		
M8	2,500	−4:				

APPENDIX A8

AN AVERAGE INTERSTELLAR EXTINCTION CURVE (from S5)

	$\lambda(\mu m)$	$\lambda^{-1}(\mu m^{-1})$	$E(\lambda - V)/E(B - V)$	$A_\lambda/E(B - V)$
	∞	0	−3.10	0.00
L	3.4	0.29	−2.94	0.16
K	2.2	0.45	−2.72	0.38
J	1.25	0.80	−2.23	0.87
I	0.90	1.11	−1.60	1.50
R	0.70	1.43	−0.78	2.32
V	0.55	1.82	0	3.10
B	0.44	2.27	1.00	4.10
	0.40	2.50	1.30	4.40
	0.344	2.91	1.80	4.90
	0.274	3.65	3.10	6.20
	0.250	4.00	4.19	7.29
	0.240	4.17	4.90	8.00
	0.230	4.35	5.77	8.87
	0.219	4.57	6.57	9.67
	0.210	4.76	6.23	9.33
	0.200	5.00	5.52	8.62
	0.190	5.26	4.90	8.00
	0.180	5.56	4.65	7.75
	0.170	5.88	4.77	7.87
	0.160	6.25	5.02	8.12
	0.149	6.71	5.05	8.15
	0.139	7.18	5.39	8.49
	0.125	8.00	6.55	9.65
	0.118	8.50	7.45	10.55
	0.111	9.00	8.45	11.55
	0.105	9.50	9.80	12.90
	0.100	10.00	11.30	14.40

APPENDIX A9

CALIBRATING GALACTIC CEPHEIDS*

(a) from S2

Star	*Cluster*	*P(days)*	μ_0	$E(B-V)$	At mean light $\langle B\rangle_0 - \langle V\rangle_0$	$M_{\langle V\rangle_0}$	$M_{\langle B\rangle_0}$
Su Cas	reflection neb.	1.95	7.5	0.33	0.38	−2.54	−2.16
EV Sct	N6664	3.09	11.03	0.58	0.57	−2.62	−2.05
CE Cas b	N7790	4.48	12.53	0.555	0.565	−3.205	−2.64
CF Cas	N7790	4.87	12.53	0.555	0.655	−3.075	−2.42
CE Cas a	N7790	5.14	12.53	0.555	0.645	−3.275	−2.63
UY Per	h + χ Per	5.36	11.90	0.98	0.58	−3.54	−2.96
VY Per	h + χ Per	5.53	11.90	1.06	0.55	−3.91	−3.36
U Sgr	M25	6.74	8.98	0.55	0.55	−3.93	−3.37
DL Cas	M129	8.00	11.28	0.50	0.70	−3.84	−3.14
S Nor	N6097	9.75	9.76	0.23	0.73	−4.03	−3.30
VX Per	h + χ Per	10.89	11.90	0.58	0.64	−4.34	−3.70
SZ Cas	h + χ Per	13.62	11.90	0.88	0.61	−4.71	−4.10
RS Pup	Pup OB III	41.38	11.30	0.55:	0.89:	−5.95	−5.06
(b) additional, from B2 (and references therein)							
CV Mon	Mon OB2	5.38	11.01	0.76	0.64	−3.35	−2.71
CS Vel	Ru 79	5.90	12.25	0.76	0.66	−3.05	−2.39
V367 Sct	N6649	6.29	11.15	1.37	0.49	−3.82	−3.33
TW Nor	Lynga 6	10.79	10.8	1.34	0.80	−3.53	−2.73
VY Car	Car OB1	18.93	11.57	0.28	0.92	−4.97	−4.05
T Mon	Mon OB2	27.02	11.01	0.23	0.96	−5.50	−4.56
SV Vul	Vul OB1	45.04	11.81		1.06:	−6.00	−4.96

: = uncertain.
* based on $\mu_{\text{Hya}} = 3.03$

APPENDIX A10

HISTORICAL SUPERNOVAE IN OUR GALAXY (from C2, T1)

A.D.	*Radio remnant*	*Duration*	*d(kpc)*	*Ref.*	*Type*	$m_V(max)$	A_V	$M_V^0(max)$
185	G315.4 − 2.3	20 months	2	C				
386 ?	11.2 − 0.3	3 months	≥5	B				
393	348.5 + 0.1	8 months	≥6.7	B				
	or 348.7 + 0.3							
1006	327.6 + 14.5	several years	1.3	C	I	−8	0.6	−19.15
1054	184.6 − 5.8	22 months	2.0	D	II	−4.8	1.55	−17.85
	(Crab)		2.2	A				
1181	130.7 + 3.1	6 months	8.2	D	II	0	3.3	−17.67
	(3C58)							
1572	120.1 + 1.4	16 months	≥6	B				
	(Tycho)		4	D	I	−4.4	2.1	−19.51
1604	4.5 + 6.8	12 months	4	D	I	−3.5	3.47	−19.98
	(Kepler)							
1667 ?	(Cas A)		2.8	D	II?	>0	4.3	>-16.5

A, proper motions; B, HI absorption; C, optical methods; D, from T1.

APPENDIX A11

WELL-OBSERVED TYPE II SUPERNOVAE IN EXTERNAL GALAXIES (from V2)

SN	*NGC*	*B(max)*
1923a	5236	12.4:
1926a	4303	14.5
1936a	4273	15.0
1937a	4157	15.55
1937f	3184	13.75
1940a	5907	14.7
1941a	4559	13.4
1948b	6946	14.4
1957a	2841	14.4
1959d	7331	13.8
1961u	3938	14.3
1965l	3631	14.65:
1966b	4688	14.9
1966e	4189	14.7 (from T2)
1968l	5236	11.75
1969l	1058	13.15
1970g	5457(M101)	11.75
1972q	4254	15.6
1973r	3627	15.0
1975t	3756	15.6
1979c	4321(M100)	12.0 (from B7, P1)
1980k	6946	11.6 (from B1)

APPENDIX A12

GALACTIC NOVAE (from V1, B3, D1)

Nova	*Year*	$M_{pg}(max)$ *(B3)*	*(V1)*	log t_2 *(B3)*	log t_3 *(V1)*
CP Pup	1942	-9.3 ± 0.3	-9.4	0.45:	0.80
V603 Aql	1918	-8.7 ± 0.2	-8.7	0.60	0.99
V446 Her	1960	-8.7 ± 0.2	-8.8	0.71	1.08
CP Lac	1936	-8.6 ± 0.2	-8.4	0.85	1.06
GK Per	1901	-8.3 ± 0.2	-8.5	0.98	1.23
V476 Cyg	1920	-8.8 ± 0.2	-8.5	1.04	1.27
T Sco	1860	-8.6 ± 0.2		1.18	1.32 (D1)
DK Lac	1950	-8.0 ± 0.4	-8.3	1.16	1.45
EU Sct	1949	-8.5 ± 0.4	-8.2	1.32	1.61
DN Gem	1912	-7.0 ± 0.4	-6.7	1.40	1.63
BT Mon	1939	-7.6 ± 0.8		1.45:	1.62 (D1)
V533 Her	1963	-6.6 ± 0.4	-7.3	1.50	1.58
LV Vul	1968		-7.9		1.66
FH Ser	1970		-7.2		1.78
DQ Her	1934	-6.5 ± 0.2	-6.2	1.90	2.00
T Aur	1891	-6.1 ± 0.3	-5.7	2.00	2.01
RR Pic	1925	-7.1 ± 0.4	-7.1	2.00	2.22

: uncertain values. Details of a further 18 Galactic novae are given in D1.
For each nova, the absolute photographic magnitude at maximum light $M_{pg}(max)$, the times t_2, t_3 (in days) for the magnitude to drop by 2 or 3 magnitudes, the absolute photographic magnitude 15 days after maximum light M_{15}, the absolute visual magnitude at maximum light $M_V(max)$, and the distance d (parsec), are given.

Nova	*Year*	M_{15} *(V1)*	$M_V(max)$ *(D1)*	*d(pc)* *(D1)*
CP Pup	1942	−4.9	−11.5	1500
V603 Aql	1918	−5.1	−9.6	330
V446 Her	1960		−8.5	790 ± 170
CP Lac	1936	−5.1	−9.6	1000 ± 100
GK Per	1901	−5.2	−9.2	525
V476 Cyg	1920	−5.6	−9.2	1650 ± 50
T Sco	1860		−9.2	1200
DK Lac	1950	−6.5	−7.2	1500 ± 200
EU Sct	1949	−6.2	−7.0:	5060 ± 1700
DN Gem	1912	−4.5	−5.3	450 ± 70
BT Mon	1939		−6.3:	1000 ± 200
V533 Her	1963		−6.7	680 ± 250
LV Vul	1968		−6.3	820 ± 50
FH Ser	1970		−6.9	650
DQ Her	1934	−6.1	−5.9	260
T Aur	1891	−5.3	−6.7	600
RR Pic	1925	−5.5	−6.9	400

APPENDIX A13

STRÖMGREN RADIUS OF HII REGIONS

(from S6)

Spectral type	T_e (K)	R/R_0	N_u (s^{-1})	$r_s(n_e n_H)^{1/3}$ $(pc\ cm^{-2})$
O5	47,000	13.8	51×10^{48}	110
O6	42,000	11.5	17.4	77
O7	38,500	9.6	7.2	57
O8	36,500	8.5	3.9	47
O9	34,500	7.9	2.1	38
B0	30,900	7.6	0.43	22
B1	22,600	6.2	0.0033	4.4

For each spectral type, the stellar radius in nits of the sun's radius, R/R_0, the number of ionizing photons emitted per second, N_u, and the Strömgren radius, r_s, are given. The latter depends on the density of electrons, n_e, and on the total density of hydrogen atoms, n_H.

APPENDIX A14

GLOBULAR-CLUSTER SYSTEMS IN 37 GALAXIES (from H1, H2, B4)

	Type	$(m-M)_V$	E_{B-V}	M_V	N_{obs}	N_t
Local Group						
M31	S	24.6	0.11	−21.1	300:	450
Galaxy	S	—	—	−20:	131	180
LMC	I	18.7	0.06	−18.5	17	23
SMC	I	19.1	0.04	−16.8	10	15
M33	S	24.5	0.03	−18.9	6	20
Fornax	dSph	21.0	0.02	−13.6	6	6
NGC 147	dE	24.6	0.15	−14.9	4	4
NGC 185	dE	24.6	0.15	−15.2	6:	7:
NGC 205	dE	24.6	0.11	−16.4	8	8
NGC 6822	I	24.2	0.30	−15.7	0	0
IC 1613	I	24.5	0.03	−14.8	0	0
WLM	I	26.2	0.06	−16.0	1:	1:

	Type	M_V	N_{obs}	N_t
Virgo System[a]				
NGC 4216	S	−21.0	21	520
NGC 4340	E	−19.9	26	650
NGC 4374 (M84)	E	−21.6	98	2500
NGC 4406 (M86)	E	−21.7	108	2600
NGC 4472 (M49)	E	−22.5	1700	4200
NGC 4486 (M87)	E	−22.3	6000	15000
NGC 4526	E	−21.3	87	2200
NGC 4564	E	−20.0	35	900
NGC 4569 (M90)	S	−21.4	32	800
NGC 4594 (M104)	S	−22.6	290	2800
NGC 4596	E	−20.4	82	2000
NGC 4621 (M59)	E	−21.1	63	1600
NGC 4636	E	−21.3	143	3600
NGC 4649 (M60)	E	−22.1	170	4200
NGC 4697	E	−21.6	72	1800

	Type	M_V	N_t
Other Groups			
NGC 891	S	−21.7	≤100
NGC 3226	E	−19.6	420
NGC 3377	E	−19.8	235
NGC 3379	E	−20.7	290
NGC 3607	E	−20.9	800
NGC 4278	E	−20.3	1075
NGC 4565	S	−22.8	100
NGC 5128	E	−21.5	1230:
NGC 5813	E	−21.3	2500
NGC 5846	E	−22.8	100

[a] $(m-M)_V = 30.9$, $E_{B-V} = 0.02$ assumed.
For each galaxy, the type, the absolute visual magnitude M_V, the observed number of globular clusters N_{obs}, and the estimated total number N_t, are given.

APPENDIX A15

BRIGHTEST CLUSTER GALAXIES WITH $V_0 \leq 10{,}000$ km/s

Cluster	*Brightest galaxy*	A_v	$V_c^{BM,R}$	μ_0	d	d_V	V_0	V_c	H_c
	N1209	0.07	10.53	33.35	47	62	2510	2324	50
Antila	N3258	0.40	10.66	33.48	50	42	2826	3330	67
A1060	N3311	0.27	10.99	33.81	58	48	3697	4225	73
	N3357	0.37	9.68	32.50	32	25	2527	3019	95
	N4373	0.33	9.80	32.62	33	26	2996	3443	103
	N4936	0.22	10.75	33.57	52	40	3074	3510	68
	I4296	0.26	10.39	33.21	44	35	3648	4031	92
	I4329	0.22	10.74	33.56	52	41	4320	4698	91
	N5898	0.25	9.52	32.34	29	24	2260	2483	85
	I4797	0.30	10.06	32.88	38	45	2653	2583	69
Not E	N6769	0.23	10.61	33.43	49	56	3953	3892	80
Tel	N6861	0.18	10.24	33.06	41	51	2733	2549	62
Pavo	N6876	0.17	10.44	33.26	45	54	3785	3759	84
Indus I	N7014	0.12	11.64	34.46	78	90	4931	4690	60
Indus II	N7079	0.11	9.64	32.46	31	44	2256	2021	65
	N7144	0.09	9.84	32.66	34	48	1919	1662	49
	N7196	0.09	10.38	33.20	44	57	2844	2601	60
	N7213	0.08	9.54	32.36	30	44	1900	1627	55
	N80	0.14	11.96	34.78	90	106	6268	5773	64
	N128	0.07	10.53	33.35	47	65	4657	4176	89
3C31	N383	0.20	11.63	34.49	73	101	5264	4835	62
	N194	0.07	11.77	34.55	88	92	5298	4826	58
3C40	N547	0.06	12.29	35.11	105	123	5395	4997	48
	N741	0.08	11.44	34.26	71	88	5637	5273	74
	N1600	0.17	11.25	34.07	65	76	4797	4781	73
	N2832	0.05	12.03	34.85	93	83	6000	6350	68
	N3158	0.0	12.03	34.85	93	80	7008	7343	79
	N5044	0.11	10.67	33.49	50	35	2600	5038	61
	N5077	0.10	10.74	33.56	52	36	2515	2946	57
	N5353	0.0	10.05	32.87	37	24	2286	2484	66
	N5486	0.08	9.77	32.59	33	22	1808	2021	61
	N7242	0.45	12.05	34.87	94	105	6122	5627	60
	N7385	0.12	12.10	34.92	96	112	7738	7192	75

Cluster	*Brightest galaxy*	A_v	$V_c^{BM,R}$	μ_0	d	d_V	V_0	V_c	H_c
Perseus	N1275	0.58	11.19	34.01	63	73	5430	5210	82
A1213		0.0	13.54	36.36	187	171	8610	9014	48
3C66		0.42	12.27	35.10	105	115	6450	6146	59
3C465		0.17	13.19	36.01	159	173	9030	8516	54
3C338	N6166	0.07	12.90	35.72	140	131	9090	9039	65
A569		0.26	11.99	34.81	92	89	5790	5903	64
A634		0.11	12.80	35.62	133	127	7980	8096	61
A1257		0.0	13.85	36.67	216	201	10170	10515	49
A1314		0.0	13.02	35.84	148	135	10050	10260	69
A1318		0.0	12.19	35.01	100	88	5670	5876	59
A1367		0.0	12.13	34.95	98	80	6120	6575	67
A2162		0.07	12.66	35.48	125	115	9540	9565	77
A2197		0.07	12.64	35.46	124	115	9660	9631	78
A2666		0.17	12.47	35.29	115	129	8190	7691	67
0131-36		0.04	12.62	35.44	123	139	8940	8714	71
1400-33		0.27	10.99	33.81	58	48	4140	4491	78

Notes: (1) The visual magnitude, corrected for the Bautz-Morgan and richness effects and for interstellar extinction, $V_c^{BM,R}$, and the recession velocity, corrected for the sun's motion round the Galaxy, V_0 are taken from S1

(2) A_v from Fisher-Tully formula (Box 2.5)

(3) d is distance from our Galaxy, d_V is distance from Virgo (μ_0 is calculated from the calibration of Table 4.11)

(4) V_0 is recession velocity corrected for Galactic rotation (300 km/s to (l,b) = (90°,0°), V_C is velocity corrected for our Galaxy's absolute motion in space (546 km/s to (L,B) = (122,−35)), $H_c = V_c/d$.

APPENDIX A16

SOURCES OF DISTANCES TO GALAXIES BY SECONDARY METHODS

(only references containing ~100 or more distances are listed)

Sandage & Tammann 1975(S3)	distances to 114 nearby galaxies using HII region diameters and luminosity class of spirals
Sandage & Tammann 1975(S4)	distances to 97 remote ScI galaxies using luminosity class
de Vaucouleurs 1979(V2)	distances to 458 spirals, using diameter, absolute magnitude and luminosity class of galaxies
de Vaucouleurs & Peters 1981(V4)	selected subset of 200 of the above
de Vaucouleurs et al 1981(V3)	distances to 200 spirals using Tully-Fisher method in the blue
Fisher & Tully 1981(F1)	21 cm line width and corrected blue magnitudes for 1171 galaxies
Aaronson et al 1982(A1)	a catalog of infrared magnitudes and HI velocity widths for 308 nearby galaxies

APPENDIX A17

THEORY OF ERRORS

1. *Internal and external errors*

Suppose n independent measurements, x_i, $i = 1 \ldots n$, are made of a physical quantity x. Then the best estimate of x is the *mean* of the x_i

$$\bar{x} = \frac{1}{n} \sum_{i=1}^{n} x_i. \tag{1}$$

The errors in the x_i are assumed to have a Gaussian distribution about this mean $\left(\text{i.e. } p(x) = \frac{1}{\sqrt{2\pi}\sigma} \exp -(x-\bar{x})^2/2\sigma^2\right)$ and the best estimate of the *dispersion* is given by

$$\sigma^2 = \frac{1}{n-1} \sum_{i=1}^{n} (x_i - \bar{x})^2. \tag{2}$$

(Note Bessel's correction, replacing $1/n$ by $1/(n-1)$.)

The standard error of the mean is

$$\sigma/\sqrt{n} \tag{3}$$

and the standard error of the dispersion is

$$\sigma/\sqrt{2(n-1)}. \tag{4}$$

If we know that some observations are more reliable than others we may assign a weight w_i to the observations x_i. The *weighted mean* is then

$$\bar{x}_w = \sum_{i=1}^{n} w_i x_i \Big/ \sum_{i=1}^{n} w_i \tag{5}$$

and the standard error of this weighted mean is

$$\left\{ \frac{\sum_{i=1}^{n} w_i (x_i - \bar{x})^2}{(n-1) \sum_{i=1}^{n} w_i} \right\}^{1/2} \tag{6}$$

Note that if the weights are all the same, (5) and (6) reduce to (1) and (3).

Suppose now we have N different physical quantities $u, v, w, \ldots$ each with dispersion $\sigma_1, \sigma_2, \sigma_3, \ldots$, respectively. Then the standard error of the linear sum $k_1 u + k_2 v + k_3 W + \ldots$ is

$$\left(\sum_{j=1}^{N} k_j^2 \sigma_j^2 \right)^{1/2} \Big/ \sum_{j=1}^{N} k_j. \tag{7}$$

This can be applied to the combination of n observations x_i of a quantity x each with different errors σ_i. It is usual in this case to assign the weight

$$w_i = 1/\sigma_i^2 \tag{8}$$

and then (5) gives the weighted mean

$$\bar{x}_w = \left(\sum_{i=1}^{n} x_i/\sigma_i^2 \right) \Big/ \left(\sum_{i=1}^{n} 1/\sigma_i^2 \right) \tag{9}$$

By (6), the standard error of this mean is

$$\sigma_{\text{ext}} = \left\{ \frac{\sum_{i=1}^{n} (x_i - \bar{x})^2/\sigma_i^2}{(n-1) \sum_{i=1}^{n} 1/\sigma_i^2} \right\}^{1/2} \tag{10}$$

the *external* error. On the other hand eqn (7) with $k_i = w_i = 1/\sigma_i^2$ gives the error estimate

$$\sigma_{\text{int}} = \left(\sum_{i=1}^{n} 1/\sigma_i^2 \right)^{-1/2}, \tag{11}$$

the *internal* error.

These two error estimates ought to agree and the quantity $Z = \sigma_{\text{ext}}/\sigma_{\text{int}}$ has an expected value of 1 with standard error $1/\sqrt{2(n-1)}$. If Z differs from 1 by several multiples of this error, then *systematic errors* are probably present, causing the errors in the x_i to be larger than the estimated value σ_i.

Now suppose we wish to combine N different sets of observations, each set consisting of n_i ($i = 1 \ldots N$) observations x_{ij}, $j = 1 \ldots n_i$, with error σ_i, and being assigned a weight w_i.

The mean for the ith set

$$\bar{x}_i = \sum_{j=1}^{n_i} x_{ij}/n_i \tag{12}$$

and the standard error in this mean is given by

$$\bar{\sigma}_i^2 = \frac{\sum_{j=1}^{n_i} (x_{ij} - \bar{x}_i)^2}{n_i(n_i - 1)} = \sigma_i^2/n_i \tag{13}$$

$$\text{where } \sigma_i^2 = \sum_{i=1}^{n_i} (x_{ij} - \bar{x}_i)^2/(n_i - 1). \tag{14}$$

The combined weighted mean of the N mean values $\bar{x}_i$ is

$$\bar{x} = \sum_{i=1}^{N} n_i w_i \bar{x}_i \Big/ \sum_{i=1}^{N} n_i w_i = \sum_{i=1}^{N} w_i \sum_{j=1}^{n_i} x_{ij} \Big/ \sum_{i=1}^{N} n_i w_i \tag{15}$$

i.e. the same as the weighted mean for the whole data set.

The internal error of this combine weighted mean, using weights $w_i/\bar{\sigma}_i^2$,

$$\sigma_{\text{int}} = \frac{\left(\sum_{i=1}^{N} w_i^2/\bar{\sigma}_i^2\right)^{1/2}}{\left(\sum_{i=1}^{N} w_i/\bar{\sigma}_i^2\right)} = \frac{\left(\sum_{i=1}^{N} n_i w_i^2/\sigma_i^2\right)^{1/2}}{\left(\sum_{i=1}^{N} n_i w_i/\sigma_i^2\right)}, \text{ by (13)}, \tag{16}$$

i.e. the same as the internal error for the weighted mean of the whole data set.

The external error of the combined weighted mean is

$$\sigma_{\text{ext}} = \left\{\sum_{i=1}^{N} (\bar{x}_i - \bar{x})^2 n_i w_i \Big/ (N-1) \sum_{i=1}^{N} n_i w_i\right\}^{1/2} \tag{17}$$

and this differs from the external error for the weighted mean of the whole data set

$$\left\{\sum_{i=1}^{N} w_i \sum_{j=1}^{n_i} (x_{ij} - \bar{x})^2 \Big/ \sum_{i=1}^{N} n_i w_i \left(\sum_{i=1}^{N} n_i - 1\right)\right\}^{1/2} \tag{18}$$

2. *Linear regression*

Often the quantity we wish to know, y, is related to an observable, x, by a simple linear relation

$$y = ax + b. \tag{19}$$

The determination of the constants a and b, i.e. the *calibration* of this relation can be made from n measurements of the y_i corresponding to x_i. Then

$$a = (\overline{xy} - \bar{x}\bar{y})/(\overline{x^2} - \bar{x}^2) \tag{20}$$

$$b = (\bar{y}\overline{x^2} - \bar{x}\overline{xy})/(\overline{x^2} - \bar{x}^2) \tag{21}$$

where $\bar{z}$ means $\frac{1}{n}\sum_{i=1}^{n} z_i$,
with standard errors given by

$$\sigma_a^2 = \sigma_b^2/\sigma_x^2 = \frac{\sum_{i=1}^{n}(y_i - ax_i - b)^2}{(n-2)n(\overline{x^2} - \bar{x}^2)} = \frac{\sigma_y^2(1-r^2)}{(n-2)\sigma_x^2} \tag{22}$$

$$\text{where } \sigma_x^2 = \overline{x^2} - \bar{x}^2,\ \sigma_y^2 = \overline{y^2} - \bar{y}^2 \tag{23}$$

and the *correlation coefficient*

$$r = (\overline{xy} - \bar{x}\bar{y})/\sigma_x\sigma_y. \tag{24}$$

In using a calibration relation of the form (19) to determine y, there will be a *random error*

$$\sigma = \left\{\sum_{i=1}^{n}(y_i - ax_i - b)^2/(n-2)\right\}^{1/2} \tag{25}$$

reflecting the intrinsic scatter in the $y(x)$ relation, which can be reduced by a factor $\sqrt{m}$ by making m independent observations. There will also be a *zero point error* due to the uncertainties in the parameters a and b

$$\sigma_{zp}^2 = \left\{\frac{\sigma^2}{n} + \sigma_a^2\sigma_x^2\right\}^{1/2} = \{2\sigma^2/n\}^{1/2}. \tag{26}$$

3. *Principle of maximum likelihood*

Consider n independent observations $x_1, x_2, \ldots x_n$, from which we wish to determine a parameter θ. Let $p(x_i \mid \theta)$, $i = 1 \ldots n$, be the probability of obtaining the result x_i for a particular value of θ Then the joint probability of the observations, called the likelihood function, is

$$L(\mathbf{x} \mid \theta) = p(x_1 \mid \theta)p(x_2 \mid \theta) \ldots p(x_n \mid \theta).$$

The *principle of maximum likelihood* tells us that we should choose the value of θ which makes L largest. If we are trying to determine several parameters $\theta_1, \theta_2, \ldots \theta_r$, then θ above should be thought of as standing for the vector $(\theta_1, \theta_2, \ldots \theta_r)$.

For the particular case of normal (Gaussian) probability distributions, the principle of maximum likelihood becomes equivalent to the method of least squares.

4. *Malmquist effect*

Let us suppose a population of galaxies, say, uniformly distributed through space, have a Gaussian luminosity function with mean $\overline{M}_0$, dispersion σ_0, so the probability that a galaxy has absolute magnitude in the range $M, M + dM$ is

$$p(M)dM = \frac{1}{\sqrt{2\pi}\sigma_0} \exp\{-(M - \overline{M}_0)^2/2\sigma_0^2\}.$$

If we observe a sample of these galaxies which is complete down to a fixed limiting magnitude m, then for those galaxies in the absolute magnitude range $M, M + dM$, we will see all of them out to a distance $d(pc)$, where

$$m = M + 5 \log_{10} d - 5$$

$$\text{or } d = 10^{1+0.2(m-M)}\ pc,$$

and the number of galaxies visible will be proportional to d^3. If we now measure the average absolute magnitude for this magnitude-limited sample, we will obtain

$$\overline{M} = \frac{\int p(M)\, M\, d^3\, dM}{\int p(M)\, d^3\, dM} = \frac{\int \exp\left\{-\frac{(M - \overline{M}_0)^2}{2\sigma_0^2} - 0.6 \log_e 10\, M\right\} M\, dM}{\int \exp\left\{\quad\right\} dM}$$

$$= \overline{M}_0 - 0.6 \log_e 10\, \sigma_0^2.$$

This shifting of the mean absolute magnitudes towards brighter magnitudes by $1.38\, \sigma_0^2$ is called the *Malmquist effect*. Clearly, for a small dispersion, say $\sigma_0 = 0.1$, the effect is negligible, whereas for a large dispersion, say $\sigma_0 = 1$, the effect is very significant and can lead to serious biassing of distance estimates.

References: B5, K2, M2

REFERENCES FOR APPENDIXES

[A1] Aaronson M. et al, "A catalog of infrared magnitudes and HI velocity widths for nearby galaxies," *Astrophys. J. Supp.* 1982, **50,** pp. 241-262.

[B1] Barbon R., Ciatti F., and Rosino L., "Two bright supernovae in NGC 6946 and NGC 4536," *Astron. Astrophys.* 1982, **116,** pp. 35-42.

[B2] van den Bergh S., *The distance scale within the Local Group,* IAU Colloquium No. 37, "Décalage vers le rouge et expansion de l'Univers," eds C. Balkowski and B.E. Westerlund (Editions du CNRS, Paris), 1977, pp. 13-22.

[B3] van den Bergh S., *The extragalactic distance scale, Stars and Stellar Systems IX "Galaxies and the Universe,"* eds A. Sandage, M. Sandage and J. Kristian (Chicago University Press), 1975, pp. 509-593.

[B4] van den Bergh S. and Harris W.E., "Globular clusters in galaxies beyond the Local Group, II The edge-on spirals NGC 891 and NGC 4565," *Astron. J.* 1982, **87,** pp. 494-499.

[B5] Bevington, P.R., *Data Reduction and Error Analysis for the Physical Sciences,* (McGraw-Hill), 1969.

[B6] Blaauw A., *The calibration of luminosity criteria,* IAU Symposium No. 54, "Problems of calibration of absolute magnitudes and temperatures of stars," eds. B. Hauck and B.E. Westerlund, (Reidel) 1973, pp. 47-56.

[B7] Branch D., Falk S.W., McCall M.L., Rybski P., Uomoto A.K. and Wills B.J., "The Type II supernova 1979c in M100 and the distance to the Virgo cluster," *Astrophys. J.* 1981, **244,** pp. 780-804.

[C1] Carney B.W., "The ages and distances of 8 globular clusters," *Astrophys. J. Supp.* 1980, **42,** pp. 481-500.

[C2] Clark D.H. and Stephenson F.R., *Historical Supernovae* (Pergamon), 1977.

[C3] Crampton D., IAU Symposium No. 54, *Problems of calibration of absolute magnitudes and temperatures of stars,* eds. B. Hauck and B.E. Westerlund, (Reidel), 1973, pp. 120-123.

[C4] Crawford D.L., IAU Symposium No. 54, *Problems of calibration of absolute magnitudes and temperatures of stars,* eds. B. Hauck and B.E. Westerlund, (Reidel), 1973, pp. 95-113.

[C5] Crawford D.L., "Empirical calibration of the ubvy, system, II The B-type stars," *Astron. J.* 1978, **83,** pp. 48-63.

[D1] Duerbeck H.W., "Light curves, absolute magnitudes and physical properties of Galactic novae," *Publ. astr. Soc. Pacific,* 1981, **93,** pp. 165-175.

[F1] Fisher J.R. and Tully R.B., "Neutral hydrogen observations of a large sample of galaxies, *Astrophys. J. Supp.* 1981, **47,** pp. 139-200.

[H1] Harris W.E. and van den Bergh S., "Globular clusters in galaxies beyond the Local Group, I New cluster systems in selected northern ellipticals," *Astron. J.* 1981, **86,** pp. 1627-1642.

[H2] Harris W.E. and Racine R., "Globular clusters in galaxies," *Ann. Rev. Astron. Astrophys.* 1979, **17,** pp. 241-274.

[K1] van der Kamp P., "The nearby stars," *Ann. Rev. Astron. Astrophys.* 1971, **9,** pp. 103-126.

[K2] Kendall M.G., Stuart A., and Ord J.K., *Advanced Theory of Statistics Vol II,* (4th Edition, Charles Griffen, London), 1983.

[M1] Mermilliod J.-C., The present data situation for stars in open clusters. IAU Symposium No. 85, *Star clusters,* ed. J.E. Hesser (Reidel), 1980, pp. 129–133.

[M2] Meyer S.L., *Data Analysis for Scientists and Engineers,* (John Wiley), 1975.

[M3] Mihalas D. and Binney J., Galactic astronomy (Freeman), 1981.

[P1] Panagia N., et al., "Coordinated Optical, Ultraviolet, Radio and X-ray Observations of Supernova 1979c in M100," *Mon. Not. R. Astr. Soc.* 1980, **192,** pp. 861–879.

[S1] Sandage A., "The redshift-distance relation VIII: magnitudes and redshifts of southern galaxies in groups: a further mapping of the local velocity field and an estimate of q_0," *Astrophys. J.* 1975, **202,** pp. 563–582.

[S2] Sandage A. and Tammann G. A., "The double cepheid CE Cassiopeiae in NGC 7790: tests of the theory of the instability strip and the calibration of the period-luminosity-colour relation," *Astrophys. J.,* 1969, **157,** pp. 683–708.

[S3] Sandage A. and Tammann G. A., "Steps towards the Hubble constant, V The Hubble constant from nearby galaxies and the regularity of the local velocity field," *Astrophys. J.* 1975, **196,** pp. 313–328.

[S4] Sandage A. and Tammann G.A., 1975, "Step towards the Hubble constant, VI The Hubble constant determined from redshifts and magnitudes of remote ScI galaxies: the value of q_0," *Astrophys. J.* 1975, **197,** pp. 265–280.

[S5] Savage B.D. and Mathis J.S., "Observed properties of interstellar dust," *Ann. Rev. Astron. Astrophys.* 1979, **17,** 73–111.

[S6] Spitzer L., *Physical processes in the interstellar medium* (John Wiley), 1978, p. 110.

[T1] Tammann G.A., in *'Supernovae',* ed. D.N. Schramm (Reidel), 1977, pp. 95–106.

[T2] Tammann G.A., Sandage A. and Yahil A., *The determination of cosmological parameters, Physical cosmology,* eds. R. Balian, J. Audouze and D.N. Schramm (North Holland), pp. 53–125.

[V1] de Vaucouleurs G., "The extragalactic distance scale, I A review of distance indicators: zero points and errors of primary indicators," *Astrophys. J.* 1978, **223,** pp. 351–363.

[V2] de Vaucouleurs G., "The extragalactic distance scale, VI Distances of 458 spiral galaxies from tertiary indicators," *Astrophys. J.* 1979, **227,** pp. 729–755.

[V3] de Vaucouleurs G., Peters W.L., Bottinelli L., Gouguenheim G. and Paturel G., "Hubble ratio and solar motion from 300 spirals having distances derived from H1 line widths," *Astrophys. J.* 1981, **248,** pp. 408–422.

[V4] de Vaucouleurs G. and Peters W.L., "Hubble ratio and solar motion from 200 spiral galaxies having distances derived from the luminosity index," *Astrophys. J.* 1981, **248,** pp. 395–407.

[W1] Walborn N.R., "Spectral classification of OB stars in both hemispheres and the absolute magnitude calibration," *Astron. J.* 1972, **77,** pp. 312–318.

[W2] Walborn N.R., "The space distribution of O stars in the solar neighbourhood," *Astron. J.* 1973, **78,** pp. 1067–1073.

ACKNOWLEDGEMENTS FOR FIGURES

I thank the following for permission to reproduce figures:

Figures on p. 7 and 11, Fig. 2.34 Yerkes Observatory

Figure on p. 8 Goodsell Observatory, Northfield, Minnesota

Figure on p. 11 Mary Lea Shane Archive of the Lick Observatory, Santa Cruz, California

Figs 1.3, 1.4, 2.48 Dover Publications

Figs 1.1, 1.2, 2.15, 2.19, 2.27b, 2.29, 2.35, 2.36, 2.50, 2.51, 3.4, 3.7b, 3.8, 3.15, 3.17b, 4.1, 4.2, 4.4, 4.5, 4.7, 4.8, 4.9, 4.11, 4.12, 4.13, 4.15, 5.10, 6.5, 6.7, 6.8 Astrophysical Journal

Figs 1.4, 2.20, 2.46, 2.47, 2.49, 3.6, 3.14c,d,e Mount Palomar Observatory

Figures on p. 14 and 15 Mount Wilson and Las Campanas Observatories, Carnegie Institution of Washington

Figs 1.6, 2.2, 2.3, 2.52, 3.7a, 4.3, 6.6 Reidel Publishing Co, Dordrecht, Netherlands

Figs 1.6, 2.52 D. Schramm; Fig 2.6 D.M. Popper

Figs 2.6, 2.25, 2.35, 2.39, 2.42 reproduced, with permission, from *Annual Review of Astronomy and Astrophysics,* © 1967, 1976, 1979, 1980 by Annual Reviews Inc.

Figs 2.4, 2.12, 2.13, 2.28, 2.33, 2.44 J. Binney

Fig 2.8 P. Malin and Mount Palomar Observatory, © 1960 National Geographic Society—Palomar Sky Survey. Reproduced by permission of the California Institute of Technology

Fig 2.9, 2.11 A. Heck and Vistas in Astronomy

Fig 2.15 O.C. Wilson; Fig 2.17 Lowell Observatory

Fig 2.18 I. Iben and G. Cayrel de Strobel; Fig 2.19 W.E. Harris

Fig. 2.21 W.A. Fowler; Fig 2.22 J.L. Greenstein

Figs 2.23, 2.31 and 2.32 R.J. Tayler; Fig 2.25 I. Iben

Fig 2.26, 2.27a Yale University Press

Figs 2.27b, 2.30, 2.51, 3.1, 4.1, 4.4c, 4.5, 4.10 A. Sandage

Fig 2.29 R. Kippenhahn; Fig 2.34 A.D. Code; Fig 2.35, 2.39 B.D. Savage

Fig 2.36 B.T. Lynds; Figs 2.37, 2.38 C. Heiles

Figs 2.37, 2.38, 2.45, 3.9 Astronomy and Astrophysics

Figure on p. 93 J.R. Fisher; Fig 2.42 W.B. Burton

Fig 2.45 Y.M. Georgelin; Figs 2.50, 3.12, 2.13 S. van den Bergh

Fig 2.53 D. Kunth and Editions Frontières, Dreux, France

Figs 3.1, 3.2, 3.3, 3.5, 3.10, 3.14a, 6.11, 6.13 Royal Astronomical Society

Fig 3.2 J.F. Dean; Fig 3.3 W.L. Martin; Figs 3.4, 4.11 G.A. Tammann

Fig 3.5 B.F. Madore; Fig 3.7a L. Rossino; Fig 3.7b W.D. Arnett

Fig 3.8 D. Branch; Fig 3.9 R. Barbon; Fig 3.11 H.Y. Chiu

Fig 3.11a J.B. Whiteoak; Figs 3.12, 3.13 University of Chicago Press

Figs 3.14b, 3.17a photography by Photolabs, Royal Observatory Edinburgh, from an original negative by the UK 1.2 m. Schmidt telescope, © Royal Observatory Edinburgh

Fig 3.15 D.S. Matthewson; Fig 3.17b R.B. Tully

Fig 4.2 R.C. Kennicutt Jr; Fig 4.3 R. Racine

Fig 4.4a,b R.M. Humphries; Fig 4.7 M.S. Roberts

Fig 4.8 M. Aaronson; Fig 4.9, 4.12 G. de Vaucouleurs

Fig 4.13 I. King; Fig 4.14 E. Holmberg and Arkiv fur Astronomi

Fig 4.15 P. Schecter; Figs 5.3, 5.4 Oxford University Press

Fig 5.5 B.E.J. Pagel and Royal Society of London

Fig 5.10 and 5.11 G.F. Smoot; Fig 6.5 J. Kristian

Fig 6.6 W.A. Baum; Fig 6.7 G. Bruzual; Fig 6.8 M. Davis

Fig 6.11, 6.13 M.J. Rees; Fig 6.12 B.J. Carr, reprinted by permission from *Nature* Vol 278, pp. 605–612, © 1978 MacMillan Journals Ltd.

NAME INDEX

SUBJECT INDEX